THE SCIENCE OF FLUIDS

THE SCIENCE OF FLUIDS

IRVING MICHELSON
ILLINOIS INSTITUTE OF TECHNOLOGY

GEORGE E. MAYCOCK

VAN NOSTRAND REINHOLD COMPANY
NEW YORK / CINCINNATI / TORONTO / LONDON / MELBOURNE

Van Nostrand Reinhold Company Regional Offices:
Cincinnati, New York, Chicago, Millbrae, Dallas

Van Nostrand Reinhold Company Foreign Offices:
London, Toronto, Melbourne

Copyright © 1970 by Irving Michelson
Library of Congress Catalog Card Number 75-76809

All rights reserved. No part of this work covered by the copyright hereon may be reproduced or used in any form or by any means — graphic, electronic, or mechanical, including photocopying, recording, taping, or information storage and retrieval systems — without written permission of the publisher. Manufactured in the United States of America.

Published by Van Nostrand Reinhold Company
450 West 33rd Street, New York, N.Y. 10001

Published simultaneously in Canada by
D. Van Nostrand Company (Canada), Ltd.

10 9 8 7 6 5 4 3 2 1

Preface

The aim of this book is to provide an introduction to classical fluid mechanics with emphasis on fundamental physical principles. Based on courses presented by the writer, it is intended to be primarily instructional in character. For the reader who has completed a first course in Calculus of the type offered to undergraduates in engineering and science, its function is more nearly that of a work-book of programmed study than a comprehensive treatise. Employing a consistent analytical approach expressing the three conservation principles of mechanics, the development advances in easy stages from the elementary and particular toward the important and interesting broad concepts of the physical sciences. In-depth exercises within the text form an essential part of the work.

A secondary aim is to strengthen the student's facility both in the analytical formulation and in the mathematical treatment of physical phenomena. Although it is as true in mechanics as in the other engineering sciences that everything depends on the insight with which ideas are handled before they reach the mathematical stage, it is the author's experience that students are often hindered by their timidity to employ mathematics already learned. Both in verifying text derivations and in solving exercise problems, the reader is introduced to much of the applied mathematics employed in related and more advanced studies. A further benefit of the analytical viewpoint is in facilitating the retracing of a reasoning newly encountered, simulating the sensation of invention or discovery and thereby affording a glimmer of the creative experience itself.

The arrangement of chapters follows a time-proven pattern: starting with some banal but clearly comprehended phenomena in fluids at rest, step-by-step generalizations and extensions lead directly to problems of genuine interest and even to the forefront of particular areas of scientific investigation. Within the several chapters examples are presented that are drawn both from the traditional branches of fluid technology and also from those areas that are likely to be of increasing technical importance in the years ahead. Large-scale effects in the earth itself, in the oceans, in the atmosphere, and in the cosmic heavens at large are thus examined both for their own sake and in order to demonstrate the near-universal validity of the underlying fluid dynamic principles. Roughly one-third of the topics considered are of this type and not to be found in the numerous standardized introductory texts on fluid mechanics.

Connections with other engineering sciences and different branches of physics are also shown in numerous places. Analogies and basic interpretations are furnished by thermodynamics, acoustics, electricity and magnetism, heat conduction, and the kinetic theory of gases, for example. The art of mathematical approximation is developed and applied in order to obtain the idealized equations that show most clearly the unity of different classes of continuum phenoma. Interdisciplinary discussions frequently present notational difficulties, however. The same algebraic symbol has different meanings in different branches of study, so that a little care is required if confusion is to be avoided. A flagrant case is presented by Rankine's temperature scale, by Reynolds' dimensionless parameter of viscous flow, by the thermodynamic gas constant, and by an occasional radial distance that are all denoted by R.

Finally, a word must be said concerning the omission of descriptions of experimental methods and instrumentation. This field of interest (increasingly specialized and notorious for rapid obsolescence) depends heavily on principles not treated in this book. The increasing tendency to eliminate laboratory sessions from the introductory courses in fluid dynamics, moreover, must also serve to justify the omission.

I. M.

Contents

Preface   v

1 FLUID STATICS   1

1. Introduction: Characteristic Fluid Property   *1*
2. Pascal's Principle; Properties and Units of Pressure   *3*
3. Uniform Gravity: Hydrostatics and Aerostatics   *6*
4. Compressibility: The Liquid-Gas Distinction   *12*
5. Pressure Distribution in the Earth's Atmosphere   *16*
6. Large Fluid Masses: Planetary and Stellar Statics   *18*
7. Pressure at Interior Point of Arbitrary Large Fluid Mass — Poisson's Equation   *22*
8. Quasi-statics: Relative Equilibrium in a Rotating Reference Frame   *24*
9. Rotational Flattening of the Earth — and of the Solar System   *26*

S1. Archimedes' Principle, Hydrostatic Pressure Force Formulae   *31*
S2. Static Stability of the Atmosphere   *32*

2 PRINCIPLES OF FLUID MOTION   35

1. Fluid Velocity; Streamlines and Stream Surfaces   *35*
2. Conservation of Fluid Mass — Continuity Equation   *39*
3. Momentum Conservation — Bernoulli's Theorem   *45*
4. Euler's Equations of Fluid Motion   *48*
5. Conservation of Energy and Summary of Flow Equations   *51*
6. Flow Condition at a Solid Boundary Surface   *54*
7. Lagrangian Form of the Equations of Fluid Motion   *56*
8. Streamline Coordinates   *60*

3 ELEMENTARY FRICTIONLESS FLOWS   64

1. Natural Convection — Chimney Flow Dynamics   *64*

2. Deductions from Bernoulli's Integral: Torricelli's Theorem, Pitot Tube 66
3. Reaction Thrust — Rocket Engine Propulsion 69
4. Simplified Momentum Theory of Propellers: Propulsive Efficiency 72
5. Motion in a Rotating Reference Frame — Geostrophic Winds 76
6. Impact of Liquid Jet Striking a Solid Surface 79
7. Pressure Recovery in Expanding Flow Passage — Gradual vs. Sudden Area Change 82
S1. Overall Propulsive Efficiency — Evaluation by Newton's Iteration Method 86

4 IRROTATIONAL FLOW—VELOCITY POTENTIAL 88

1. Fluid Rotation: Vorticity 88
2. Flow Circulation and Irrotationality — Kelvin's Theorem 91
3. Vorticity in Nonbarotropic Fluid — Sea Breeze 95
4. Nonconservative Forces — Vorticity at Edge of Lifting Surface 97
5. Bernoulli's Integral in Irrotational Flow: Velocity Potential 100
6. Minimal Energy Property of Irrotational Flow 102
7. Examples of Flow Defined by Elementary Velocity Potential Functions 104
8. Irrotationality Condition Using Streamline Coordinates and Hodograph Variables 106

5 TWO-DIMENSIONAL PERFECT FLUID MOTIONS 109

1. Quantity of Flow — The Stream Function 109
2. Simple Flows Represented by Stream Function 112
3. Superposition of Elementary Flows — Circular Cylinder in Uniform Motion 116
4. Irrotational Flow with Circulation: Aerodynamic Lift 118
5. Methods for Determining Irrotational Flows 120
6. Linearized Analysis of Uniform Flow Past a Wavy Wall 122
7. Kinetic Energy of Flow Determined at Boundary Points: Apparent Mass 127
8. Reduction to Steady Flow — Galilean Invariance 129

6 IRROTATIONAL FLOW IN THREE DIMENSIONS 133

1. Elementary and Derivative Flows in Spherical Coordinates 133

2 Ocean Tide Height, Elementary Calculation *136*
3 Euler Equations in Polar Form *139*
4 Axial Symmetry — Sphere Flow *143*
5 Vortex Dynamics — Rectilinear Vortices *145*
6 Extension to Curved Vortex Lines — Ring Vortices *151*
7 Ground Effect — Momentary Enhancement of Airplane Wing Lift Force *153*
8 Starting Resistance *155*

7 SURFACE GRAVITY WAVES 157

1 Gravity and Inertia Forces Compared: Froude Number *157*
2 Waves in Shallow Water *160*
3 Waves in Water of Arbitrary Depth *163*
4 Waves in Deep Water: Dispersion *167*
5 Oscillation of a Spherical Liquid Mass *168*
6 Particle Paths in Progressive Wave Motion *170*

8 COMPRESSIBILITY 172

1 Acoustic Propagation: Sound Waves *172*
2 Energy Conservation in Compressible Fluid *176*
3 Normal Shock Waves in Ideal Gas *178*
4 Mach Number: The Dynamic Measure of Fluid Compressibility *180*
5 Heat Shock; Heat Pulse *182*
6 Linearized Small-Disturbance Flow in Two Dimensions *184*
7 Supersonic Flow Past Wavy Wall *186*
8 Stars Heated by Radiative Energy Loss *189*
9 Acoustic Dispersion: Gravitational Collapse and the Birth of Galaxies *192*
S1 Prandtl-Glauert Compressibility Correction in Subsonic Flow *195*
S2 Application of the Hodograph Equations — Chaplygin's Approximation *197*
S3 Entropy Jump Across Normal Shock Wave, Nonideal Gas *202*

9 VISCOSITY FLUID FRICTIONS 206

1 Newton's Friction Law: Pipe Flow Resistance *206*
2 Viscous Stresses: The Navier-Stokes Equations *210*

3 Swirling Flow: The Ideally Stirred Coffee Cup *213*
4 The Frictional Wind *216*
5 Anomalous Positive Frictional Acceleration in One-Dimensional Flow: Frictional Cooling *218*
6 Viscous Boundary Layer; Blasius' Calculation *219*
7 Nonzero Pressure Gradient — Integral of Momentum *224*
8 Viscous Dissipation of Flow Energy *228*
9 Decay of a Plane Sound Wave *229*
10 Viscous Dissipation Related to Deformation of Fluid Element *231*
S1 Precise Formulas for Viscous Stresses *234*
S2 General Equations of Unsteady, Viscous, Compressible Fluid Motion *237*
S3 Conditions Within and Thickness of Shock Waves *240*

10 TURBULENCE 244

1 Transition to Turbulent Flow; Mean Values and Fluctuations *244*
2 Equations of Turbulent Fluid Motion — Reynolds Stresses *248*
3 Correlation of Velocities: Mixing Length Hypothesis *250*
4 Universal Resistance Law for Smooth Pipes, Large Reynolds Number *252*
5 Turbulent Flat Plate Boundary Layer; Logarithmic Formula *254*
6 One-Seventh Power Law — Momentum Integral for Turbulent Flow *256*

11 THE MOLECULAR BASIS OF FLUID MECHANICS 258

1 Identification of Pressure and Temperature *258*
2 Flow Velocity Expressed in Terms of Molecular Speeds; Boltzmann's Equation *261*
3 Deduction of Flow Equations from Boltzmann's Equation *263*
4 Boltzmann's Law and the Maxwellian Velocity Distribution *265*
5 Finite Molecular Diameter: Real-Fluid Properties *269*
6 Virial Theorem — Molecular Basis of Ideal Gas Equation of State *272*
7 Intermolecular Forces — The Second Virial Coefficient and van der Waals' Volume Correction *275*

APPENDIX

A   Green's Theorem — Transformations of Gauss and Stokes *279*
B   On Partial Differential Equations — Gravitational Potential *282*
C   Properties of Jacobian Determinants *287*

Index *293*

THE SCIENCE OF FLUIDS

I
Fluid Statics

1 INTRODUCTION: CHARACTERISTIC FLUID PROPERTY

FLUID MECHANICS deals with the motions of fluids and with the forces exerted on solid bodies in contact with fluids. In common with other branches of classical mechanics its fundamental principles are *Newton's laws of motion*, the *indestructibility of matter*, and the *conservation of energy*. The objectives of fluid mechanics are to explain and predict observable phenomena in Nature and in fluid devices and equipment of all kinds. Because of the broad validity of Newton's laws, a great variety of physical effects falls within the province of fluid mechanics, ranging from ancient questions of waterway hydraulics to the most esoteric studies of solar system exploration and cosmic dynamics. One purpose of this book is to display the great power of the analytical approach at the level of the calculus for providing a coherent understanding of the widest possible scope of physical problems.

A useful definition of the fluid state is based on the most conspicuous characteristic of all fluids, the extreme ease with which a given volume of fluid can be deformed. This feature is expressed in technical language by the statement that, in contrast to solids that support both normal and tangential internal stresses, a fluid at rest can sustain only *normal* stresses (Fig. I-1): it yields to even the smallest tangential force. The normal stress in a fluid at rest (i.e., without relative motion between its parts), counted positive in the compressive sense, is termed the fluid *pressure*, denoted by the symbol p, and is identified with the corresponding thermodynamic variable. The stress at any point of a fluid-filled region, representing the quotient of a force and an area in the limit of vanishing area, is a useful concept that has both physical and mathematical

fluid motion and fluid forces

fundamental principles are three:
 (i) mass conservation
 (ii) momentum conservation
(iii) energy conservation

wide range of physical applications old and new

definition of fluid: normal stresses only in fluid at rest

fluid pressure: the compressive normal stress

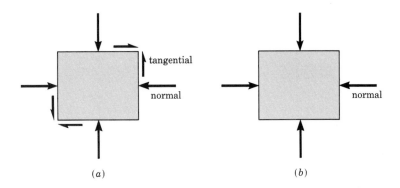

Fig. I-1 Contrasting equilibrium stress states, solids vs fluids; (a) solid: normal and tangential stresses; (b) fluid: normal stresses only.

homogeneous and continuous

meaning only when the fluid medium is regarded as *continuous* and *homogeneous*, i.e., such that the smallest parts considered are identical in character with the fluid in bulk. For only when the granular nature of the constituent molecules is ignored can the required limits be formed that lead to the common derivative expressions so convenient throughout all the following developments. The pressure then becomes a strictly "local" property in the sense that its value is defined at each point of the fluid, and the fluid "particle" can be referred to as the ultimately small element of fluid at any point. Justification for this idealization is found in the fact that the macroscopic scale of ordinary fluid phenomena is vastly greater than the molecular dimensions and intermolecular distances in virtually all cases. A great convenience results from considering the fluid as a *continuous medium* in this manner, since the pressure and all other flow quantities vary smoothly from one point to another, permitting the unrestricted use of derivatives and differentials whenever they are needed.

fluids: liquids and gases

The term "fluid" comprises both *liquids* and *gases*, the most familiar examples of which are water and air. The two corresponding branches of fluid mechanics are called *hydrodynamics* and *aerodynamics*, the former relating to water as well as other liquids, the latter to air and other gases. Another customary division of the subject depends on the practical importance of fluid friction in any given case. "Perfect fluids" are treated as if all tangential stresses caused by friction can be ignored. The designation "real fluids" refers to the cases in which friction must properly be taken into account.

fluid friction, perfect fluids and real fluids

The preceding categories and classifications within fluid mechanics provide a framework for indicating a traditional arrangement of topics in the study of the subject. Among the simplest flows are those of fluids like water in which volume changes and friction are both insignificant; therefore perfect fluids are examined first. Volume changes (or fluid *compressibility*) have the smallest effect in low-speed flows and are logically treated next. Fluid friction (or *viscosity*) introduces fundamentally different flow characteristics and difficulties that require special methods of analysis and measurement. The most general formulation of classical fluid mechanics is completed by the inclusion of viscous effects.

_{divisions of the study}

A useful preliminary to the study of perfect fluid flows is provided by the mechanics of fluid at rest (or *fluid statics*) since in both cases tangential stresses are absent and even inertia effects do not appear in fluid statics. Problems of fluid statics are also of considerable intrinsic interest and theoretical importance as well, as the entire development that follows can be seen as a sequence of successive generalizations of the mechanics of fluid in equilibrium.

_{statics: pressure forces related to fluid *density*}

From the outset it is essential to specify the condition (or *state*) of a fluid. In addition to the pressure, one more thermodynamic variable is necessary in general, for this purpose. It is a familiar fact that, when either the temperature or the specific volume is determined, together with the pressure, the other is determined by the equation of state of the substance. In fluid mechanics it is customary to employ the *density* (mass per unit volume) in place of the specific volume, and for liquids this quantity is taken to be constant. Other variables will also be introduced as needed, as well as certain physical properties of fluids that influence fluid motions.

_{fluids are thermodynamic substances}

2 PASCAL'S PRINCIPLE; PROPERTIES AND UNITS OF PRESSURE

The pressure is regarded as known when a relationship has been established that furnishes a one-to-one correspondence between each fluid particle and the pressure associated with it at every instant of time. When the purpose is to emphasize that the pressure p is represented as an explicit function of particle position and time, the "functional" notation will often be employed; it is written as

$$p = p(x, y, z, t) \tag{2.1}$$

for example, where the right side of (2.1) stands for a certain function of the Cartesian coordinates x, y, z, and the time t, all referred to as *independent variables*. Then each set of values $(x, y,$

4 FLUID STATICS

pressure is the dependent variable, coordinates and time are independent

z, t) determines a definite value of the pressure, here termed the *dependent variable* of the function $p(x, y, z, t)$.

The form of (2.1) implies that precisely one value of the pressure is assigned to each fluid particle at any instant of time. Although this is entirely consistent with the thermodynamic usage, in which a single value of each thermodynamic variable is associated with the substance at every point, the same property may seem less obvious when the pressure is introduced as an internal stress. Specifically it might be thought that different values of the normal force per unit area could coexist at a single point, depending on the orientation in space of the surface element on which the stress is considered to act. It is therefore necessary to show that this possibility is ruled out by the definition of the pressure taken in conjunction with Newton's laws, demonstrating thereby that the pressure is a scalar quantity (i.e., fully specified by a single magnitude at each point of space and instant of time).

For this purpose the equilibrium of a small fluid volume element in the form of a triangular prism of unit depth having horizontal end surfaces of dimensions dh_1, dh_2, and dh_3, as shown in Fig. I-2 is considered. The corresponding pressures, all co-planar, are provisionally denoted p_1, p_2, p_3, and it is required to show that they are identical with each other. The three horizontal forces on the sides of the prism are then $p_1\,dh_1$, $p_2\,dh_2$, and $p_3\,dh_3$, each perpendicular to one side of the prism. The fact that the volume element is in equilibrium means, according to Newton's second law, that the vector diagram of the forces forms a closed figure,

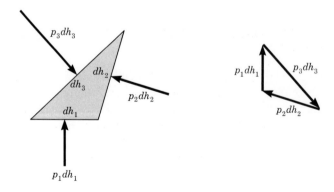

Fig. I-2 Equlibrium of fluid element: hydrostatic pressure independent of direction (pressure is a scalar quantity); (a) fluid element under pressure; (b) triangle of forces.

here a triangle. But, since each side of the force triangle is perpendicular to one side of the prism, the force triangle is geometrically similar to the prism surfaces having dimensions dh_1, dh_2, and dh_3. Comparison of the two triangles then shows that the pressures p_1, p_2, and p_3 must be equal to each other, so that, when their common value is denoted by p,

$$p_1 = p_2 = p_3 = p$$

pressure is a scalar quantity

This result is known as Pascal's principle.

An additional requirement for the equilibrium of the elementary fluid volume is that the moments of the external forces, in this case the pressures exerted by surrounding fluid, must be zero when referred to any axis. That this condition is also satisfied is now demonstrated by simply noting that the distributed pressures can be replaced by the resulting concentrated forces coincident with the normal bisectors of the triangle. When it is recalled that all three normal bisectors of an arbitrary triangle meet at a common point, the vanishing of moments about axes through this point (and all other points as well) follows at once.

Exercise 1 The pressure in a two-dimensional fluid motion is independent of the time; it is given by

$$p(x, y) = p_0 - 36(x^2 + y^2)$$

where p_0 is a positive constant. (a) Determine the coordinates x, y of the point where the pressure is a maximum. (b) At what distance from the origin is the pressure equal to one-half of the maximum value? (c) Take $p_0 = 1200$ and plot the pressure variation $p(x)$ in the range $-4 \leq x \leq +4$ when y takes each of the values 0, 2, and 4.

Exercise 2 The pressure variation on a circular cylinder $r = a$ is given as a function of the plane polar coordinate angle θ by the function

$$p(\theta) = p_0 + \rho \frac{U^2}{2} \{1 - 4 \sin^2 \theta\}$$

(a) Indicate the maximum and minimum values of the pressure on a sketch of a normal section of the cylinder. (b) Evaluate the resultant pressure force F, per unit length of cylinder, on the upper half of its surface for which $0 \leq \theta \leq \pi$, by summing (integrating) components of pressure normal to the $\theta = 0$ axis:

$$F = \int_0^\pi p(\theta) \sin\theta \, a \, d\theta$$

(The symbols ρ and U denote fluid density and flow speed, respectively, treated as constants.)

Exercise 3 The coordinates (x, y) of the vertices of an arbitrary triangle are given by $(0, 0)$, $(a, 0)$, and (b, c). Find the point of intersection of the three normal bisectors as the simultaneous solution of the equations of these lines. As a check on your result, verify the reduction to the trivial case of the right triangle obtained by setting $b = a$ or $b = 0$, so that coordinates of the common intersection point are $(a/2, c/2)$.

fundamental units: length L, mass M, and time T: dimensions

Among the various quantities and concepts encountered in mechanics three are by common consent taken as basic and independent: *length*, *mass*, and *time*. The corresponding fundamental units L, M, and T serve, so to speak, as Nature's independent variables by permitting other quantities to be given in terms of them. Thus velocity is a ratio of length to time, force is mass multiplied by acceleration, pressure is force per unit area, and so on. The "dimensions" of pressure are shown as

$$\text{Pressure:} \quad \text{force} \div \text{area} \quad M L^{-1} T^{-2}$$

British and metric systems: units of force: dimensions MLT^{-2}

In the British and metric systems, respectively, the units of force are the pound and the dyne, related to the corresponding mass, length, and time units by the equations that express Newton's second law of motion:

$$1 \text{ pound} = 1 \text{ slug} \frac{\text{ft}}{\text{sec}^2} \qquad 1 \text{ dyne} = 1 \text{ gm} \frac{\text{cm}}{\text{sec}^2}$$

It follows that the fundamental units of pressure in the two systems are slug/ft-sec² and gm/cm-sec², although it is more usual to express them in terms of force units as pound/ft² and dyne/cm². When a conversion of the British length scale is made, the former expression becomes pound/in², written psi. The value of the International Standard Atmosphere sea level pressure, at temperature $T_0 = 15°C$ (59°F), is indicated for future reference as

$$p_0 = 14.691 \text{ psi} = 1.0132 \times 10^6 \frac{\text{dyne}}{\text{cm}^2}$$

3 UNIFORM GRAVITY: HYDROSTATICS AND AEROSTATICS

balance of pressure and gravity forces

The pressure can be calculated at any internal or boundary point of a fluid whenever an adequate statement of external conditions is given. In fluid statics the vanishing of all velocities and accelerations in a particular frame of reference means that pressure forces are balanced everywhere by any other forces that may be acting. Gravitational forces are by far the most common and hence of greatest practical importance. The reason is that every fluid particle is attracted by all other masses of the universe in accordance with *Newton's law of gravitation:*

$$F = k^2 \frac{m_1 m_2}{r^2} \tag{3.1}$$

The meaning of (3.1) is that between each pair of particles of masses m_1 and m_2 a force of mutual attraction is present; it is proportional to the product of their masses and inversely proportional to the square of their separation distance. In terrestrial applications where the Earth's attraction is predominant, a uniform value of the gravity force is adopted; pressure calculations will be shown to be simplified accordingly.

The coefficient k^2 in (3.1), called the universal gravitational constant, has been determined experimentally to have the numerical value

$$k^2 = 6.670 \times 10^{-8} \text{ cm}^3 \text{ gm}^{-1} \text{ sec}^{-2}$$

universal constant of gravitation

in the cgs system of units. The Earth's attraction on each unit of mass at sea level is denoted by g and treated as a constant and is expressible in units of acceleration. (It actually varies slightly from one point to another on the Earth's surface.) The commonly accepted value is

$$g = 981 \text{ cm sec}^{-2} = 32.2 \text{ ft sec}^{-2} \tag{3.2}$$

the Earth's gravity at sea level

and it is supposed that the force is everywhere directed downward toward the Earth's center. An accurate determination of g at any point according to (3.1) would involve the sum of a very large number of terms, one for each mass particle m_2 of the Earth that attracts unit "test mass" $m_1 = 1$ situated at the point in question. Such a calculation is clearly so arduous as to be utterly impractical. A crude approximation sufficient for many purposes that depends on the roughly spherical form of the Earth will be shown below.

Exercise 4 Calculate the gravity force per unit mass at sea level directly from (3.1) by considering a test mass of 1 gm ($m_1 = 1$) and supposing that the entire mass of the Earth E is concentrated at its center. * (Take $E = 5.97 \times 10^{27}$ gm and the Earth's radius as $a = 6.378 \times 10^8$ cm.) Compare with (3.2) and suggest how the calculation might be improved.

Exercise 5 Use the same assumptions to evaluate the Earth's gravity force at height h above sea level by setting $r = a + h$ when $h = 10$ m, when $h = 10$ m^6, and when $h = a$. Give the justification for considering g to be constant in ordinary terrestrial applications.

*The basis for this assumption is given in Appendix B, Eq. (B. 10).

8 FLUID STATICS

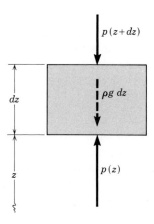

Fig. I-3 Gravity and pressure forces on fluid element in static equilibrium.

The balance of pressure and gravity forces is expressed by equating to zero the sum of their components in each of three mutually orthogonal directions. If the z-axis is taken to be directed upward opposite to gravity, the absence of gravity force components in the horizontal x- and y-directions is seen to imply the absence of pressure variations in these directions, so that the pressure may be written as a function of z only, i.e., $p = p(z)$. A right-cylindrical fluid volume element of altitude dz, extending from z to $z + dz$ (Fig. I-3) is now considered. The pressure on its lateral surfaces is directed horizontally, so that the summation of vertical forces must include, in addition to the gravity force, only the pressure $p(z)$ exerted by fluid beneath the element and the pressure $p(z + dz)$ by the fluid above it. By taking the cross section of the cylindrical element to be unity, the sum of these forces is found to be

$$p(z) - p(z + dz)$$

in the upward direction. From the assumed continuity of the pressure as a function of position, it follows that we can write with sufficient accuracy

$$p(z + dz) = p(z) + \frac{dp(z)}{dz} dz \qquad (3.3)$$

when dz is small (the right-hand side of (3.3) is recognized as the first two terms of the Taylor series expansion of the pressure, neglecting a remainder that vanishes like the square of dz). The resultant vertical pressure force on the element is thus seen to be

$$-\frac{dp}{dz} dz$$

infinitesimal analysis; the fluid "element"

When ρ denotes the density of the fluid (mass per unit volume), the mass of fluid within the element of volume $dz\,1$ is seen to be $\rho\,dz\,1$, while the Earth's gravitational attraction force acting on it is simply g times this value: $-\rho g\,dz\,1$, where the negative sign again indicates downward direction. The equilibrium condition now is

$$-\left\{\frac{dp}{dz} + \rho g\right\} dz = 0$$

or finally

$$\frac{dp}{dz} = -\rho g \qquad (3.4)$$

fundamental equation of hydro- and aero- statics

since the thickness dz is nonzero. Equation (3.4), relating the pressure variation in the vertical direction to the density and the gravity constant, is referred to as the fundamental equation of fluid statics. The pressure itself is obtained from (3.4) by means of single integration with respect to the z-coordinate when the density is known as a function of that variable. It is particularly to be noted that (3.4) applies both to liquids and to gases, since the derivation of this equation has placed no restriction on the density and this quantity may therefore be either a constant or a variable.

Although (3.4) contains the crucial statement of fluid equilibrium from which pressure calculations are made below, a formally complete expression is obtained by supplementing (3.4) with the statements referring to possible variations in the horizontal x- and y-directions. From what has gone before it is clear that these constitute the first two of the following set of three equations:

$$\left.\begin{array}{l} \dfrac{\partial p}{\partial x} = 0 \\[6pt] \dfrac{\partial p}{\partial y} = 0 \\[6pt] \dfrac{\partial p}{\partial z} = -\rho g \end{array}\right\} \quad (3.4a)$$

three-dimensional form, gravity acting in $-z$ direction

where the partial derivative notation is adopted simply to emphasize the general three-dimensional character of the pressure variations (it will be seen that the set (3.4a) provides a suitable basis for extension to the discussion of arbitrary fluid motions).

Hydrostatic pressure is computed from (3.4) by noting that the liquid density is constant in time and constant from one point to another in space. Integration is thus immediate; it gives

$$p(z) = -\rho g z + \text{const} \qquad (3.5)$$

hydrostatics: ρ = const and g = const

where the additive constant is readily evaluated by specifying one level at which the pressure is taken as zero. When water stands to a height H in an open vessel, for example, it is natural to take the origin of the z-coordinate at the bottom of the vessel, so that the "free surface" is situated at $z = H$ and the pressure is taken there as zero by writing

$$p(z) = \rho g\{H - z\} \tag{3.6}$$

hydrostatic pressure is proportional to depth below free surface

The interpretation of (3.6) is immediate, indicating that the hydrostatic pressure increases linearly with depth $H - z$ below the free surface. An equivalent reading of (3.6) is that the pressure at level z is simply the weight of the fluid column of unit cross section above z. As an explicit formula for the pressure, (3.6) is employed directly for calculating forces in a manner analogous to Exercise 2. Simplified computation techniques based on the linearity of the pressure-depth relationship (3.6) are demonstrated in the Supplement S1 at the end of this chapter.

Exercise 6 Atmospheric pressure is, of course, *not* absolute zero in value, and (3.6) must therefore be understood to give the *excess* of the pressure over the atmospheric value (i.e., the "gauge" pressure). Indicate the justification for regarding (3.6) as a good approximation for the absolute pressure at substantial depths below the free surface of a water-filled vessel open to the atmosphere. HINT: Recall that the density of water is of the order of one thousand times greater than that of air at standard sea level conditions.

aerostatics: $\rho \neq$ const and $g =$ const

In *aerostatic* calculations based on (3.4) the density appears as a variable and the pressure cannot be evaluated until the density is specified in some manner. The most direct procedure would depend on substituting in (3.4) a definite function of z, for example,

$$\rho = \rho(z) \tag{3.7}$$

but the density is neither measured directly nor, consequently, quoted in the form (3.7) as a rule. More commonly air is regarded as an ideal gas obeying the familiar thermodynamic equation of state

ideal gas equation of state

$$p = \rho R T \tag{3.8}$$

and a supplementary statement is given concerning the (absolute) temperature T. Then (3.4), combined with (3.8), is seen to be suitable for evaluating the pressure when, for example, the temperature is a given function $T(z)$.

The simplest statement of this type is that the temperature is a constant, for example, $T = T_0$. Then the density can be eliminated from (3.4) and (3.8) to obtain

$$\frac{dp}{dz} = -\frac{p}{RT_0} g \tag{3.9}$$

The variables p and z are easily separated from each other in (3.9); the resulting integration yields

$$p(z) = p_0 \exp\left\{\frac{-gz}{RT_0}\right\} \qquad (3.10)$$

pressure diminishes exponentially with height

where p_0 is the pressure at level $z = 0$. Equation (3.10), known as Halley's law, is useful when limited thicknesses of the atmosphere are considered.

Exercise 7 An open tank of water is filled to depth H, where the pressure is taken as zero. Calculate the hydrostatic pressure force F on unit breadth of one of its vertical walls (Fig. 1-4), and locate the line of action of the equivalent concentrated pressure force.

SOLUTION Writing the element of force acting on infinitesimal area dA as $dF = p(z)\,dA$, obtain the total force by integration as

$$F = \int_A p(z)\,dA \qquad (3.11)$$

where it is convenient to express dA as $dz\,1$ so that integration limits are 0 and H. According to (3.6) then

$$F = \rho g \int_0^H (H - z)\,dz = \rho g \frac{H^2}{2} \qquad (3.12)$$

Recall that the line of action of the equivalent concentrated force is specified by the coordinate value $z = l$, where the force moment Fl equals the moment of the distributed pressures.
Hence

$$Fl = \rho g \int_0^H z(H - z)\,dz = \rho g \frac{H^3}{6} \qquad l = \frac{H}{3} \qquad (3.13)$$

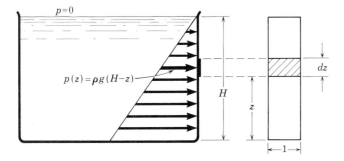

Fig. I-4 Hydrostatic pressure variation in an open tank.

Exercise 8 A circular pipe 5 ft in diameter is placed with its axis horizontal and one end capped by a vertical flat plate. Determine the hydrostatic force on the plate if the pipe is filled to a depth of 4 ft and the remaining space is open to the atmosphere. Take the specific weight of water as $\rho g = 62.4$ lb ft^{-3}, i.e., water density $\rho = 1.94$ slug ft^{-3}.

4 COMPRESSIBILITY: THE LIQUID-GAS DISTINCTION

When the pressure of air at standard conditions is increased by 1 psi while the temperature remains constant, the air density increases by about 7 percent. The corresponding pressure increase in water is accompanied by a substantially smaller increase of the density, the two cases differing by a factor of roughly 200. The *compressibility* of air and other gases is on the whole so much greater than that of water and other liquids that the reference to liquids as "incompressible fluids" is at least useful despite its inexactness. The increased importance of the concept of fluid compressibility with relationship to high-speed flows justifies a closer examination of the question at this point, which leads to the introduction of a useful quantitative parameter of compressibility.

> no actual fluid is either an ideal gas or a fully incompressible liquid

The concepts of pressure and density as already introduced are sufficient by themselves for indicating the degree to which any substance is compressible in the ordinary sense of the word. Thus the incremental density increase accompanying a small change in pressure might be considered in the limit of vanishing pressure variation, leading to the formation of a derivative. According to customary usage, in fact, the *fluid compressibility* is defined by this ratio divided by the density,

$$K = \frac{1}{\rho}\frac{\partial \rho}{\partial p} \tag{4.1}$$

the dimensionality of the parameter K being somewhat reduced by the introduction of the factor $1/\rho$. The partial derivative notation again emphasizes that the density of a thermodynamic substance is not a function of the pressure alone but depends on *two* thermodynamic variables, such as the pressure and the absolute temperature T. In order to be more specific therefore, the term *isothermal compressibility* is used when the derivative in (4.1) is understood to be evaluated at constant temperature:

> compressibility K depends on detailed thermodynamic behavior of a substance

$$K_T = \frac{1}{\rho}\left(\frac{\partial \rho}{\partial p}\right)_T \tag{4.1a}$$

where the meaning of the subscript on the derivative is that the temperature is treated as a constant in the differentiation with

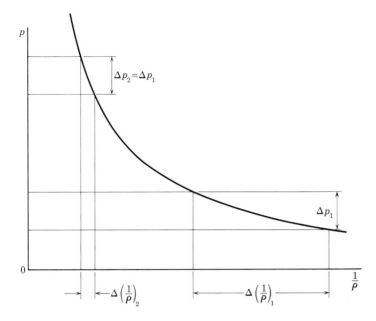

Fig. I-5 Ideal gas isothermal compressiblity varies inversely as pressure.

respect to pressure. The parameter (4.1a) can evidently be evaluated unambiguously when the characteristic relationship between the three variables shown is given; it is called the thermodynamic equation of state.

For an ideal gas, substitution from (3.8) into (4.1a) gives

$$K_T = \frac{1}{p} \quad \text{(ideal gas)} \qquad (4.2)$$

compressibility is greatest at *low* pressures

which shows that the isothermal compressibility is the greater when the pressure is smaller — the greatest change in volume of a toy balloon occurs at the beginning of the inflation process. When the pressure is elevated, however, the fractional compression is much smaller, as seen by examining the isotherms that appear as hyperbolae when plotted in a plane where the Cartesian coordinates are taken as $1/\rho$ and p, as in Fig. I-5.

Exercise 9 A metal container of internal volume 1,000 ft³ is filled with water at atmospheric pressure. How much more water must be pumped in to increase the pressure 100 psi? Allowing for the expansion of the vessel, assume that the equivalent fluid compressibility K of the fluid is 4×10^{-6} in² lb⁻¹. HINT: Approximate the derivative in (4.1) by a ratio of differentials.

14 FLUID STATICS

Because liquids behave very differently from ideal gases, a suitable representation of the compressibility (4.1a) in liquids therefore requires the introduction of a more accurate equation of state. The equation of van der Waals

non-ideal gas: van der Waals' equation of state

$$(p + a\rho^2)(1 - \rho b) = \rho RT \qquad (4.3)$$

is commonly adopted because, despite its simplicity, it brings out all the essential mechanical properties of substances that possess both liquid and gaseous states. The numerically small constants a and b account for the facts that the molecules themselves occupy a finite volume and are subject to intermolecular forces even when not in collision with each other. The form of the isotherms (Fig. I-6) shows that the ideal gas equation (3.8), recoverable from (4.3) by setting $a = b = 0$, is an approximation to (4.3) when both pressures and densities are small. Of particular interest with regard to the isotherms shown is the occurrence of both negative and positive slopes in the region closest to the origin of coordinates. These are separated from the isotherms corresponding to higher temperatures by a unique isotherm identified with the "critical temperature" T_C and having negative slopes everywhere except at one point, where it is zero. The critical state defined in this manner provides a reference standard not to be found in ideal gases, and

critical state

Fig. I–6 van der Waals' isotherms — small compressibility (liquid) when $T \gg T_C$ and $p \ll p_C$.

permits the recognition of the region of small compressibility (i.e., liquid state) as belonging to the part of the plane that represents high pressures and high densities:

$$p > p_C \qquad \frac{1}{\rho} < \frac{1}{\rho_C}$$

The inequalities allow the liquid state to be identified as a continuation of the gaseous state represented by the isotherms in those regions of the plane where they are similar in form to the ideal gas case shown in Fig. I-5. It will be seen that the constants a and b can be evaluated on the basis of the geometrical properties already mentioned; thus they can be expressed in terms of the *critical values* of the pressure and the density at the uniquely defined point on the critical isotherm. Laboratory determinations of the critical values for different gases then permit numerical evaluation of the constants and hence physical characterization of the liquid state in definite terms. In Chapter XI it is also shown that an elementary mathematical model of the real gas also leads to evaluation of the constants a and b, confirming that van der Waals substance *incompressibility* ($K_T = 0$) corresponds to close packing of the molecules in conformity with accepted notions of the molecular structure of liquids.

liquid behavior: large pressure and large density values

Exercise 10 Determine the critical pressure p_C and critical density ρ_C of a van der Waals substance in terms of the constants a, b, and R by noting that the critical point represents the only state for which the vanishing isotherm slope coincides with a point of inflection of the isotherm, i.e., where

$$\left(\frac{\partial p}{\partial \rho}\right)_T = 0 \qquad \left(\frac{\partial^2 p}{\partial \rho^2}\right)_T = 0 \qquad (4.4)$$

Substitute for $p(\rho, T)$ from (4.3) and solve the resulting pair of algebraic equations (4.4) for p_C and ρ_C.

Exercise 11 Obtain the general expression for the isothermal compressibility of a van der Waals substance according to the definition (4.1a). Verify that the result reduces to (4.2) when $a = b = 0$, and also indicate the conditions for strictly liquid behavior in the sense of vanishing compressibility $K_T = 0$. HINT: Recall that $(\partial \rho / \partial p)_T = 1/(\partial p/\partial \rho)_T$.
  Ans. $K_T = 0$ when $\rho = 1/b = 3\rho_C$.

to specify the liquid behavior

It will be seen in Chapter VIII that the fundamental parameter of high-speed flow dynamics is closely related to the *isentropic* compressibility K_S obtained by considering the density variation with pressure while the *entropy s* is held constant:

$$K_S = \frac{1}{\rho}\left(\frac{\partial \rho}{\partial p}\right)_S \tag{4.1b}$$

*Exercise 12 Show that for an ideal gas the ratio of isothermal and isentropic compressibilities is a constant, equal to the ratio γ of specific heats c_p and c_v; that is,

$$K_T = \gamma K_S, \quad \text{where} \quad \gamma = \frac{c_p}{c_v}$$

(the same result may be established for an arbitrary substance as well). HINT: Recall that

$$T\,ds = c_v dT + p\,d\left(\frac{1}{\rho}\right)$$

5 PRESSURE DISTRIBUTION IN THE EARTH'S ATMOSPHERE

The chief components of ordinary air are nitrogen and oxygen, for which the critical temperature and pressure are $-147°C$, 35 atmospheres (atm), and $-118.8°C$, 49.7 atm, respectively. It follows that, at least within the lower parts of the atmosphere, air may be reasonably treated as an ideal gas. The properties of a horizontally stratified atmosphere at rest are then subject to the static equilibrium requirement (3.4.) as well as to the ideal gas equation of state (3.8):

$$\frac{dp}{dz} = -\rho g \qquad p = \rho RT \tag{5.1}$$

As mentioned earlier, the appearance of the three state variables p, ρ, and T in the set of two equations (5.1) indicates the necessity for one additional statement about these quantities. When regions of appreciable thickness are considered, the constant-temperature assumption that led to Halley's law (3.10) should be replaced by a more realistic condition. Of nearly equal simplicity but far greater physical significance is the assumption of temperature decreasing uniformly with altitude; that is,

linear temperature lapse rate α = const

$$T(z) = T_0 - \alpha z \tag{5.2}$$

*Exercises and discussions denoted by an asterisk, involving principles drawn from related fields like thermodynamics and mathematics, may be omitted in a first reading.

where T_0 is the sea level temperature corresponding to $z = 0$ and α is a constant *temperature lapse rate*.

Exercise 13 Eliminate the density from the pair of equations (5.1) and use (5.2) to effect the separation of pressure and position variables in the differential equation

$$\frac{dp}{p} = -\frac{g}{R}\frac{dz}{T_0 - \alpha z}$$

Observe that each side of the preceding equation is the differential of a logarithm term and establish the pressure variation as

$$p(z) = p_0 \left\{ 1 - \frac{\alpha z}{T_0} \right\}^{+g/R\alpha} \tag{5.3}$$

Also determine the density variation $\rho(z)$ and hence the "height" H of the atmosphere, i.e., the value of z for which the density becomes zero, $\rho(H) = 0$.

finite "height" of atmosphere

Exercise 14 Standard Atmosphere sea level conditions indicated in Section 2 are supplemented with a specification of the standard lapse rate

$$\alpha = 0.003566°\text{F ft}^{-1}$$

standard atmosphere lapse rate α

Recall that the static pressure at any altitude measures the weight of the unit column of fluid above that level, and determine the height $z_{1/2}$ beneath which half of the atmosphere's mass is found. Take the gas constant for air as

$$R = 1718 \text{ ft}^2 \text{ sec}^{-2} \text{ (°R)}^{-1} = 2.884 \times 10^6 \text{ cm}^2 \text{ sec}^{-2} \text{ (°K)}^{-1}$$

gas constant for air

and note that the absolute temperature T_0 at sea level is obtained from the Fahrenheit (or Celsius) value by the addition of a constant:

$$59°\text{F} = 459.7 + 59 = 518.7°\text{R} \quad \text{(Rankine)}$$
$$15°\text{C} = 273.1 + 15 = 288.1°\text{K} \quad \text{(Kelvin)}$$

standard atmosphere temperature at sea level, and density

Also evaluate the sea level air density ρ_0 by substituting the preceding numerical values in the ideal gas equation of state, and compare with the Standard Atmosphere value $\rho_0 = 0.002377$ slug ft^{-3}.

Exercise 15 Show that the density variation according to Halley's law is of the same form as the pressure variation; that is,

$$\rho(z) = \rho_0 \exp\left\{-\frac{gz}{RT_0}\right\} \tag{5.4}$$

in contrast to the uniform lapse rate atmosphere corresponding to (5.2) and (5.3). Note that in this case there is no finite "height" H at which the density is zero, and calculate $z_{1/2}$ and compare with the value obtained in the preceding exercise. As a practical estimate of the effective upper limit of the atmosphere according to Halley's law calculate also $z_{9/10}$.

18 FLUID STATICS

margin note: barotropic fluid, polytropic index n

*Exercise 16 (a) Determine the pressure as a function of elevation in the *barotropic* atmosphere model characterized by the pressure-density relationship

$$\frac{p}{p_0} = \left(\frac{\rho}{\rho_0}\right)^n \tag{5.5}$$

where the *polytropic index n* is constant, $n \neq 1$, and the relationship (5.5) replaces (5.2). (b) Evaluate the height of the *adiabatic* atmosphere corresponding to taking n as the ratio of specific heats of an ideal diatomic gas, $n = \gamma = 1.4$, using standard sea level conditions. (c) Show why Halley's law is *not* recovered by setting $n = 1$ in the result of part (a).

margin note: adiabatic atmosphere, $n = 1.4$

6 LARGE FLUID MASSES: PLANETARY AND STELLAR STATICS

It is clear from the findings of Exercise 5 that variations of the Earth's gravitational attraction should not be neglected over distances comparable with the Earth's dimensions. As a final example of the balance of gravity and pressure forces, conditions within the Earth itself are now considered, due allowance being made for variations of both density and gravity. An extension of (3.4) is obtained that provides the basis for studying the mechanics of all the large quasi-spherical masses occurring in Nature (notably the planets and the stars). A basic difference from the preceding calculations is that in the present case the pressure at *interior* points of the attracting body is examined, whereas in ordinary hydro- and aerostatics the gravity force has been treated as purely external to the fluid element under consideration.

margin note: planetary and stellar statics $\rho \neq$ const and $g \neq$ const

Because of the more or less spherical form of the free surface it is apparent that spherical coordinates are the most appropriate, and a tempting idealization is to regard the configuration as spherically symmetrical with regard to all its properties, at least as a first approximation. Then the pressure and the gravity force will each depend only on r, the distance from the center, so the pressure can be expressed as a function $p(r)$ of a single variable. Unlike the argument leading to (3.4), however, a thin spherical shell of radius r and thickness dr is not an appropriate fluid element, because the sum of pressure (or gravity) forces in any direction vanishes on each surface separately. The difficulty is easily removed, however, by considering a thin *hemispherical* shell element, the axis of which defines a preferred direction along which the different forces have nonvanishing resultants (Fig. I-7)

A spherical mass of radius a is considered, and the pressure $p(r)$ at an arbitrary interior point $r < a$ has the same value at all points

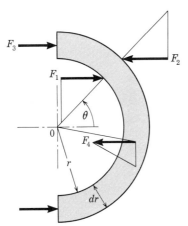

Fig. I-7 Equilibrium of pressure and gravity forces on hemispherical shell

of the hemispherical surface of radius *r* passing through it. The force elements are everywhere normal to the surface, hence differently directed at different points of the surface. It is a consequence of the axial symmetry of a hemispherical surface that all force components normal to the axis have a zero sum when integrated over the surface, and it is sufficient to calculate the component of pressure forces parallel to the axis. This is obtained as $p(r)\cos\theta$, where θ is the polar angle, times the surface on which it acts. All points on the circle of radius $r\sin\theta$, perimeter $2\pi r \sin\theta$, experience the same pressure force component, and the area is given by multiplying by the factor $r\,d\theta$ representing the arc length element on the hemisphere surface. The element dF_1 of pressure force on the surface *r* is then

$$dF_1 = p(r)\cos\theta\, 2\pi r \sin\theta\, r\, d\theta$$

and summation over values of θ in the interval $(0, \pi/2)$ gives

$$F_1 = 2\pi \int_0^{\pi/2} r^2 p(r) \sin\theta \cos\theta\, d\theta = \pi r^2 p \qquad (6.1)$$

when it is noted that

$$\int_0^{\pi/2} \sin\theta \cos\theta\, d\theta = \tfrac{1}{2}$$

The opposing pressure force on the outer surface of the shell of

pressure *within* the attracting mass itself

choice of a fluid element

force balance

thickness dr, at distance $r + dr$ from the center, is of the same form as (6.1) and is obtained from it by replacing r by $r + dr$ in the preceding expressions. It is readily shown that, to the first order of small terms dr, the affected terms can be written

$$r^2 p + \frac{d}{dr}(r^2 p)\, dr$$

so that the axial component of pressure forces on the outer surface, denoted F_2, is obtained as

$$F_2 = -\left\{\pi r^2 p + \frac{d}{dr}(\pi r^2 p)\, dr\right\} \qquad (6.2)$$

where the sign reversal corresponds to the opposing direction of axial force components.

The pressure $p(r)$ also acts on the edge of the hemispherical shell, the differential surface area of which is $2\pi r\, dr$; hence the force is of the same order as the sum of (6.1) and (6.2). When this force is denoted F_3, we obtain

$$F_3 = p 2\pi r\, dr \qquad (6.3)$$

and the sum of forces (6.1–6.3) gives the resultant hydrostatic pressure on the hemisphere element. (In similiar manner we obtained the balance of forces on a rectangular element in section 3 above.) The total pressure force is now

$$-\pi r^2 \frac{dp}{dr}\, dr \qquad (6.4)$$

When the gravitational attraction on unit mass at distance r from the sphere center is denoted $f(r)$, its component parallel to the symmetry axis of the hemispherical shell is $f(r)\cos\theta$ as before, and the summation proceeds in the same manner as for the pressure forces.

Exercise 17 Show that the axial component of the elementary gravity force on the hemispherical shell is

$$dF_4 = -f(r)\cos\theta\, \rho r\, d\theta\, dr\, 2\pi r \sin\theta \qquad (6.5)$$

Exercise 18 An elementary result of potential theory shows that the attraction of all parts of a sphere on unit mass situated at distance r from its center can be written as

$$f(r) = \frac{k^2 M(r)}{r^2}$$

where $M(r)$ is the mass of the fluid interior to the sphere passing through r, (cf. (3.1) and (B.1) of Appendix B). Recall that

$$M(r) = \int_0^r \rho(r) 4\pi r^2 \, dr$$

and show that the force F_4 is given by

$$F_4 = -k^2 \rho \pi \, dr \left\{ \int_0^r \rho(r) 4\pi r^2 \, dr \right\} \quad (6.6)$$

Adding (6.4) and (6.6) and equating their sum to zero expresses the condition of static equilibrium in the form

$$0 = r^2 \frac{dp}{dr} + 4\pi k^2 \rho \int_0^r \rho r^2 \, dr \quad (6.7)$$

a form analogous to (3.4), which resembles (6.7) still more closely when the gravity term g in (3.4) is written in more basic form as in (3.1).

The slight awkwardness of an equation like (6.7) containing both a derivative and an integral with respect to the same variable is removed by recalling the reciprocal nature of the two operations. By isolating the integral and differentiating with respect to its upper limit, the integrand is recovered and, when the same step is applied to the first term of (6.7) multiplied by the reciprocal of the coefficient of the integral (recall that the density may be dependent on r), the result is

$$\frac{1}{r^2} \frac{d}{dr}\left(\frac{r^2}{\rho} \frac{dp}{dr}\right) + 4\pi k^2 \rho = 0 \quad (6.8)$$

fundamental equation of planetary and stellar statics ($k^2 = 6.670 \times 10^{-8}$ cm^3 gm^{-1} sec^{-2})

Equation (6.8) is taken as the final form of the fundamental equation of planetary and stellar statics. It may also be derived from the balance of forces on fluid contained both in the spherical shell and in a central cone of small solid angle.

second-order equation for pressure in terms of density and radial distance r

Exercise 19 Note that, when the density is constant, (6.8) contains the sum of two terms each of which is constant. Prove that the pressure is a quadratic function of the radial distance r in this case.

Exercise 20 Show that the mean density of the Earth is approximately $\rho = 5.5$ gm cm^{-3}, and use the result of the preceding exercise to estimate the central pressure $p(0)$ in a liquid sphere of mass and radius equal to the Earth values given in Section 3. Compare the result with the accepted value 3.9×10^6 atm (1 atm = 1.01×10^6 dyne cm^{-2}).

Ans. $p(0) = \frac{1}{2}\rho g a$.

*7 PRESSURE AT INTERIOR POINT OF ARBITRARY LARGE FLUID MASS — POISSON'S EQUATION

A more fundamental development of (6.8) proceeds by expressing the balance of pressure and gravity forces in terms of the gravity potential U (see Appendix B) in vector form as

$$\frac{1}{\rho} \nabla p = -\nabla U \qquad (7.1)$$

where the components of gravity force are represented as the corresponding components of the gradient $-\nabla U$ of the negative of the potential U, in Cartesian form (3.4a) or in any other coordinate system. When it is shown that the divergence $\nabla \cdot (-\nabla U)$ of the gravity force, written $-\nabla^2 U$, is given by

equation satisfied by gravitational potential at interior point of fluid mass

$$\nabla^2 U = 4\pi k^2 \rho \qquad (7.2)$$

where ρ is the density of the fluid at the point where $\nabla^2 U$ is evaluated, it follows from (7.1) that

reduces to (6.8), derivation does not require hemispherical shell elements

$$\nabla \cdot \left(\frac{1}{\rho} \nabla p\right) + 4\pi k^2 \rho = 0 \qquad (7.3)$$

without any assumption of radial symmetry such as was employed in obtaining (6.8). It is readily established that the first term of (6.8) is exactly the expression shown in (7.3) when this symmetry is present, of course.

to establish Poisson's equation (7.2)

To show that the gravity potential U satisfies the partial differential equation (7.2), known as Poisson's equation, at an interior point of the attracting mass, and hence to establish (7.3), the entire volume V of fluid is regarded as decomposed in two parts. Surrounding the point at which the pressure is to be evaluated, taken as the origin from which distances r are now measured, is a small sphere of radius ϵ, entirely within V; the second part encloses the first and is bounded by surfaces S and S_ϵ where S_ϵ is the boundary surface of the small spherical volume. An essential property of both bounding surfaces is that the surface integral of the normal component of the vector $\mathbf{e}_r/r^2$ is given by the same constant 4π in each case:

decompose the fluid volume into two parts

$$\int_{S_\epsilon} \frac{\mathbf{e}_r}{r^2} \cdot \mathbf{n} \, dS_\epsilon = \int_S \frac{\mathbf{e}_r}{r^2} \cdot \mathbf{n} \, dS = 4\pi \qquad (7.4)$$

no matter what form the surface S may have.

The first integral of (7.4) is evaluated by noting that $r = \epsilon$ at each point of the surface, and that the surface element dS_ϵ can be written as $\epsilon^2 \, d\Omega$, where $d\Omega$ is the solid angle subtended by the element from $r = 0$. The radial unit vector $\mathbf{e}_r$ and the outward unit normal vector n are parallel on S_ϵ, and it follows that the value of the integral is 4π because the surface area of the sphere is $4\pi\epsilon^2$, the total solid angle around any point being 4π. That the second integral of (7.4) has the same value follows by considering the volume $V - V_\epsilon$ contained between the two surfaces and noting that the integral of the divergence of $\mathbf{e}_r/r^2$ throughout this volume vanishes:

evaluation of integrals

$$0 = \int_{V-V_\epsilon} \nabla \cdot \left(\frac{\mathbf{e}_r}{r^2}\right) dV = \int_{S+S_\epsilon} \frac{1}{r^2} \mathbf{e}_r \cdot \mathbf{n} \, dS \quad (7.5)$$

as seen, for example, by the evaluation in Cartesian form as

$$\frac{\partial}{\partial x}\left(\frac{x}{r^3}\right) + \frac{\partial}{\partial y}\left(\frac{y}{r^3}\right) + \frac{\partial}{\partial z}\left(\frac{z}{r^3}\right) = 0$$

where $r^2 = x^2 + y^2 + z^2$ and the Cartesian components of $\mathbf{e}_r$ are obtained as terms of the form x/r. It is clear that $r \neq 0$ everywhere in the volume $V - V_\epsilon$ because of the removal of the small sphere. Green's theorem (A.6) permits the volume integral to be expressed in terms of surface integrals over the boundaries S and S_ϵ [the outward normal on the latter is now directed toward the sphere center, opposite to the case of (7.4)]. The vanishing of the sum of two surface integrals permits one of them to be expressed as the negative of the other, hence (7.4).

Considering next the gravitational potential of fluid of density ρ within the sphere of radius ϵ,

$$U = -k^2 \int_{V_\epsilon} \frac{\rho}{r} dV_\epsilon \quad (7.6)$$

which is convergent at $r = 0$ (notice $dV_\epsilon = 4\pi r^2 \, dr$ vanishes more rapidly than r as $r \to 0$), we see that the force per unit mass is given as

$$\mathbf{f} = -\nabla U = -k^2 \int_{V_\epsilon} \rho \frac{\mathbf{e}_r}{r^2} dV_\epsilon$$

By integrating the normal component of $\mathbf{f}$ on the spherical surface S_ϵ,

$$\int_{S_\epsilon} \mathbf{f} \cdot \mathbf{n} \, dS_\epsilon = -k^2 \int_{V_\epsilon} \rho \int_{S_\epsilon} \frac{\mathbf{e}_r}{r^2} \cdot \mathbf{n} \, dS_\epsilon \, dV$$
$$= -k^2 \rho 4\pi V_\epsilon \quad (7.7)$$

according to (7.4) and the mean-value theorem of the integral calculus (i.e., the value of the integral is given by an appropriately chosen mean value of the integrand, which is the density, multiplied by the volume).

By using Green's theorem to express the first integral of (7.4) as the volume integral of the divergence of the force vector **f** expressed in terms of U,

$$\int_{S_\epsilon} \mathbf{f} \cdot \mathbf{n} \, dS_\epsilon = \int_{V_\epsilon} \boldsymbol{\nabla} \cdot (-\boldsymbol{\nabla} U) \, dV_\epsilon = (-\nabla^2 U) V_\epsilon \qquad (7.8)$$

Finally, equating the right sides of the two preceding equations gives exactly (7.2), establishing (7.3) as the equation of static equilibrium at any point inside a large fluid mass held together by self-gravitation and prevented by pressure forces from collapsing entirely into a point.

8 QUASI-STATICS: RELATIVE EQUILIBRIUM IN A ROTATING REFERENCE FRAME

When a fluid is said to be at rest, the absence of motion is always understood with reference to a particular set of coordinate axes or reference frame. Thus a tank of water appears to be stationary to an observer at a fixed point, but it is realized that the Earth's rotational and orbital motions entail nonzero accelerations of the observed and of the observer as well. A logical difficulty is presented, because on the one hand the laws of motion (and hence of statics also) are valid only in "inertial" reference frames having zero acceleration, but no reference frames of this kind are available in which these laws might be tested. Since practical questions of mechanics always deal with relative motion, the purpose of this section is to furnish a criterion for neglecting relative motion and a modification of the equations (3.4a) of fluid statics when its effects are significant. The discussion is limited to uniform rotations both for the sake of simplicity and on account of their great physical importance.

A particle is at rest in the x, y, z coordinate system if each of its coordinates is constant in time. It is supposed that Newton's laws are valid in another coordinate frame, X, Y, Z, with respect to which the first system is in uniform rotation. Gravity and pressure terms are balanced in the X, Y, Z system, and the required equations of relative equilibrium are then obtained by direct transformation.

Exercise 21 Note that uniform rotation ω about the Z-axis is expressed by the coordinate transformation equations

$$\begin{aligned} X &= x \cos \omega t - y \sin \omega t \\ Y &= x \sin \omega t + y \cos \omega t \\ Z &= z \end{aligned} \qquad (8.1)$$

I. 8 / QUASI-STATICS: RELATIVE EQUILIBRIUM IN A ROTATING REFERENCE FRAME

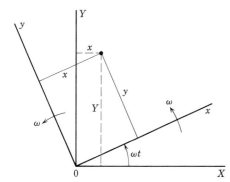

Fig. I-8 Rotating coordinate axes x,y referred to stationary coordinates X,Y.

where ω is constant (see Fig. I-8), and show that the acceleration components obtained by differentiating (8.1) are given, respectively, by

$$-\omega^2 X, \quad -\omega^2 Y, \quad 0 \tag{8.2}$$

when x, y, and z are constant.

Exercise 22 Invert the transformation equations (8.1) to express x, y, and z in terms of X, Y, and Z, and note that the components of acceleration, $\ddot{X}$, $\ddot{Y}$, $\ddot{Z}$, transform in the same manner as the coordinates. Show that the components of acceleration in the rotating coordinate system are

$$-\omega^2 x, \quad -\omega^2 y, \quad 0 \tag{8.2a}$$

components of centripetal acceleration

Exercise 23 Regarding $\partial p/\partial X$ and $\partial p/\partial Y$ as the X- and Y-components of a vector, use the chain rule of differentiation and the inverse transformation found in the preceding exercise to show that

$$\frac{\partial p}{\partial x} = \frac{\partial p}{\partial X}\cos \omega t + \frac{\partial p}{\partial Y}\sin \omega t$$

Show that the equations of static equilibrium in the stationary coordinate system analogous to (3.4a) account for effects of rotation when the terms (8.2) are included, and that these equations are generalized as

$$\left.\begin{aligned}
-\omega^2 x + \frac{1}{\rho}\frac{\partial p}{\partial x} &= 0 \\
-\omega^2 y + \frac{1}{\rho}\frac{\partial p}{\partial y} &= 0 \\
\frac{1}{\rho}\frac{\partial p}{\partial z} &= -g
\end{aligned}\right\} \tag{8.3}$$

extension of (3.4a): enter the "centrifugal" forces

It is seen that the components of the pressure derivatives in (8.3) do in fact determine a vector termed the pressure gradient. The new terms in (8.3) are recognized also as components of the centripetal

26   FLUID STATICS

acceleration vector, and when their signs are reversed they are commonly referred to as centrifugal forces.

Each of the equations (8.3) is seen to be obtained as a derivative of a sum of three terms. Specifically, by writing

gravitational force related to gravitational potential energy

$$U = gz \qquad (8.4)$$

and considering incompressible fluid (ρ = const), the x-, y-, and z-derivatives of the quantity on the left side of the equation

three forms of energy (per unit mass)

$$-\frac{\omega^2}{2}(x^2 + y^2) + \frac{p}{\rho} + U = \text{const} \qquad (8.5)$$

correspond exactly to the separate equations of (8.3). It follows that all the statements expressed by these three equations are contained in the single equation, or *integral*, (8.5), which has the advantage of referring to the pressure directly without the necessity for integration. Moreover, when the quantity U introduced in (8.4) is recognized as the potential energy of unit mass in the gravity force field g, the remaining terms of (8.5) are also seen to represent energies; the equation therefore expresses the constancy of the sum of three separate types of energy. Energy integrals like (8.5) are found for more general conditions of fluid motion and are always important because they furnish overall, or in-the-large, statements that are more comprehensive than the separate differential equations like (8.3) which refer to conditions at a point only, conditions comprising in-the-small or local statements.

an integral contains information of numerous separate equations

utility of potential functions

It may be noted that the existence of the integral (8.5) depends on finding a gravity *force potential U*. In this regard an important property of gravitational forces can be seen directly from the gravitation law (3.1) by showing that the component forces are always expressible as partial derivatives of a gravitational potential which is determined directly from the form of the attracting body or bodies.

From (8.5) it is plain that effects of relative motion may be neglected in general when the first term on the left is negligible compared with either of the others. A more delicate criterion is given by (8.3).

9 ROTATIONAL FLATTENING OF THE EARTH — AND OF THE SOLAR SYSTEM

an example to show effects of rotation — and the utility of an integral

The spherical symmetry of a large fluid mass is modified by rotation, and the equations (8.3) of quasi-static equilibrium permit evaluation of the rotational distortion in approximate form when the speed of rotation, and hence the effective centrifugal forces, are sufficiently small. The calculation is carried out in detail for

I. 9 / ROTATIONAL FLATTENING OF THE EARTH — AND OF THE SOLAR SYSTEM 27

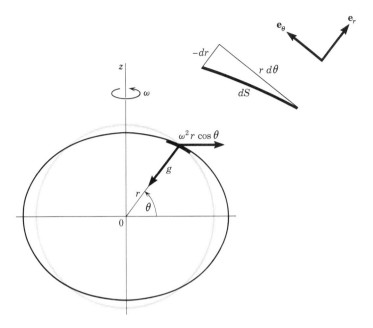

Fig. I-9 Rotational flattening of spherical fluid mass.

conditions corresponding to the Earth's nearly uniform rotation. Identical conclusions are also demonstrated more directly on the basis of the integral (8.5).

When the Earth's rotational speed is ω, parallel to the z-axis, centripetal acceleration components appear in the x- and y-directions, as in the first two equations of (8.3). The Earth's gravitational attraction force then can be taken as directed toward the center of the Earth from each point of the surface, with components in each coordinate direction. By assuming symmetry about the rotation axis, it is possible to determine the form of the resulting slightly flattened *spheroid* by finding any of its meridian curves, and only conditions within the x, z plane are examined here. In anticipation of the greater convenience of plane polar coordinates r and θ, where r is the distance from the center of the Earth and θ is the latitude angle measured from the equatorial plane $z = 0$ (Fig. I-9, the equations are written as

$$\left. \begin{array}{r} -\omega^2 r \cos\theta + \dfrac{1}{\rho}\dfrac{\partial p}{\partial x} = -g \cos\theta \\[1em] +\dfrac{1}{\rho}\dfrac{\partial p}{\partial z} = -g \sin\theta \end{array} \right\} \quad (9.1)$$

force balance: centrifugal plus pressure plus gravity forces

According to the fundamental property of fluids, viz., to support only normal stresses in the absence of relative motion, the meridian curve $r(\theta)$ is determined by observing that the pressure gradient terms in (9.1) specify the remaining forces with regard to direction. Specifically, by equating the pressure gradient terms to the gravity force components and the effective centrifugal force, i.e., by rearranging terms to put the equations (9.1) in the form

$$\frac{1}{\rho}\frac{\partial p}{\partial x} = \omega^2 r \cos\theta - g\cos\theta$$
$$\frac{1}{\rho}\frac{\partial p}{\partial z} = -g\sin\theta \qquad (9.2)$$

the force components on the right side of (9.2) are conveniently treated by resolving in the radial and transverse directions. In vector notation, the force **F** is referred to unit vectors $\mathbf{e}_r$ and $\mathbf{e}_\theta$ in the respective coordinate directions by writing

$$\mathbf{F} = \mathbf{e}_r\{-g + \omega^2 r \cos^2\theta\} + \mathbf{e}_\theta\{-\omega^2 r \cos\theta \sin\theta\}$$

The direction of the force **F** is also the direction of the pressure gradient force, according to (9.2), and the condition of normal stress means that the tangential component of **F** at the free surface is zero. If the unit tangent vector to the surface is introduced as $\boldsymbol{\tau}$, this condition is expressed by the vanishing of the scalar product of the vectors **F** and $\boldsymbol{\tau}$, written

fundamental property of fluid: normal stress at fluid boundary

$$\mathbf{F}\cdot\boldsymbol{\tau} = 0 \qquad (9.3)$$

while the unit vector $\boldsymbol{\tau}$ is referred to the base vectors $\mathbf{e}_r$ and $\mathbf{e}_\theta$ by resolving the components of the arc length ds of the meridian curve:

$$\boldsymbol{\tau}\,ds = dr\,\mathbf{e}_r + r\,d\theta\,\mathbf{e}_\theta \qquad (9.4)$$

The condition (9.3) then becomes the first-order differential equation

a first-order differential equation — variables not separable

$$dr\{-g + \omega^2 r \cos^2\theta\} + r\,d\theta\{-\omega^2 r \cos\theta \sin\theta\} = 0 \qquad (9.3a)$$

the integration of which gives the meridian profile $r(\theta)$ and hence the form of the spheroid. A difficulty is observed in the fact that the variables are not separable, as the equation of atmospheric statics was found to be; it was immediately integrated in the form (5.3). A physical argument leads to an approximation of (9.3a), however, which can be directly integrated. For this purpose it is noted that the gravity acceleration g is known, as well as the Earth's rotation speed ω, and the distance r which differs only slightly from the value earlier identified as a. Comparison of the coefficients of dr in (9.3a) then establishes that the first term is

approximation

much greater than the second in magnitude, and, if this term is ignored when added to the first, it is found that

$$-\frac{1}{r^2} dr = \frac{\omega^2}{g} \sin\theta \cos\theta \, d\theta \qquad (9.5)$$

where each side of the equation is recognized as an exact differential. By identifying the equatorial radius as a, i.e., $r(0) = a$, integration of (9.5) gives r as a function of θ;

$$\frac{1}{r} - \frac{1}{a} = \frac{\omega^2}{2g} \sin^2\theta \qquad (9.6)$$

integration gives meridian profile $r(\theta)$

When the approximation

$$\frac{\omega^2 a}{g} \ll 1$$

is used again, (8.6) is written in final form as

$$r(\theta) = a\left\{1 - \frac{\omega^2 a}{2g} \sin^2\theta\right\} \qquad (9.6a)$$

which shows that the equatorial radius a is a maximum value and that the distance r to the surface at any other place is slightly less. The equatorial bulge then exhibits both axial symmetry and symmetry about the equatorial plane.

Exercise 24 Denote the polar radius of the Earth by c, i.e., $r(\pi/2) = c$, and evaluate the standard flattening parameter

$$\frac{a-c}{c} \qquad (9.7)$$

by taking for a and g the numerical values given in Section 3 and for the square of the Earth's rotational speed (in radian measure)

$$\omega^2 = 0.52885 \times 10^{-8} \quad \sec^{-2} \qquad (9.8)$$

Which assumptions of the present calculation are likely to contribute to the error?

Exercise 25 Show that the component gravity forces in (9.2) in the x- and z-directions are obtained as the negative of the respective partial derivatives of the gravitational potential

$$U = -\frac{k^2 E}{(x^2 + z^2)^{1/2}} \qquad (9.9)$$

potential of the gravity force

The meridian curve (9.6a) has been obtained by integration of the equations (9.1) of quasi-statics, and it is reasonable to ask whether the same result could have been obtained more directly from the integral corresponding to (8.5), which must embody both of the equations (9.1). That this is actually the case will be demonstrated for the specific evaluation of the flattening parameter (9.7). For this purpose it is only necessary to note that (9.9) is the appropriate form of the gravitational potential U in (8.5) and that the sum of three terms in that equation must be the same at the equator, where $r = a$, and at the pole, where $r = c$, while the surface pressure is the same at both places (and may be taken as zero). Hence

application of integral at equator and pole

$$-\frac{\omega^2 a^2}{2} + \frac{p}{\rho} - \frac{k^2 E}{a} = \frac{p}{\rho} - \frac{k^2 E}{c} \qquad (9.10)$$

and when

$$\frac{k^2 E}{a^2}$$

the integral (8.5) gives the flattening directly, without regard to detailed force balance

is recognized as the sea level gravity constant g, (9.10) reduces to the result of Exercise 24. The use of the integral (8.5) thereby permits bypassing the calculations leading to (9.6); the integral, in fact, simply represents the equivalent of the statements contained in (9.2) for the totality of fluid particles.

The integral (8.5) is of course also valid when the relative importance of the respective terms differ from that in the case just examined. Thus it is held that the planets of the solar system were deposited as cosmic dust particles that once formed a hot cloud at low pressures. When pressure is ignored completely in (8.5), a balance of centrifugal kinetic energy and gravitational potential energy is exhibited in which the latter can be represented in the same manner as (9.9), referring in this case to the solar mass. According to the Kant-Laplace solar nebula hypothesis of cosmogonic origins, now widely accepted in modified form, the fact that all the planets travel in orbits that nearly coincide with a single plane, called the ecliptic, is explained by the conditions implicit in the form of the integral (8.5).

Exercise 26 Take the speed ω as constant, corresponding to uniform rotation of the dust cloud, and again consider conditions in the meridian plane x, z in order to show that equilibrium conditions in the absence of pressure are satisfied only in the $z = 0$ plane at all points at distance

equilibrium in rotating coordinate system — only at particular points

$$\left(\frac{k^2 S}{\omega^2}\right)^{1/3} \qquad (9.11)$$

from the axis of rotation passing through the Sun, the mass of which is denoted S. The successive stages of cooling and contraction are thought to correspond to different distances of the several planets. HINT: The x- and z-derivatives of the integral corresponding to (8.5) specify the equilibrium locations where particles might collect and accumulate.

S1 ARCHIMEDES' PRINCIPLE, HYDROSTATIC PRESSURE FORCE FORMULAE

If we interpret hydrostatic pressure in terms of fluid weight the pressure increasing linearly with depth, as expressed by (3.6) and depicted in Fig. I-4, we are led to simplified practical calculations. In the first place the fact that the pressure difference between two different levels equals the weight of the unit column of liquid of the same height means that the liquid column experiences a buoyant force equal to its weight. The same is true of a solid body that displaces the liquid, and it follows that a submerged body of arbitrary form is buoyed upward by a force equal to the weight of the fluid it displaces (Archimedes' celebrated principle), even if the fluid is not of uniform density. The detailed argument can be completed by the reader; the result is employed on pg. 33.

buoyancy: consequence of greater hydrostatic pressures at greater depths

When the density is constant, the hydrostatic force on a plane surface of arbitrary form and area A is obtained from (3.6) as the sum of two separate integrals of the form (3.11). The integral of dA multiplied by constants is termed the zero moment of the area, and the remaining term that contains the first power of the distance z multiplied by dA is called the first moment of the area relative to the level $z = 0$. The latter integral is recalled to define the coordinate z_c of the centroid of the area A, so that F becomes

$$F = \rho g \int_A (H - z) \, dA = \rho g (H - z_c) A = p_c A \qquad (S1.1)$$

where p_c is the pressure at the centroid, by (3.6). The last form of (S1.1) therefore gives the total hydrostatic force as the product of the area A and the pressure p_c at its centroid. Whenever the area in question is of a simple form for which the centroid location is known, therefore, the force evaluation is effected without recourse to integration. One example of this result is provided by (3.12).

an elementary geometrical interpretation: force determined by pressure at centroid

Analogous interpretation of the general expression for the hydrostatic couple corresponding to (3.13), that is,

$$Fl = \rho g \int_A z(H - z) \, dA \qquad (S1.2)$$

gives first moment of the area, and a second moment. Second moments of a plane area are known as the moments of inertia of the area, and it is clear that the hydrostatic couple as well as the location of the effective resultant force is now expressible in terms of a particular moment of inertia of the area subject to hydrostatic pressures.

The required formulae are found most easily if distances are now measured from the centroid level z_c, i.e., by the coordinate ζ defined as $\zeta = z - z_c$. Substitution and reduction of (S1.2) now give

$$Fl = \rho g[z_c(H - z_c)(A - I)] \qquad \text{where} \quad I \equiv \int_A \zeta^2 \, dA$$

another geometric interpretation: line of action of resultant force is displaced downward from centroid an amount depending on moment of inertia

so that I is the moment of inertia of the area A about the horizontal axis passing through its centroid. The first term on the right in the preceding equation contains the force F as one factor, and it follows at once that the *line of action* of the resultant force is displaced below the centroid a distance given by the formula

$$z_c - l = \frac{I}{(H - z_c)A} \qquad (S1.3)$$

The distance in question is therefore determined by the moment of inertia I, the area A, and the depth $H - z_c$ of its centroid below the free surface. The geometric forms of the results (S1.1) and (S1.3) are clearly consequences of the linearity already noted in (3.6).

S2 STATIC STABILITY OF THE ATMOSPHERE

Static equilibrium can occur only when forces are properly balanced, but the fact that forces are so balanced by no means guarantees that the equilibrium configuration in question does in fact occur. Just as a pencil does not remain standing on its point on a smooth horizontal surface, there are also atmospheric stratifications consistent with the fundamental aerostatic equations (5.1) that have never been observed. The reason is that in actual physical systems small extraneous forces are always present that are not taken into account and which upset a strict equilibrium condition. The manner in which a system responds to small disturbing influences is characterized by its *stability*, and an easy introduction to this often delicate subject is furnished by a slight extension of aerostatics as developed in Section 5.

equilibrium does not guarantee stability

Equilibrium in this case involves only gravity and buoyant (pressure) forces, and attention is focused on the possible changes of these two forces when the equilibrium is assumed to be disturbed in

I S2 / STATIC STABILITY OF THE ATMOSPHERE

some arbitrary and unspecified manner. It is assumed for simplicity that the disturbance, such as might result from a gust or from uneven heating of the air, leads to a displacement of the air in a small region without imparting to it any significant velocity. The stratification is then said to possess *static stability* in a positive sense if the modified forces are such as to create a tendency to restore the initial configuration (*instability* is understood when the opposite tendency appears, *neutral stability* when no tendency in the one direction or the other occurs). By considering the gravity force to be constant, the question of static stability depends only on the manner in which the buoyant force is altered as a small volume of fluid is displaced from its equilibrium location.

It is supposed that during any vertical displacement of a "parcel" of air its pressure remains equal to that of the undisturbed atmosphere at the level it occupies. When it is also assumed that the pressure drop accompanying an increase of altitude occurs adiabatically, so that the pressure-density variation of the parcel is given by (5.5) (as before, $n = \gamma = 1.4$). The resulting density of the displaced air is in general different from that of its surroundings, and it is clear that the sign of this difference determines whether the undisturbed air is in a stable stratification or unstable. Specifically, when a displacement from level z to new level $z + dz$ (upward when $dz > 0$) is considered, it is found that (5.1) and (5.2) give for the atmospheric density at level $z + dz$

static stability: to find forces that arise when equilibrium is disturbed

$$\rho(z + dz) \doteq \rho(z)\left\{1 - \left[\frac{\rho g}{p} - \frac{\alpha}{T}\right]dz\right\} \quad (S2.1)$$

to the first order in dz. The buoyant force on the parcel, according to Archimedes' principle, is then proportional to the quantity (S2.1).

The most direct assessment of the stability consists in comparing the density (S2.1) with the value corresponding to the displaced parcel; when the latter exceeds the former, the weight exceeds the buoyant force and the resultant downward force is an indication of positive stability. The density of the displaced air parcel is now obtained from (5.5) by noting that the density of the lifted parcel, denoted by $\rho'(z + dz)$, is

compare buoyant and gravity forces

$$\frac{\rho'(z + dz)}{\rho(z)} = \left(\frac{p'(z + dz)}{p(z)}\right)^{1/\gamma} \doteq \left(1 - \frac{\rho g}{p}dz\right)^{1/\gamma} \quad (S2.2)$$

because of (5.1). When the first two terms of the series expansion of (S2.2) are retained, the result is

$$\rho'(z + dz) \doteq \rho(z)\left\{1 - \frac{\rho g\, dz}{\gamma p}\right\} \quad (S2.3)$$

According to what has been said above, the condition for stability is expressed as the inequality

$$\rho'(z + dz) > \rho(z + dz)$$

which provides an upper limit for the lapse rate α as

$$\alpha < \frac{\gamma - 1}{\gamma} \frac{g}{R} \qquad (S2.4)$$

criterion of stability: lapse rate less than a certain value

From (S2.4) it appears that Halley's isothermal atmosphere, like other models in which the lapse rate is less than the amount indicated by the expression on the right side of (S2.4), is statically stable. It is left for the reader to verify whether the "standard atmosphere" also shares this property. From (S2.1) and (S2.3) the reader will also find no difficulty in expressing the values of the buoyant and gravity forces on elementary mass δm lifted through distance dz (whether positive or negative).

The instability of stratified layers having strong temperature gradients (i.e., large values of the lapse rate α) is also noted from (S2.4). In general such conditions are not found in the atmosphere, but there are notable exceptions. One of them is in the immediate vicinity of the Earth's surface, where the impervious character of this boundary surface inhibits vertical motions. Tornadoes represent the violent consequence of short-lived configurations that are statically unstable, although the precise conditions favorable to their formation are not fully specified on the basis of the arguments presented above.

II
Principles of Fluid Motion

1 FLUID VELOCITY; STREAMLINES AND STREAM SURFACES

The motion of a fluid is fully known when the *velocity* of each of its particles can be specified. Then the location of an individual particle is given by position coordinates through integration of the velocity, while forces and pressures are related to accelerations in the form of velocity derivatives. It is useful to regard the vector velocity as the basic variable of fluid motion, and to develop related concepts that facilitate geometric and analytic expressions of the basic laws of fluid flow.

fluid velocity: motion of fluid particles

Fluid velocity is specified at every point of a moving fluid by adopting the compact vector-functional notation written as

$$\mathbf{q} = \mathbf{q}(\mathbf{r}, t) \tag{1.1}$$

velocity field

where $\mathbf{q}$ denotes the velocity vector at the point having position vector $\mathbf{r}$ measured from a definite point of reference, and the scalar time variable is written as t. In terms of Cartesian coordinates x, y, z and their associated unit vectors $\mathbf{i}, \mathbf{j}, \mathbf{k}$, for example,

$$\mathbf{r} = x\mathbf{i} + y\mathbf{j} + z\mathbf{k} \tag{1.2}$$

so that the assignment of four numbers as values of x, y, z, and t, treated as independent variables, determines the velocity $\mathbf{q}$ at each location and instant of time. When coordinate axes are not in motion, the unit vectors are independent of time; hence (1.1) is obtained from (1.2) by differentiation:

from position to velocity to acceleration by time differentiations

$$\begin{aligned}\mathbf{q} &= \frac{d\mathbf{r}}{dt} = \frac{dx}{dt}\mathbf{i} + \frac{dy}{dt}\mathbf{j} + \frac{dz}{dt}\mathbf{k} \\ &\equiv u\mathbf{i} + v\mathbf{j} + w\mathbf{k}\end{aligned} \tag{1.3}$$

where the *Cartesian velocity components u, v, w* appearing in (1.3) are defined as the time derivatives of each of the respective coordinates *x, y,* and *z* regarded as functions of time. Another differentiation of the velocity (1.3) gives the acceleration, each of the Cartesian components again appearing as a coefficient of the corresponding unit vector. The *field of velocities* is visualized by regarding a velocity vector as a directed arrow at each point of the flow according to (1.1) at every instant *t*.

When a definite function $\mathbf{q}(\mathbf{r}, t)$ is known for a region of flow determined by a range of values of the position vector $\mathbf{r}$, the velocity vectors at various points can be sketched to give a graphic representation of the fluid motion. If the vectors are connected at closely spaced points, the flow directions are shown by the family of curves that are tangent to the velocities at all points. The *streamlines* found in this manner are characterized by the property of being parallel everywhere to the local velocity vectors, and the form of the streamlines can be determined directly from this property once the field of velocities (1.1) is known (Fig. II-1). It is readily seen that when a flow is *steady*, so that the velocities are independent of time, $\partial \mathbf{q}/\partial t = 0$, the streamline at any point is identical with the *path* or trajectory of the fluid particles passing through the point.

<small>streamlines are everywhere parallel to local velocity vector — streamlines never intersect each other</small>

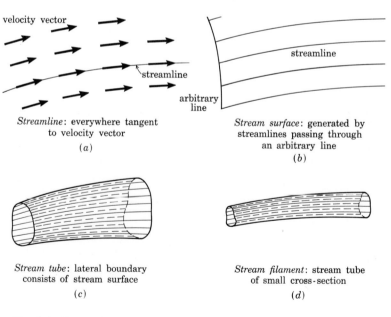

Fig. II-1 Illustrating streamline, stream surface, stream tube, and stream filament.

II. 1 / FLUID VELOCITY; STREAMLINES AND STREAM SURFACES

Exercise 1 The Cartesian velocity components of a given steady flow are

$$u = 2x^2y \qquad v = -2xy^2 \qquad w = 0 \qquad (1.4)$$

(a) Determine the coordinates $x(t)$, $y(t)$, $z(t)$ as functions of time for the particle situated at $x = y = 1$, $z = 0$ at time $t = 0$. Sketch the particle trajectory, indicating the velocity by an arrow for each of the instants $t = 0.1, 0.2$, and 0.3.

(b) Find the x-component of particle acceleration by twice differentiating the function $x(t)$ found in part (a). Compare the result with the expression obtained from the velocity (1.4) by a single differentiation, noting that

$$\frac{du}{dt} = \frac{\partial u}{\partial x}\frac{\partial x}{\partial t} + \frac{\partial u}{\partial y}\frac{\partial y}{\partial t} = u\frac{\partial u}{\partial x} + v\frac{\partial u}{\partial y} \qquad (1.5)$$

according to (1.3). HINT: Find the equation of the streamlines from the defining property

$$\left.\frac{dy}{dx}\right|_{\text{S.L.}} = \frac{v}{u} \qquad (1.6)$$

equation of streamline determined by the velocity field

taking the values of u and v given by (1.4). Integrate (1.6) to find the equation of the streamline as

$$xy = 1 \qquad z = 0 \qquad (1.7)$$

Use (1.7) in the first of (1.4) and integrate according to (1.3) to find $x(t)$, etc. so that

$$x = e^{2t} \qquad y = e^{-2t} \qquad z = 0 \qquad (1.4a)$$

Exercise 2 Repeat part (a) of the preceding Exercise for the particle situated at $(x_0, y_0, 0)$ at time $t = 0$, obtaining thereby instead of (1.4a) functions $x(t)$ and $y(t)$ dependent on the parameters x_0 and y_0:

$$x = x(t; x_0, y_0) \qquad y = y(t; x_0, y_0) \qquad (1.8)$$

that is, find the explicit form of the functions indicated by (1.8). The representation of fluid motion by means of equations of the form of (1.8) is referred to as the *Lagrangian description*, while the equations (1.4) that express velocities as functions of position and time are known as the *Eulerian* form. Establish the equivalence of the two forms by evaluating the velocity components (1.4) from (1.8) and eliminating the Lagrangian parameters x_0, y_0. HINT: Use the equations of streamlines, corresponding to (1.7).

Lagrangian description: particle coordinates as functions of time
Eulerian description (more commonly used): velocity as function of position and time

Exercise 3 (a) Consider a velocity vector **q** in the x, y plane, and show that its radial and transverse components u_r and u_θ are respectively proportional to the line element components dr and $r\,d\theta$ of the streamline tangent to **q**:

$$u_r : dr = u_\theta : r\,d\theta \quad \text{or} \quad \frac{u_r}{u_\theta} = \left.\frac{dr}{r\,d\theta}\right|_{\text{S.L.}} \qquad (1.9)$$

equation of streamlines using polar coordinates

(b) Verify the result (1.9) directly from (1.6) by using the transformations of coordinates and velocities written as

transformation of polar coordinates and velocity components to Cartesian form

$$x = r\cos\theta \qquad u = u_r \cos\theta - u_\theta \sin\theta$$
$$y = r\sin\theta \qquad v = u_r \sin\theta + u_\theta \cos\theta \qquad (1.10)$$

Hence the polar coordinate form of the equation of streamlines can be obtained directly from (1.9) when the velocity components are given in polar form as functions of the variables r and θ.

The last expression on the right side of (1.5) expresses the "total" time derivative of the velocity u in terms of the Eulerian variables x and y as shown in (1.4). Derivatives of this type are of such frequent occurrence as to justify a special compact notation. When the total time derivative of an arbitrary flow property $f(x, y, t)$ is intended, for example, this is written

$$\frac{Df}{Dt} = \frac{\partial f}{\partial t} + u\frac{\partial f}{\partial x} + v\frac{\partial f}{\partial y} \qquad (1.11)$$

reducing to (1.5) for steady flow when f signifies the velocity component appearing there. The first term on the right in (1.11) gives the *local* or *instantaneous* variation at position x, y different from zero only in unsteady flow. The remaining terms, called the *convective* portion of the derivative, represent the effects of non-uniformity of property f viewed as a function of x and y at a fixed time. Thus, in the example in which $f = u$, the derivative represents the acceleration in the x-direction, and the convective contribution is nonzero in steady flow in regions containing velocity gradients.

total derivative, Eulerian derivative, substantial derivative ...

The total derivative, also variously termed the Eulerian derivative, substantial derivative, streamline derivative, or particle derivative, can likewise be written for a three-dimensional flow as

$$\frac{Df}{Dt} = \frac{\partial f}{\partial t} + u\frac{\partial f}{\partial x} + v\frac{\partial f}{\partial y} + w\frac{\partial f}{\partial z} \qquad (1.11a)$$

in Cartesian form. Different discussions that follow show that f can be taken to represent particle coordinate, velocity or acceleration component, fluid pressure or density, or other flow quantity. Both scalar and vector quantities are differentiated in this manner, and it is easy to see that the physical meaning is retained in different coordinate systems.

Since there is a well-determined streamline direction at every point of the space occupied by fluid in motion, a definite streamline is seen to pass through each point of an arbitrary line element drawn within the fluid. The one-parameter family of lines thus determined then forms a surface, called a *stream surface* (an exception occurs if the initial line element coincides with a streamline, when the stream surface degenerates into the streamline itself). Because two streamlines cannot intersect each other, it follows that the stream surface creates a separation between fluid on each of its two sides. If the the initial line element is extended to form a closed curve, the associated stream surface that it generates assumes a tubular form, separating the fluid into distinct interior and exterior regions, like the walls of an actual pipe or tube, giving rise to a *stream tube*. In steady flow the form of the stream tube is permanent, and fluid particles inside the tube at one point remain inside it throughout the motion. When a stream tube has an especially small cross section, it may be referred to as a *stream filament;* this corresponds to consideration of fluid in closest proximity to a single streamline (Fig. II-1).

stream surface, composed of streamlines

stream tube, stream filament

It is often convenient to consider a limited length, or section, of a stream tube; precise statements about the flow behavior are then formulated by comparing "inlet" conditions at one of its ends with the "outflow" at the other. Such discussions are simplified by the fact that no fluid crosses the lateral boundary (composed of streamlines) at any of its points. The fluid contained within the control volume bounded by the lateral stream tube surface and by the end sections is isolated in a manner that lends itself easily to dynamical analysis.

utility of stream tube concept: localize mass flux, momentum flux, etc.

2 CONSERVATION OF FLUID MASS — CONTINUITY EQUATION

The fact that streamlines do not intersect each other is related to the indestructibility of matter. Expressed as an equation, the same principle takes the form of a restriction on the possible velocities of a fluid motion. Introducing the other restrictions imposed by Newton's second law of motion leads to the possibility of fully and uniquely specifying flow fields in terms of gross external data. The combination of these two principles, written in this and succeeding sections as statements of conservation of mass and momentum, thereby represents the entire physical basis of hydrodynamics. The stream tube concept lends itself easily to the development of the underlying equations and also to the expression of the principle of energy conservation that is required for the extension to aerodynamic flows of variable density.

Just as two different bodies do not occupy the same place at any given time, the idea of a continuous fluid implies that every point of a fluid-filled region is occupied by precisely one fluid particle. In

40 PRINCIPLES OF FLUID MOTION

indestructibility of matter: relationship between inward mass flux and outward, in steady motion

a stream tube segment in steady flow it follows that the total mass of fluid contained in an arbitrary segment of the stream tube remains constant in time. Said differently, the *mass flux* (i.e., mass flow per unit time) across the inlet section equals the mass flux across the outlet section of the stream tube segment. It is this statement of the principle of fluid mass conservation that furnishes the first of the fundamental equations of fluid mechanics.

Specifically, where the speed of flow is q, all the fluid particles within a distance q units of length flow past the point in question in unit time (Fig. II-2). The volume of fluid that passes through a cross section S drawn normal to the flow, in unit time, is given by the product qS, and multiplying by the fluid density ρ (mass per unit volume) then gives the mass flux as $\rho q S$. With the understanding that this product represents the outflow, while the same terms written with subscript 1 represent flow conditions at the inlet section, the *equation of continuity* becomes

$$\rho q S = \rho_1 q_1 S_1 \tag{2.1}$$

continuity equation in stream tube form for steady flow

in "stream tube" form where the product of density and velocity represents an average taken over all points of the cross section in each case. The narrower the stream tube, the smaller is the difference between the average and local values, and the distinction vanishes completely when the tube is so narrow as to be regarded as a stream filament.

The constancy of mass flux expressed by (2.1) also permits this equation to be written as

$$\rho q S = \text{const} \tag{2.1a}$$

for all sections of the stream tube. A still simpler form suffices when the fluid is a liquid and the density is the same at all points, since in this case

$$q S = \text{const} \tag{2.1b}$$

Although the preceding equations were obtained for steady flow, it is evident that (2.1b) is also valid for *unsteady* flow, because in the case of liquid the total fluid mass within the stream tube segment remains the same regardless of whether the flow is steady or un-

ρ = const — valid for steady flow and unsteady also

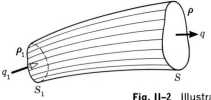

Fig. II-2 Illustrating Eq. (2.1); fluid continuity.

steady. The interpretation in each instance is that the flow speed varies inversely with stream tube cross section, increasing where the section is narrower, decreasing where less constricted.

The stream tube form of the continuity equation is most simple from the mathematical standpoint, but practically useful only in particular cases. Thus, when (2.1b), for example, is applied to flow in a pipe of variable cross section, the average flow speed can be specified at once in terms of given pipe dimensions. For "external" flows this is not possible, however, because the variation of stream tube cross section is not known in advance, nor even the direction of its central axis. It will now be seen in what manner (2.1) can be extended to permit the stream tube dimensions to be eliminated from the continuity equation and the uncertainty as to flow direction to be removed.

<small>limitation of utility of (2.1)</small>

It is convenient to introduce the unit normal (outward-directed) vector **n** on any point of a closed surface such as the stream tube segment shown in Fig. II-2. The vector **n** is of length unity at each point of the surface, whereas its direction varies from point to point on the curved lateral boundary surface of the stream tube segment. On the plane end sections S_1 and S its directions are respectively opposite and parallel to the average velocities q_1 and q when the density of the fluid is constant. According to the definition of the scalar, or inner, product of two vectors $\mathbf{n} \cdot \mathbf{q}$ as the product of their lengths multiplied by the cosine of the included angle,

<small>arbitrary fluid volume element, outward flux expressed in vector form</small>

$$\mathbf{q}_1 \cdot \mathbf{n} S_1 = -q_1 S_1 \quad \text{and} \quad \mathbf{q} \cdot \mathbf{n} S = qS$$

are the *outward* volume flux expressions at the inlet and outlet, respectively, of the stream tube segment (minus sign indicating that inward-directed flow is subtracted from the outward when the total outward flux is computed). Corresponding terms vanish for all points of the lateral boundary surface, because the velocity and normal vectors there are mutually perpendicular and the cosine of the included angle is therefore zero. It follows that, by bringing both terms to one side of the equation, (2.1) expresses the fact that the total outward mass flux across all the boundary surfaces of the stream tube segment is zero. The same is true if the restriction that the surfaces S_1 and S are normal to the average velocities is dropped since any increase in area that might result would be offset by the change in the value of the cosine that enters into the scalar product evaluation. Finally, it is seen at once that the volume of fluid need not be that of a stream tube segment, so long as outward flux is computed at each point of the boundary surface.

If we limit ourselves to the incompressible case (2.1b) for simplicity, equation (2.1b) can be written as

steady-flow continuity: total outward mass flux equals zero

$$\Sigma \mathbf{q} \cdot \mathbf{n} S = 0 \qquad (2.2)$$

and (2.2) also applies to an arbitrary volume at each plane surface boundary element S on which the vector $\mathbf{n}$ is drawn in the outward normal sense. The summation implies that the product $\mathbf{q} \cdot \mathbf{n} S$ is formed separately for each facet of the surface, and the same meaning is contained in the "integral" form of the continuity equation, written as

integral form of (2.2)

$$\int_S \mathbf{q} \cdot \mathbf{n} \, dS = 0 \qquad (2.2a)$$

in which it is not even necessary to regard the surface as composed of plane facets. The meanings of (2.2) and (2.2a) are depicted in Fig. II-3, where the included angle between the normal and the velocity vector is denoted θ. The surface S is of arbitrary form, and special choices of the fluid volume and its boundary surface S now lead to particularly useful forms of the continuity equation.

In the most important case of Cartesian coordinates x, y, z an infinitesimal volume element of dimensions dx, dy, and dz is taken, so that the six rectangular boundary surfaces are given by terms like $dy\,dz$, $dz\,dx$, etc. The direction normal to the elementary area $dy\,dz$ is parallel to the x-axis, and the normal velocity component is simply $u = dx/dt$. The surface in question is shown in end projection in Fig. II-4 as AB, regarded as the surface element on which the normal vector $\mathbf{n}$ opposes the direction of the velocity component u, so that $\mathbf{n} \cdot \mathbf{q} = -u$. If the surface element shown in projection as CD is at distance dx from AB, the normal flow component there is $u + (\partial u/\partial x) \cdot dx$, so that the total outward volume flux across both surfaces is given as the sum of the two terms

Cartesian element: outward flux evaluated in terms of u, v, and w

$$-u\,dy\,dz \quad \text{and} \quad \left\{ u + \frac{\partial u}{\partial x} dx \right\} dy\,dz \qquad (2.3\text{a, b})$$

That is,

$$\frac{\partial u}{\partial x} dx\,dy\,dz \qquad (2.3c)$$

The same argument applied to the remaining four surfaces gives flux terms like (2.3a) and (2.3b) with sums like (2.3c), so that the vanishing of the total outward mass flux is expressed by the equation

$$\rho \left\{ \frac{\partial u}{\partial x} + \frac{\partial v}{\partial y} + \frac{\partial w}{\partial z} \right\} dx\,dy\,dz = 0 \qquad (2.3d)$$

II. 2 / CONSERVATION OF FLUID MASS — CONTINUITY EQUATION

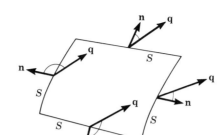

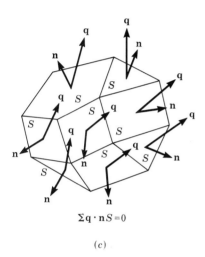

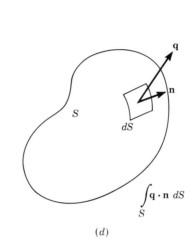

Fig. II-3 Conservation of (incompressible) fluid mass implies zero net outward flux through any closed surface S drawn in the fluid.

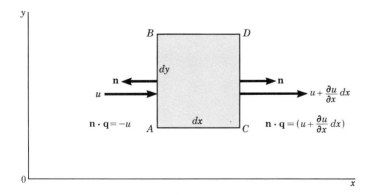

Fig. II-4 Infinitesimal fluid element; x-component of velocity and its differential variation.

Both the density ρ and the volume $dx\,dy\,dz$ are nonzero, and the Cartesian form of the continuity equation for an incompressible fluid becomes

continuity equation in Cartesian form

$$\frac{\partial u}{\partial x} + \frac{\partial v}{\partial y} + \frac{\partial w}{\partial z} = 0 \qquad (2.4)$$

The two-dimensional counterpart of (2.4) corresponds, e.g., to $w = 0$ or to $\partial/\partial z \equiv 0$, giving

two-dimensional form

$$\frac{\partial u}{\partial x} + \frac{\partial v}{\partial y} = 0 \qquad (2.4a)$$

The continuity equation in Cartesian form (2.4) or (2.4a) provides a relationship between the velocity components at any point. Although it places no restriction on the absolute values of the components u and v in (2.4a), for example, at any given point, it does indicate that any nonuniformity of one of them in the *vicinity* of that point is exactly offset by a nonuniformity of the other. Specifically, if $\partial u/\partial x > 0$ at a point in the flow, so that the outward flux through a surface CD exceeds the inward flow through a nearby parallel surface AB, the constancy of the fluid mass in the volume element is assured by $\partial v/\partial y < 0$, in accord with (2.4a). A similar interpretation is evident for three-dimensional flow and the terms of (2.4).

continuity equation imposes restrictions on the velocity derivatives

Exercise 4 Obtain the equation of continuity of an incompressible fluid in cylindrical coordinates r, θ, z by evaluating the integral (2.2a) on the boundary surface consisting of the nearly rectangular areas $dr\,r\,d\theta$, $dr\,dz$, and $r\,d\theta\,dz$, respectively perpendicular to the velocity components u_r, u_θ, and u_z. Verify that

cylindrical coordinates

$$\frac{1}{r}\frac{\partial}{\partial r}(ru_r) + \frac{1}{r}\frac{\partial u_\theta}{\partial \theta} + \frac{\partial u_z}{\partial z} = 0 \qquad (2.5)$$

Exercise 5 When spherical polar coordinates r, θ, ω are adopted ($\theta \sim$ polar angle, $\omega \sim$ longitude), the elementary volume is $(dr)(r\,d\theta)(r\sin\theta\,d\omega)$. Show that the continuity equation for incompressible flow is

spherical polar coordinates

$$\frac{1}{r^2}\frac{\partial}{\partial r}(r^2 u_r) + \frac{1}{r\sin\theta}\frac{\partial}{\partial \theta}(\sin\theta\,u_\theta) + \frac{1}{r\sin\theta}\frac{\partial u_\omega}{\partial \omega} = 0 \qquad (2.6)$$

Exercise 6 Show that the fluid acceleration in steady radial flow of incompressible fluid in three dimensions varies as the inverse fifth power of the distance from the center. HINT: The acceleration is equal to $u_r(\partial u_r/\partial r)$.

3 MOMENTUM CONSERVATION — BERNOULLI'S THEOREM

That the principle of fluid mass conservation is an insufficient physical basis for the study of fluid mechanics is apparent from the fact that the continuity equation (2.4a), for example, is a single relationship between two flow quantities u and v to be determined. A similar inadequacy of the other forms of the same equation is also clear; it is still more pronounced in (2.4) or in (2.1) where there are three "unknowns" in one equation. Newton's second law of motion is now shown to furnish additional relationships (i.e., equations) just sufficient in number that perfect fluid flows become mathematically determinate in principle and in reality as well for many cases.

insufficiency of continuity equation alone

Newton's second law in its most elementary form relates the acceleration of a body to the force acting on it. A slight restatement of the same principle facilitates the application to fluid motion by emphasizing the interpretation of force as a time-rate of increase of momentum. For each individual particle of mass m and linear velocity $\mathbf{q}$ subject to force $\mathbf{F}$,

$$\mathbf{F} = m\mathbf{a} = \frac{d}{dt} m\mathbf{q}$$

the fundamental dynamical principle: Newton's second law $F = ma$

where $m\mathbf{q}$ is the linear momentum. Vector addition for all particles of fluid within an arbitrary volume then shows that *the total force in any direction acting on a definite fluid volume equals the rate of increase of its momentum in the direction of the force.* In steady flow this is the difference between the momentum flux leaving the volume at some parts of its boundary surface and the momentum flux entering at others.

For definiteness a very short length dl of an arbitrary stream tube segment is considered now; its inlet and outlet cross sections are S and $S + (dS/dl)\,dl$, written $S + dS$ for brevity, as shown in Fig. II-5. In steady flow the time variable t is absent and l is the only independent variable, the omission of which in the more concise differential notation can cause no confusion. The pressures and the flow speeds at the two locations are written p, $p + dp$ and q, $q + dq$, the differential in each case referring to the increment on the normal section $S + dS$. The balance of pressure forces on the two end faces gives

application to segment of stream tube

$$pS - (p + dp)(S + dS) \doteq -p\,dS - S\,dp = d(-pS) \quad (3.1)$$

summation of forces

when the product $dp\,dS$ of differentials is neglected compared with terms containing differentials of the *first order* like dp or dS

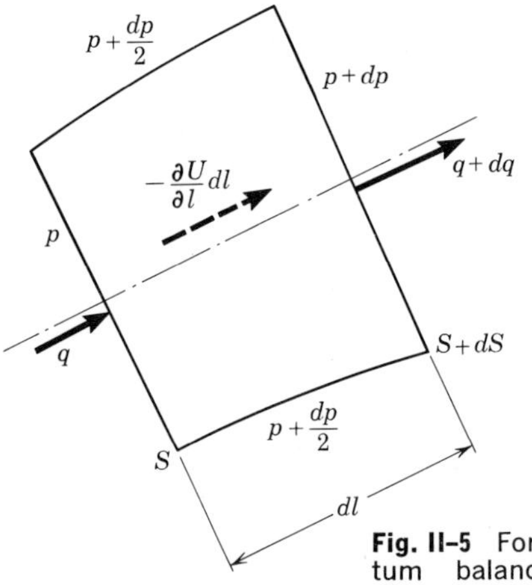

46 PRINCIPLES OF FLUID MOTION

Fig. II-5 Force and momentum balance, infinitesimal stream tube segment.

separately. The justification of this important step may be seen by referring to the more complete notation

$$dp\, dS = \frac{dp}{dl}\, dl\, \frac{dS}{dl}\, dl$$

the meaning of infinitesimal analysis: to neglect "higher order" terms

where the second power of the small quantity dl appears explicitly. In the limit of vanishing dl, to which expressions like (3.1) apply, it is evident that $(dl)^2$ vanishes to a *higher order* than the linear term dl.

In addition to the pressure force term (3.1) representing the action of the fluid external to the element on its end faces, there is also a force component parallel to the axis of the stream tube segment; it results from the pressures on the lateral boundary surface. Its projection normal to the flow direction is dS, and the force is obtained by multiplication with the average of the pressures at the end faces, viz., $p + dp/2$; the result is

$$\left(p + \frac{dp}{2}\right) dS \doteq p\, dS \qquad (3.2)$$

to the same order of accuracy as (3.1). The sum of these terms then gives for the total pressure force exerted on the fluid volume element by the surrounding fluid

$$-S\, dp \qquad (3.3)$$

The total fluid mass contained within the volume element under consideration is given by the product of the density ρ and the volume $S\, dl$. When the gravitational potential energy per unit mass is written as before as U, the gravity force per unit mass is $-dU/dl$

in the flow direction, while the total gravity force on the element is

$$-\frac{dU}{dl}\,dl\,\rho\,S = -\rho S\,dU \qquad (3.4)$$

Adding (3.3) and (3.4) gives the total external force in the flow direction as

$$-S\,dp - \rho S\,dU \qquad (3.5)$$

which must therefore be equal to the rate at which momentum in the same direction increases with time.

The momentum of one unit of mass is equal to its speed q, and it follows that the *momentum flux* across the stream tube inlet is obtained by multiplication by the mass flux (2.1a), or as

$$\rho q^2 S \qquad (3.6)$$

The outward momentum flux at the exit section of the stream tube segment is likewise given by

$$(\rho + d\rho)(q + dq)^2(S + dS) \qquad (3.7)$$

momentum flux values, inward and outward

Exercise 7 Expand the product (3.7) and arrange its terms as the sum of a zero-order term plus a first-order term plus a second-order term, etc., up to the fourth order. Use (2.1a) to show that the first-order term can be written simply as

$$\rho q S\,dq \qquad (3.8)$$

The outward momentum flux across the boundary surfaces of the elementary volume, to the first order in small quantities, is obtained as (3.8) by subtracting (3.6) from (3.7). Equating (3.8) and (3.5) and removing a common factor yields the differential expression of the principle of momentum conservation as

$$\rho q\,dq + dp + \rho\,dU = 0 \qquad (3.9)$$

In the more explicit notation that exhibits the independent variable, this is

"momentum conservation": balance of momentum flux by forces acting

$$\rho q\frac{dq}{dl} + \frac{dp}{dl} + \rho\frac{dU}{dl} = 0 \qquad (3.10)$$

Each of the terms in (3.10) can be written as a derivative with respect to l if the density ρ is taken as a constant. It therefore follows that for an *incompressible fluid* in steady motion, when only gravity and pressure forces are considered,

$$\frac{q^2}{2} + \frac{p}{\rho} + U = \frac{q_1^2}{2} + \frac{p_1}{\rho} + U_1 = \text{const} \qquad (3.11)$$

Bernoulli's theorem — valid at different points of a single streamline

Equation (3.11), known as Bernoulli's theorem, relates the kinetic energy $q^2/2$ of unit mass of fluid to its potential energy U and to the pressure-density quotient p/ρ, showing that the sum remains the same at all points along a given streamline. Bernoulli's theorem in the form (3.11) is evidently a statement concerning the conservation of energy in an incompressible fluid motion and may be compared with (I.8.5). The same relationship will be recovered in a later section by adopting an energy principle, i.e., on a basis fundamentally different from the momentum argument employed here.

constancy of energy *total (kinetic, potential, and "pressure") — deduced from momentum principle*

4 EULER'S EQUATIONS OF FLUID MOTION

The Bernoulli equation (3.11), obtained by a consideration of momentum in the flow direction, is a single relationship between the pressure and the flow speed. Taken in conjunction with the continuity equation (2.1b), it provides a pair of equations containing q, p, and S as three unknowns — still insufficient basis for determining these quantities. The inadequacy is overcome by exploiting Newton's second law more completely: as many as three independent relationships are obtained by applying the same momentum consideration to each of three mutually perpendicular directions in space. Again for simplicity, flow in the x, y plane is considered, and two momentum relationships for the velocity components u, v and the pressure p are obtained. These equations, together with the corresponding form (2.4a) of the continuity equation, form a system of equations from which incompressible flow solutions may be found.

inadequacy of streamline forms of continuity and momentum principles

It is convenient to refer again to Fig. II-4 and to calculate the flux of momentum across pairs of parallel boundary surfaces of the rectangular fluid element $ACDB$. The velocity component u is now the x-component of momentum of unit mass of the fluid, and multiplication by the mass flux $\rho u\, dy$ across the surface AB gives the momentum flux into the elementary volume as $\rho u^2\, dy$ when the dimension of the element in the direction parallel to the z-axis is taken as unity. Converting this expression to represent the *outward* momentum flux by changing the sign yields

consideration of Cartesian fluid element, momentum flux

$$x\text{-momentum flux}|_{AB} = -\rho u^2\, dy$$

for the outward flux. The corresponding term for the surface CD is

$$x\text{-momentum flux}|_{CD} = \rho u^2\, dy + \frac{\partial}{\partial x}(\rho u^2\, dy)\, dx$$

Hence the total outward flux through the pair of surfaces is

$$x\text{-momentum flux}|_{AB,CD} = \frac{\partial}{\partial x}(\rho u^2)\, dx\, dy \qquad (4.1a)$$

momentum flux analysis

Since x-momentum is also carried across the surface AC at speed v, the mass flux there being $\rho v\, dy$, the x-momentum flux $\rho uv\, dy$, added with due regard for sign to the flux across surface BD at distance dy from AC, is

$$x\text{-momentum flux}|_{AC,BD} = \frac{\partial}{\partial y}(\rho uv)\, dx\, dy \qquad (4.1b)$$

The pressure force exerted by the fluid to the left of AB on the volume element is $p\, dy$, and the force on CD is likewise $-[p + (\partial p/\partial x)\, dx]dy$; their sum is

$$x\text{-component of pressure forces}|_{AB,CD} = -\frac{\partial p}{\partial x}\, dx\, dy \qquad (4.1c)$$

summation of forces

and the gravity force on the fluid mass $\rho\, dx\, dy$ within the element is

$$x\text{-component of gravity force}|_{ABCD} = -\frac{\partial U}{\partial x}\rho\, dx\, dy \qquad (4.1d)$$

Equating the forces (4.1c) and (4.1d) to the momentum flux terms (4.1a) and (4.1b) gives the steady-flow form of *Euler's equation* for two-dimensional flow as

$$u\frac{\partial u}{\partial x} + v\frac{\partial u}{\partial y} + \frac{1}{\rho}\frac{\partial p}{\partial x} = -\frac{\partial U}{\partial x} \qquad (4.2)$$

with the aid of the continuity equation (2.4a) when the fluid is regarded as incompressible [it is easily shown that (4.2) is valid without this restriction]. Momentum parallel to the y-axis and the corresponding forces are accounted for in the same manner, so that the pair of equations

$$\left.\begin{array}{l} u\dfrac{\partial u}{\partial x} + v\dfrac{\partial u}{\partial y} + \dfrac{1}{\rho}\dfrac{\partial p}{\partial x} = -\dfrac{\partial U}{\partial x} \\[2mm] u\dfrac{\partial v}{\partial x} + v\dfrac{\partial v}{\partial y} + \dfrac{1}{\rho}\dfrac{\partial p}{\partial y} = -\dfrac{\partial U}{\partial y} \end{array}\right\} \qquad (4.3)$$

Euler equations of steady fluid motion, compressible or incompressible

represents the complete statement of Newton's dynamics for steady, two-dimensional frictionless flow subject only to conservative forces (i.e., forces derived from a potential U) like gravity.

Exercise 8 Demonstrate that equations (4.3) are also valid when the fluid is not assumed to be uniform and incompressible by noting the elementary modification of (2.4a) required to extend to the case of compressible fluid. HINT: Show that $\partial(\rho u)/\partial x + \partial(\rho v)/\partial y = 0$.

50 PRINCIPLES OF FLUID MOTION

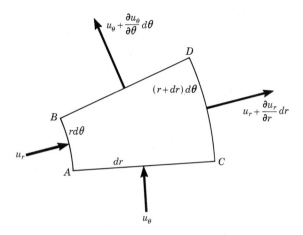

Fig. II-6 Velocity variations, infinitesimal plane polar fluid volume element.

When the gravity force is known (given, for example, by its potential U), the three equations (2.4a) and (4.3) form a system of *differential* equations for the determination of the three quantities represented by the two components of velocity, u and v, and the pressure p. As the number of equations now available is identical with the number of unknowns, the mathematical problem thus formulated is theoretically solvable on the basis of the two physical principles which have been invoked, viz., conservation of fluid mass and conservation of fluid momentum. It will be seen that in many cases flow "solutions" of practical interest are actually obtained as solutions of this set of equations.

<small>incompressible case: (4.3) and (2.4a) form a determinate system of equations</small>

<small>extension to flow in three dimensions</small>

Exercise 9 For flow in three dimensions the three velocity components u, v, and w, as well as the pressure, are considered "unknown." By what means is a fourth equation obtained that makes the mathematical problem a determinate one? Write the complete set of equations in this case, and discuss the connection with equations (8.3) of Chapter 1.

Exercise 10 The Cartesian form (4.3) of Euler's equations for steady flow can be transformed to any other coordinate system by following the rules of calculus, but the presence of partial derivatives makes the task inordinately tedious. It is both simpler and more instructive to obtain the equations by resolving momentum and force components in the different coordinate directions. Apply the latter procedure in the case of *plane polar coordinates* r, θ and the corresponding velocity components u_r and u_θ; calculate the flux of radial momentum across the four boundary surface elements AC, AB, BD, and CD shown in Fig. II-6, where both AC and BD

are of length dr, AB is of length $r\,d\theta$, and CD is of length $(r + dr)\,d\theta$. Equate the outward momentum flux to the resultant of pressure and gravity forces in the same direction to obtain the equation

$$u_r \frac{\partial u_r}{\partial r} + u_\theta \frac{\partial u_r}{r\,\partial \theta} - \frac{u_\theta{}^2}{r} + \frac{1}{\rho}\frac{\partial p}{\partial r} = -\frac{\partial U}{\partial r} \qquad (4.4a)$$

plane polar coordinates — Euler equations equivalent to (4.3)

HINT: Note that the sides AC and BD form an angle $d\theta$ with each other, so that the transverse velocity component u_θ at AC has a *radial* component $u_\theta\,(d\theta/2)$ when projected on the axis of symmetry of the element. The outward-directed momentum flux $-u_\theta \rho u_\theta\,dr$, combined with a similar term for the surface BD, leads to the third term on the left side of (4.4a). Note also that the continuity equation (2.5) is used as before.

Exercise 11 Obtain the Euler equation expressing momentum conservation in the θ-direction as

$$u_r \frac{\partial u_\theta}{\partial r} + \frac{u_\theta}{r}\frac{\partial u_\theta}{\partial \theta} + \frac{u_r u_\theta}{r} + \frac{1}{\rho r}\frac{\partial p}{\partial \theta} = -\frac{1}{r}\frac{\partial U}{\partial \theta} \qquad (4.4b)$$

by the same procedure as employed above (noting particularly the angle formed by the radial flow components on AC and BD), which leads to the third term on the left side of (4.4b).

5 CONSERVATION OF ENERGY AND SUMMARY OF FLOW EQUATIONS

Momentum and energy are concepts so totally distinct from each other that the deduction of Bernoulli's (energy) theorem (3.11) directly from Newton's second law of motion must seem surprising and enigmatic. It raises the question whether *conservation of energy*, introduced as an independent physical principle, might lead to any conclusions (relationships) different from those expressed by the mass and momentum principles. The preceding calculations now suggest both the appropriate form of the statement of principle as well as its treatment in terms of energy flux at the boundary surfaces of a conveniently chosen "control" volume.

energy conservation is an independent principle...

In the interest of simplicity alone, the steady motion of an incompressible fluid is considered first. The only energies that are essential are kinetic and potential (i.e., gravitational) in form, and their sum referred to unit mass of fluid is denoted by E (when compressible fluids are considered later, the same symbol will be used to denote energy in a less restricted sense). The principle of energy conservation is expressed by equating the net outward flux of energy E across the surfaces of the element of Fig. II-4 to the rate at which work is done to the fluid by external agencies.

Proceeding exactly as in the development of (4.1a) and (4.1b) leads to the expression for the energy flux as

$$\rho\left\{\frac{\partial(Eu)}{\partial x} + \frac{\partial(Ev)}{\partial y}\right\} dx\, dy \qquad (5.1)$$

Fluid at AB moves at speed u parallel to the direction of the pressure exerted on the element by adjacent fluid, so that work is performed on that part of the elementary surface at the rate pu per unit area. Summing for AB, CD, AC, and BD gives

...analyzed in standard manner...

$$-\left\{\frac{\partial(pu)}{\partial x} + \frac{\partial(pv)}{\partial y}\right\} dx\, dy \qquad (5.2)$$

for the total work done on the element in unit time by pressure forces.

Equating (5.1) to (5.2), with the aid of (2.4a), shows that

$$0 = \rho\left(u\frac{\partial E}{\partial x} + v\frac{\partial E}{\partial y}\right) + u\frac{\partial p}{\partial x} + v\frac{\partial p}{\partial y} \qquad (5.3)$$

which can be written in total derivative form as

$$\frac{D}{Dt}\left(E + \frac{p}{\rho}\right) = 0 \qquad (5.3a)$$

Since the energy is taken as the sum of kinetic and potential parts per unit mass,

$$E = \frac{q^2}{2} + U \qquad (5.4)$$

it is seen that (5.3a) is identical with Bernoulli's equation (3.11):

...leading to no new information, the fluid being considered imcompressible

$$\frac{q^2}{2} + \frac{p}{\rho} + U = \text{const} \qquad (5.5)$$

for each particle of fluid in its motion along a streamline. The conclusion that may be drawn is that the momentum and energy conservation principles are identical in this case. The energy principle is therefore consistent with the momentum principle but is not needed and may be ignored.

Exercise 12 Show that the substitution of (5.4) in (5.3) recovers Euler's equations (4.3), so that the principle of energy conservation for frictionless incompressible fluid places no new requirements on the flow beyond those found earlier on the basis of mass and momentum conservation.

From the form of Bernoulli's integral (5.5) as well as the observation that the pressure divided by density has the dimensions of energy per unit mass, it is clear that this equation expresses the simple constancy of fluid energy. The pressure in an incompressible fluid may therefore be regarded as a third form of flow energy, convertible to kinetic or potential form according to the balance indicated by (5.5).

The principles of conservation of fluid mass, momentum, and energy applied to the steady plane motion of a frictionless and incompressible fluid are now seen to be expressed by the three equations

$$\frac{\partial u}{\partial x} + \frac{\partial v}{\partial y} = 0 \qquad (5.6)$$

summary of the differential equations of mass and momentum conservation

$$\left. \begin{array}{l} \dfrac{Du}{Dt} + \dfrac{1}{\rho}\dfrac{\partial p}{\partial x} = -\dfrac{\partial U}{\partial x} \\[6pt] \dfrac{Dv}{Dt} + \dfrac{1}{\rho}\dfrac{\partial p}{\partial y} = -\dfrac{\partial U}{\partial y} \end{array} \right\} \qquad (5.7)$$

for the determination of the three quantities u, v, and p. It will be recalled that the equation of continuity (5.6) and Euler's equations (5.7) were obtained without regard to the energy principle which yielded Bernoulli's integral (5.5) along a streamline in exactly the same form obtained earlier on momentum considerations as (3.11). The set of equations (5.6) and (5.7) provides the complete basis for the study of two-dimensional flow of frictionless incompressible fluid [including unsteady flow, when the total derivatives of the velocity components in (5.7) are understood in the sense of (1.11a) rather than in the more limited meaning of (4.3)].

A more compact form of the two equations (5.7) may be noted here, since the vector notation permits them to be written as a single equation. Thus, by multiplying the first of (5.7) by the unit vector **i** in the x-direction, and the second by **j**, as in (1.3), the velocity vector **q** is obtained in the total derivative term when the two equations are added, so that

$$\frac{D\mathbf{q}}{Dt} + \frac{1}{\rho}\boldsymbol{\nabla} p = -\boldsymbol{\nabla} U \qquad (5.7a)$$

vector form of (5.7)

The gradient notation for the derivatives of pressure and gravitational potential is employed in (5.7a), so that, for example,

$$\boldsymbol{\nabla} p = \mathbf{i}\frac{\partial p}{\partial x} + \mathbf{j}\frac{\partial p}{\partial y}$$

6 FLOW CONDITION AT A SOLID BOUNDARY SURFACE

differential equations must be supplemented by a certain number of auxiliary conditions

The fluid mass and momentum conservation principles are expressed by (5.6) and (5.7) as a set of *differential* equations that specify the motion of each fluid particle at each instant of time. In the same manner that an ordinary differential equation possessing many solutions is made to yield a particular solution of interest by imposing an auxiliary restriction (boundary condition or initial condition), the equations of steady fluid motion do not completely determine a particular flow until suitable auxiliary conditions are specified at the physical boundary surfaces of the flow. Since a physical boundary serves as a stream surface that is not traversed by the fluid particles, it is clear that a limitation is imposed on the velocity of fluid particles at a boundary surface. For a fixed boundary the normal component of the fluid velocity is zero at the boundary surface, so that

$$\mathbf{q} \cdot \mathbf{n} = 0 \qquad (6.1)$$

where $\mathbf{n}$ again denotes the unit normal vector, here taken as directed toward the fluid from a point of the boundary surface (Fig. II-7).

When the flow is two-dimensional in the x, y plane, the velocity is given by its Cartesian components as

solid boundary forms a stream surface — zero relative normal flow

$$\mathbf{q} = u\mathbf{i} + v\mathbf{j}$$

and the normal vector $\mathbf{n}$ can be expressed in terms of the differential elements dx and dy that are the Cartesian projections of the boundary surface line element dl. For this purpose it is noted that the unit *tangent* vector $\boldsymbol{\tau}$ is given directly as

$$\boldsymbol{\tau}\, dl = dx\mathbf{i} + dy\mathbf{j} \qquad (6.2a)$$

while $\mathbf{n} = \boldsymbol{\tau} \times \mathbf{k}$, so that

$$\mathbf{n}\, dl = dy\mathbf{i} - dx\mathbf{j} \qquad (6.2)$$

and the condition (6.1) becomes

Cartesian expression

$$u\, dy - v\, dx = 0 \qquad (6.3)$$

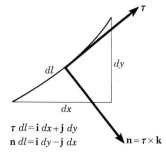

$\boldsymbol{\tau}\, dl = \mathbf{i}\, dx + \mathbf{j}\, dy$
$\mathbf{n}\, dl = \mathbf{i}\, dy - \mathbf{j}\, dx$

Fig. II-7 Illustrating oriented unit tangent and normal vectors.

in scalar form. The equation of the boundary surface can be written as

$$y = F(x) \qquad (6.4)$$

the right side of (6.4) being regarded as a given function of x. Since the equation of the boundary surface can also be written as

$$f(x, y) \equiv y - F(x) = 0 \qquad (6.4a)$$

more symmetric expression

the differentials dx and dy in (6.3) can be expressed in more symmetrical form by noting that $f(x, y) = 0$ implies

$$\frac{\partial f}{\partial x} dx + \frac{\partial f}{\partial y} dy = 0$$

Consequently (6.3) becomes

$$0 = u \frac{\partial f}{\partial x} + v \frac{\partial f}{\partial y} = \frac{Df}{Dt}$$

or simply

$$\frac{Df}{Dt} = 0 \qquad (6.3a)$$

condition at a solid boundary is a conserved (i.e., permanent) property

in total derivative form. As (6.3) expresses the vanishing of the (volume) flux $u\,dy - v\,dx$ across the boundary surface element dl, (6.3a) has the interpretation that the *property* of zero normal flux is a permanent one, i.e., the property is *conserved* in time. The interest of the boundary condition expressed in the form (6.3a) is that it is also valid for *unsteady* flow, i.e., when the boundary is in motion so that its equation becomes $f(x, y, t) = 0$ and (6.3a) takes the form

$$\frac{\partial f}{\partial t} + u \frac{\partial f}{\partial x} + v \frac{\partial f}{\partial y} = 0 \qquad (6.3b)$$

for two-dimensional flow, corresponding to (1.11). The demonstration of (6.3b) proceeds by noting that, in unsteady flow, (6.1) is replaced by the statement that the component of fluid velocity normal to the moving boundary is equal to the component of the velocity of the boundary normal to itself.

valid for unsteady flow also, the boundary surface motion being non-zero

Exercise 13 Note that the equation of the boundary surface at time t is written as

$$f(x, y, t) = 0 \qquad (6.5)$$

and at time $t + dt$ as

$$f(x + dx, y + dy, t + dt) = 0 \qquad (6.5a)$$

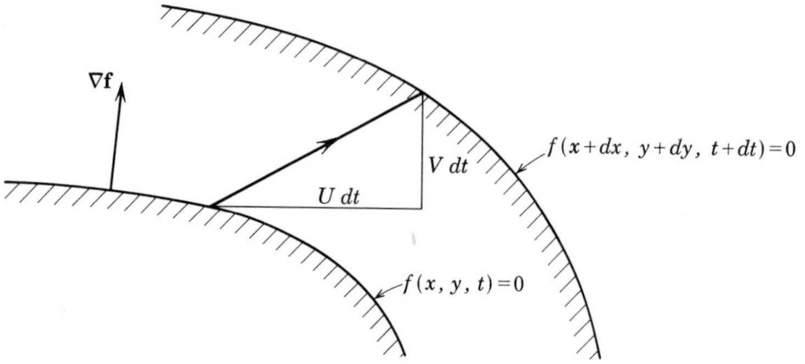

Fig. II-8 Illustrating Eqs. (6.5) and (6.5a).

so that expanding (6.5a) in ascending powers of the differentials gives, to the first order,

$$\frac{\partial f}{\partial x} U + \frac{\partial f}{\partial y} V = -\frac{\partial f}{\partial t} \qquad (6.6)$$

where U and V are the components dx/dt and dy/dt of the motion of a point on the surface (Fig. II-8). The left side of (6.6) is the scalar product of the gradient of f, ∇f, and the velocity $U\mathbf{i} + V\mathbf{j}$. Show that the unit normal vector $\mathbf{n}$ is given by

$$\mathbf{n} = \frac{\nabla f}{|\nabla f|}$$

so that the normal component of *fluid* velocity, $\mathbf{q} \cdot \mathbf{n}$, is also given by (6.6). Justify (6.3b) as the normal-flow boundary condition at a solid surface, both for steady and for unsteady flows.

7 LAGRANGIAN FORM OF THE EQUATIONS OF FLUID MOTION

nonlinearity of Euler equations: a major obstacle to mathematical solution

A serious inconvenience of the Euler equations is that they are *nonlinear* in form, and known general methods of solution apply generally only to linear equations. In contrast to the nonlinearities appearing in (4.3) as products of dependent variables and their derivatives, the acceleration terms in the Lagrangian description are entirely linear. It is therefore appropriate to examine the entire set of Lagrangian equations corresponding to (5.6) and (5.7) to determine whether they are more tractable. Although the Lagrangian equations might also be derived from first principles, the method to be employed here consists in transforming the Eulerian equations by changing the independent variables.

When x_0 and y_0 are replaced by a and b in (1.8) to conform to a more customary notation, the particle coordinates of two-dimensional flow that serve as dependent variables are written as

$$x = x(a, b, t) \qquad y = y(a, b, t) \tag{7.1}$$

with the understanding that each pair of values of a and b identifies a definite fluid particle (the one having x- and y-coordinates at time $t = 0$ respectively given by a and b, for example). Then the Cartesian velocity components (1.3) in the Eulerian system denoted by u and v are understood as

Lagrangian acceleration terms are not nonlinear

$$u = \left.\frac{\partial x}{\partial t}\right|_{a,b} = \frac{\partial(a, b, x)}{\partial(a, b, t)} \equiv \dot{x} \quad \text{and} \quad v = \dot{y} \tag{7.2}$$

The dot notation for the Lagrangian time derivative, also shown in (7.2) in the Jacobian determinant notation introduced in Appendix C, is readily extended to the accelerations as $\ddot{x}$, $\ddot{y}$. For the sake of simplicity in the transformations required the discussion is now limited to steady flow, so that $\dot{x}$ and $\dot{y}$ are functions of position coordinates x and y alone, and not the time t, in the Eulerian description. In the Lagrangian representation the same terms are regarded as functions of a and b alone, and

a useful formalism: expression of partial derivative as Jacobian determinant

$$\frac{\partial u}{\partial x} = \frac{\partial(u, y)}{\partial(x, y)} = \frac{\partial(\dot{x}, y)}{\partial(a, b)} \frac{\partial(a, b)}{\partial(x, y)} \tag{7.3}$$

basic transformation property (see Appendix C)

according to the transformation rule shown as (C.5) in Appendix C. The term (7.3) appears in the continuity equation (5.6), together with the term that is written in like manner as

$$\frac{\partial v}{\partial y} = \frac{\partial(x, \dot{y})}{\partial(a, b)} \frac{\partial(a, b)}{\partial(x, y)} \tag{7.3a}$$

Because the last factor on the right sides of (7.3) and (7.3a) is different from zero, (5.6) becomes

$$\frac{\partial(\dot{x}, y)}{\partial(a, b)} + \frac{\partial(x, \dot{y})}{\partial(a, b)} = 0 \tag{7.4}$$

which is recognized as the time derivative of the Jacobian determinant $\partial(x, y)/\partial(a, b)$. The interpretation of a and b as x- and y-coordinates at a reference time $t = 0$ means that at that instant

$$t = 0: \quad \frac{\partial x}{\partial a} = 1 \quad \frac{\partial y}{\partial b} = 1 \quad \frac{\partial(x, y)}{\partial(a, b)} = 1 \tag{7.5}$$

Finally (7.4) shows that the value of the Jacobian determinant in (7.5) is independent of time; hence the *Lagrangian form of the continuity equation* of an incompressible fluid in steady motion has been obtained as

$$\frac{\partial(x, y)}{\partial(a, b)} = 1 \tag{7.6}$$

Lagrangian form of continuity equation

which is valid also for unsteady motion [cf. discussion following (2.1b)]. When this is written in full as

$$\frac{\partial x}{\partial a}\frac{\partial y}{\partial b} - \frac{\partial x}{\partial b}\frac{\partial y}{\partial a} = 1$$

it is apparent that (7.6) is a nonlinear equation, which is less convenient than the corresponding Eulerian form (5.6).

reverse transformation recovers (2.4a)

Exercise 14 Recover the Eulerian form (2.4a) of the continuity equation by transforming the Lagrangian equation (7.6). HINT: Write (7.6) as

$$\frac{\partial(x, y, t)}{\partial(a, b, t)} = 1 \qquad (7.6a)$$

and differentiate with respect to time to obtain

$$\frac{\partial(u, y, t)}{\partial(a, b, t)} + \frac{\partial(x, v, t)}{\partial(a, b, t)} + \frac{\partial(x, y, 1)}{\partial(a, b, t)} = 0$$

before introducing Eulerian variables x, y, t.

Exercise 15 Derive the Lagrangian form of the continuity equation for a *compressible* fluid in plane motion by considering the mass of fluid contained within a region S of the x, y plane at time t

$$\int_S \rho(x, y, t)\, dx\, dy$$

and equating to the mass of the same fluid at time $t = 0$, when the density is denoted ρ_0:

$$\int_{S'} \rho_0(a, b, 0)\, da\, db$$

(cf. Fig. C-1). HINT: Apply the transformation (C-8) of Appendix C.

extension to compressible flow

$$\text{Ans.} \quad \rho_0 = \rho\, \frac{\partial(x, y)}{\partial(a, b)} \qquad (7.7)$$

Exercise 16 Use (7.2) and (7.6) to show that

$$u = \frac{\partial(a, b)}{\partial(y, t)} = \frac{\partial a}{\partial y}\frac{\partial b}{\partial t} - \frac{\partial a}{\partial t}\frac{\partial b}{\partial y} \qquad v = \frac{\partial(a, b)}{\partial(t, x)} = \frac{\partial a}{\partial t}\frac{\partial b}{\partial x} - \frac{\partial a}{\partial x}\frac{\partial b}{\partial t} \qquad (7.8)$$

for an incompressible fluid.

transformation or total derivative terms

Transformation of the Euler equations (5.7) involves expanding the acceleration components according to (1.11) and replacing each term and derivative by the corresponding Lagrangian expression. For two-dimensional flow the typical expression is

$$\frac{Du}{Dt} = \frac{\partial u}{\partial t} + u\frac{\partial u}{\partial x} + v\frac{\partial u}{\partial y} \qquad (7.9)$$

for which five separate transformations are required. An equivalent but more concise calculation starts with the Lagrangian form

$$\left.\frac{\partial u}{\partial t}\right|_{a,b} = \ddot{x} = \frac{\partial(a, b, u)}{\partial(a, b, t)} \qquad (7.10)$$

When the expression on the right side of (7.10) is transformed to Eulerian variables, it is found to be identical with the terms on the right side of (7.9), under the same conditions as before.

It is also seen from (7.3) and (7.3a) that

$$\frac{\partial f}{\partial x} = \frac{\partial(f, y)}{\partial(a, b)} \quad \text{and} \quad \frac{\partial f}{\partial y} = \frac{\partial(x, f)}{\partial(a, b)}$$

for an incompressible fluid, so that identifying f with the pressure p and with the potential U gives the first of the equations

$$\left.\begin{aligned}\ddot{x} + \frac{1}{\rho}\left(\frac{\partial p}{\partial a}\frac{\partial y}{\partial b} - \frac{\partial p}{\partial b}\frac{\partial y}{\partial a}\right) &= -\left(\frac{\partial U}{\partial a}\frac{\partial y}{\partial b} - \frac{\partial U}{\partial b}\frac{\partial y}{\partial a}\right) \\ \ddot{y} + \frac{1}{\rho}\left(\frac{\partial p}{\partial b}\frac{\partial x}{\partial a} - \frac{\partial p}{\partial a}\frac{\partial x}{\partial b}\right) &= -\left(\frac{\partial U}{\partial b}\frac{\partial x}{\partial a} - \frac{\partial U}{\partial a}\frac{\partial x}{\partial b}\right)\end{aligned}\right\} \quad (7.11)$$

Lagrangian form of equations of fluid motion

for the Lagrangian form of the dynamical equations (5.7), using (7.10) and the corresponding expression for the acceleration in the y-direction. Equations (7.11), although simpler than the Euler equations in respect to the acceleration terms which are now linear, are seen to be nonlinear because of the appearance of products of pressure and coordinate derivatives, both of these quantities now occurring as dependent variables. It follows that there is no clear and general advantage in adopting the Lagrangian equations, although there are special cases in which they are preferable.

pressure terms are now nonlinear

Exercise 17 Show that multiplication of each of the equations (7.11) by an appropriate factor and adding leads to an alternative form:

$$\left.\begin{aligned}\ddot{x}\frac{\partial x}{\partial a} + \ddot{y}\frac{\partial y}{\partial a} + \frac{1}{\rho}\frac{\partial p}{\partial a} &= -\frac{\partial U}{\partial a} \\ \ddot{x}\frac{\partial x}{\partial b} + \ddot{y}\frac{\partial y}{\partial b} + \frac{1}{\rho}\frac{\partial p}{\partial b} &= -\frac{\partial U}{\partial b}\end{aligned}\right\} \quad (7.11')$$

alternative form with linear pressure terms — accelerations nonlinear again

In analogy with (5.7) linearity of the pressure terms is secured only by reintroducing a nonlinearity in the acceleration terms.

Exercise 18 Transform each of the terms on the right side of (7.9) in order to show that their sum is $\ddot{x}$ [observe that, when (7.3) and (7.8) are used, it is only necessary to evaluate $\partial u/\partial t$ and $\partial u/\partial y$ in Lagrangian form].

$$\text{Ans:} \quad \frac{\partial u}{\partial t} = \ddot{x} + \frac{\partial \dot{x}}{\partial a}\left(v\frac{\partial x}{\partial b} - u\frac{\partial y}{\partial b}\right) - \frac{\partial \dot{x}}{\partial b}\left(v\frac{\partial x}{\partial a} - u\frac{\partial y}{\partial a}\right).$$

Exercise 19 Expand the Jacobian determinant of (7.10) to recover (7.9).

introduce velocity components as independent variables

Exercise 20 Show that the continuity equation (2.4a) remains linear when the roles of the independent variables x, y and dependent variables u, v are *interchanged* [assume that the transformation determinant $\partial(x, y)/\partial(u, v) \neq 0$].

8 STREAMLINE COORDINATES

A judicious choice of variables usually simplifies the solution of a physical problem, particularly with respect to the coordinates adopted in the description of fluid motions. Whereas Cartesian coordinates may seem appropriate in one case, the plane and spherical polar systems are distinctly preferable in others, etc. It is clear that the selection of a coordinate system involves, in addition to an element of arbitrariness, also a certain degree of artificiality: fluid flows according to natural laws, not in conformance with our coordinate preferences. A set of directions determined by the flow itself would appear to offer a unique advantage from the calculation standpoint, and the streamlines already introduced form the basis of a system of "natural coordinates."

When the discussion is limited again to plane flow, distances measured along the streamline serve as one coordinate, and the distance normal to the streamline tangent in a given direction furnishes another in a definite manner. When these coordinates are denoted s and n, respectively, and their increments by Δs and Δn, a curvilinear coordinate grid in the x, y plane as indicated in Fig. II-9 is obtained; it is simpler than the ξ, η mesh of Appendix C in that s and n are mutually orthogonal at each point. It is now possible to carry out stream tube flow analyses in greater detail than in Sections 2 and 3 where the stream tubes were of more general three-dimensional form.

streamlines can serve as coordinates

The cross-section area of a stream tube of width Δn and unit depth in the direction normal to the flow plane is now $\Delta n 1$, and the differential form of the continuity condition (2.1a) can therefore be written at once as

$$\frac{\partial}{\partial s}(\rho q\, \Delta n) = 0 \tag{8.1}$$

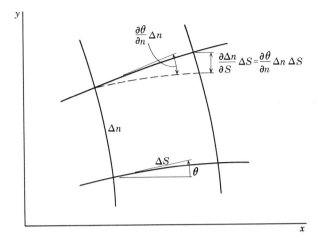

Fig. II-9 Streamline coordinates, differential variations.

When, for simplicity, the density is taken to be constant and it is noted that the stream tube width Δn varies from one point to another, it is found from the preceding development that, for an incompressible fluid,

$$q \frac{\partial \Delta n}{\partial s} + \Delta n \frac{\partial q}{\partial s} = 0$$

continuity equation in primitive streamline coordinate form

The directions of s and n being everywhere normal to each other, it is possible to express the streamwise variation of the stream tube width Δn in terms of the variable flow direction from one point to another. Thus the inclination of the flow velocity vector with respect to the x-axis is denoted by θ (this is *not* the polar angle associated with the x, y Cartesian axes); then the increment of Δn corresponds to the n-variation of the flow angle:

$$\frac{\partial \Delta n}{\partial s} \Delta s = \frac{\partial \theta}{\partial n} \Delta n \, \Delta s$$

analysis of differential variations of coordinates

The continuity equation in *streamline coordinates* s, n is now obtained from (8.1) as

$$q \frac{\partial \theta}{\partial n} + \frac{\partial q}{\partial s} = 0 \qquad (8.2)$$

When the flow speed q and direction θ are both regarded as unknowns, (8.2) is again nonlinear, with the further seeming complication that the coordinate directions s and n are also not known until the flow is determined. It will be seen later, however, that the interchange of dependent and independent variables in (8.2) in the manner of Exercise 20 leads to linearity of the continuity equation and also of the dynamical equations.

Exercise 21 Note that the relationship between Cartesian coordinates x, y and the streamline coordinates s, n is given by the differential transformation equations

$$\Delta x = \Delta s \cos\theta - \Delta n \sin\theta$$
$$\Delta y = \Delta s \sin\theta + \Delta n \cos\theta \qquad (8.3)$$

and obtain (8.2) by direct transformation of (2.4a). HINT: Express the Cartesian velocity components in terms of q and θ as

$$u = q\cos\theta \quad \text{and} \quad v = q\sin\theta$$

Exercise 22 Interchange the roles of dependent and independent variables in (8.2) to obtain

hodograph variables: velocity furnishes the independent variables

$$q\frac{\partial s}{\partial q} + \frac{\partial n}{\partial \theta} = 0 \qquad (8.4)$$

as the *linear* form of the continuity equation in *hodograph variables* q, θ. HINT: Write the derivative $\partial\theta/\partial n$ in determinant form, $\partial(s,\theta)/\partial(s,n)$, and introduce (q, θ) as new variables, to find

$$\frac{\partial\theta}{\partial n} = \frac{\partial s}{\partial q}\frac{\partial(q,\theta)}{\partial(s,n)}$$

Exercise 23 Recover (8.4) by transforming the derivatives $\partial y/\partial v$ and $\partial x/\partial u$ in the result of Exercise 20 by noting relationships obtained from (8.3), such as

$$\frac{\partial y}{\partial q} = \frac{\partial s}{\partial q}\sin\theta + \frac{\partial n}{\partial q}\cos\theta$$

The condition of momentum conservation in the steady frictionless motion of an incompressible fluid is obtained in streamline coordinates in the same manner as in Section 3. It is observed that in the present notation the force-momentum balance (3.10) contains derivatives with respect to s only; then the Bernoulli integral (3.11) for plane flow is recovered as

streamline form of momentum requirement

$$\frac{q^2}{2} + \frac{p}{\rho} + U = \text{const} \qquad (8.5)$$

The pair of equations (8.2) and (8.5) indicate that the pressure is known when the flow speed q has been determined. It will be seen that an additional relationship, taken in conjunction with (8.2), permits calculation of both the speed q and the inclination of the flow given by θ, so that (8.5) may be regarded as specifying the pressure and need not be considered in determining the form of the fluid motion.

Exercise 24 Indicate the reason why an additional independent relationship between flow speed and pressure is *not* obtainable by differentiation of (8.5) with respect to the normal coordinate n.

III
Elementary Frictionless Flows

The systematic study of fluid mechanics consists of the examination of various types of flows that are found as solutions of the equations of motion. Because of the features of the fundamental equations already mentioned, especially the nonlinearities, the most readily found solutions do not necessarily correspond to the most elementary flows from the physical standpoint. When the complete formal solution is not readily obtained, the main features of a flow process nevertheless can often be explained directly in terms of the dominant physical effects such as forces and boundary surface geometry. Several flows are so analyzed in this chapter before an orderly treatment of the basic equations in more complete form is begun.

1 NATURAL CONVECTION — CHIMNEY FLOW DYNAMICS

Aerostatic equilibrium is upset when the weight of a column of gas is less than the buoyant pressure force acting on it. This is what happens, for example, when a heated and expanded gas fills a chimney flue the pressures of which at the upper and lower ends remain at the values determined by the relatively cool and heavy atmosphere. The resultant upward force on the chimney gas, given by the excess buoyant force, thereby accounts for an outward flux of upward momentum at the top of the chimney.

If the motion is in the upward vertical z-direction, the acceleration in steady flow is

$$\frac{Dw}{Dt} = w\frac{dw}{dz}$$

III. 1 / NATURAL CONVECTION — CHIMNEY FLOW DYNAMICS

and the equation (I.5.1) of static equilibrium is replaced by *an extension of aerostatics: inertia as well as pressure and gravity forces*

$$w \frac{dw}{dz} = -\frac{1}{\rho} \frac{dp}{dz} - g \qquad (1.1)$$

where w, p, and ρ refer to conditions within the chimney gas. Recall that the pressure decreases in the upward direction, $dp/dz < 0$, and that the density ρ must be less than, for example, the value ρ' of the air outside the chimney. Then it is seen from (1.1) that the speed w increases in the upward direction, $dw/dz > 0$. Motions of this type, "driven" by an insufficiency of the gravity force acting downward, are termed "natural convection."

If the density ρ is regarded as known and the pressure gradient in (1.1) is taken as the atmospheric value, it becomes possible to infer from this equation a relationship between chimney height and the chimney draft velocity at its upper end.

For this purpose the height of the chimney is denoted by h, the atmospheric pressures at levels $z = 0$ and $z = h$ by p_0' and $p'(h)$, respectively. The pressures in the chimney gas at the same levels are $p(0) \equiv p_0$ and $p(h)$, supposed equal to the ambient values, so that (Fig. III-1)

$$p(h) = p'(h) \quad \text{and} \quad p_0 = p'_0 \qquad (1.2)$$

It follows that the vertical pressure gradients are equal inside and outside the chimney, on the average, and it is assumed that

$$\frac{dp}{dz} = \frac{dp'}{dz} \qquad (1.3)$$ *a plausible assumption*

at all levels as well. The last term, by (I.5.1), equals $-\rho'g$, and substitution in (1.1) now gives

$$w \frac{dw}{dz} = \left\{ \frac{\rho'}{\rho} - 1 \right\} g \qquad (1.4)$$

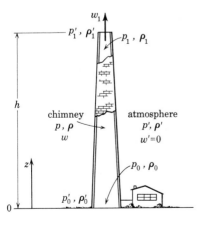

Fig. III-1 Chimney draft flow: chimney conditions p, ρ, w; atmosphere conditions p', ρ', $w' = 0$.

The coefficient of g on the right side of (1.4) is positive, and if its average value over the height h is denoted by the same symbol, the fact that the left side is the z-derivative of $w^2/2$ leads to

chimney draft proportional to $h^{1/2}$

$$w(h) = \left[2\left\{\frac{\rho'}{\rho} - 1\right\}gh\right]^{1/2} \quad (1.5)$$

by integration, taking $w(0) = 0$. Equation (1.5) indicates that the chimney draft increases as the square root of the height of the chimney, in accord with a commonly used rule of thumb for the design of smokestacks and chimneys. It follows that contamination of the air by noxious gases is reduced by using chimney column of sufficient height to impart a considerable speed to gases carried to those levels where they are unlikely to create a nuisance at the surface. An additional benefit is that air drawn into the chimney "ventilates" the flame in which the gases are heated, leading to more intensive and thorough combustion in general.

limitations of the analysis

There are several factors, not taken into account in the preceding calculation, which prevent a strict meaning from being attached to such a relationship as (1.5). The flow is not strictly one-dimensional and vertical, for example, and friction forces are by no means unimportant, so that (1.5) should be regarded as only approximate: but it is questionable whether the effort required to secure greater accuracy would be justified by the results.

a numerical example

Exercise 1 Calculate the chimney draft $w(h)$ for a chimney of height 250 ft when the density ρ corresponds to air at temperature 300°F. Use (1.5) to show that the quantity of gas removed by a chimney does *not* increase indefinitely as the temperature is increased.

2 DEDUCTIONS FROM BERNOULLI'S INTEGRAL: TORRICELLI'S THEOREM, PITOT TUBE

It is readily shown that the result (1.5), obtained by integrating the dynamical equation (1.4), is found still more directly by applying Bernoulli's integral (II.5.5) in the same manner as in the calculation of Section I.8. Among the many other flow problems where the integration of the fundamental equations can be avoided by using the Bernoulli integral at the outset, two of particular importance are now considered.

incompressible fluid: application of Bernoulli's theorem

The discharge of liquid through an orifice in a tank open to the atmosphere can be regarded as an immediate extension of the hydrostatic pressure calculations of Section I.3. The flow speed

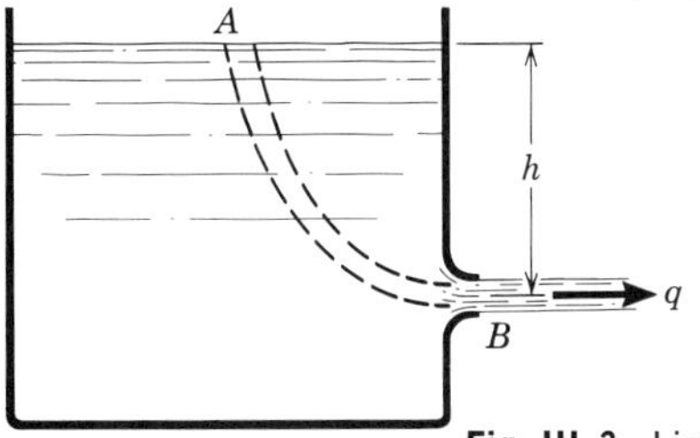

Fig. III-2 Liquid efflux through an orifice.

q of the liquid escaping at depth h below the free surface (Fig. III-2) is determined by noting that the drop in pressure from the hydrostatic value as the liquid enters the atmosphere is accompanied by an increase of kinetic energy as indicated by (II.5.5). Specifically, the escaping fluid particle at B can be considered to lie on the same streamline as a point A at the free surface where the flow speed is taken as zero (the steady state implying that the level is maintained by continually refilling the tank). Then writing the gravity potential U as gz and indicating each of the two points by a subscript gives Bernoulli's integral as

gravitational potential energy converted to fluid motion (kinetic energy)

$$\frac{q_B^2}{2} + \frac{p_0}{\rho} + gz_B = \frac{p_0}{\rho} + gz_A \qquad (2.1)$$

The difference in heights z_B and z_A corresponds to h, and it follows at once that

$$q_B = [2gh]^{1/2} \qquad (2.2)$$

which closely resembles (1.5). The fact that the flow speed q_B is identical with the speed acquired in free fall from rest through distance h, according to (2.2) is known as Torricelli's theorem, in honor of the student of Galileo who is credited with the discovery nearly four hundred years ago. The interpretation of Torricelli's theorem is simplified by the fact that the fluid velocities are everywhere very small until a fluid particle reaches the orifice. Then a fluid particle at A descends with negligible kinetic energy, the incremental loss of potential energy $g\,\Delta z$ being offset by increased pressure in accordance with Bernoulli's theorem (II.3.11), where U again takes the form gz. The accumulated "pressure energy" ultimately increases to the amount gh, per unit mass, as the particle approaches the level z_B. The sudden reduction of pressure as the fluid passes through the orifice is accompanied by the conversion of this energy to kinetic form, again in accordance with (II.3.11). A freely falling particle experiences the same conversion of potential to kinetic energy, without the intermediary pressure

Torricelli's theorem: flow speed equals free-fall value

form, so that the final velocities in the two cases are identical. A noteworthy feature of the flow in question is that the two energy conversion processes occur without internal losses of a frictional character such as are found in different types of energy conversion. It may also be mentioned that the area of the jet of fluid narrows as it leaves the orifice, the amount of area contraction depending on the pressure difference and other effects.

A second example is provided by flow at essentially constant level, so that the potential energy remains constant. In passing around a blunt obstacle, fluid particles move to one side or another of a *stagnation point* where the velocity at the boundary surface is zero. The *stagnation pressure* p_1 there provides a measure of the relative speed q of the flow far ahead of the body, where the pressure p is regarded as given. According to Bernoulli's theorem again,

kinetic energy conversion to pressure form: Pitot's airspeed indicator

$$q = \left[2 \frac{p_1 - p}{\rho} \right]^{1/2} \tag{2.3}$$

In the impact tube (or *Pitot tube*) arrangement indicated schematically in Fig. III-3 the free-stream pressure p is sensed at an opening in the side of the tube where the flow speed is closely given by q, the pressure difference being measured directly by the differential height h of the two columns of an attached manometer. Because of the appearance of the density in (2.3), altitude corrections are required in practice on Pitot tubes installed on airplanes that fly at appreciable elevations.

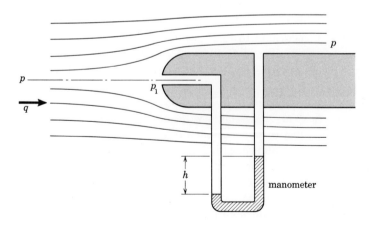

Fig. III-3 Pitot tube airspeed indicator.

3 REACTION THRUST — ROCKET ENGINE PROPULSION

When a rocket engine expels its combustion products rearward at high speed, the reaction to the momentum flux of the exhaust jet appears as forward thrust on the rocket itself. The main features of rocket performance are found by expressing the rocket acceleration directly in terms of the flux of momentum in the exhaust jet.

The required relationship is obtained by considering the motion of the rocket and its exhaust at two successive instants, t and $t + dt$, when the rocket speed increases from V to $V + dV$. If the rocket mass changes from M to $M + dM$ in the same interval of time ($dM < 0$ expresses the *reduction* of mass corresponding to the amount of expelled fuel and oxidizer in solid or liquid chemical propulsion systems), conservation of total mass is implied by denoting the expelled mass $-dM$.

When air resistance and gravity force on the rocket are both ignored, momentum conservation is assured by equating the momentum MV at the instant t to the sum of the rocket momentum and the increment of exhaust momentum at time $t + dt$. If these are written as $(M + dM)(V + dV)$ and $-dM(V - c)$, respectively, where c is the exhaust speed relative to the rocket (Fig. III-4), then

mass and momentum conservation — gross analysis

$$MV = (M + dM)(V + dV) - dM(V - c) \qquad (3.1)$$

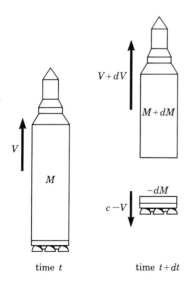

Fig. III-4 Rocket mass and speed at times t and $t + dt$.

The right side of (3.1) contains terms of zero, first, and second order in the differentials, and the left side is equal to the zero-order term on the right. When the remaining first-order terms are retained and those of second order are neglected, (3.1) gives

$$c\, dM = -M\, dV \qquad (3.2)$$

as a differential equation relating instantaneous values of rocket mass and velocity. If c is taken as constant, separation of variables and integration are possible and they lead to

$$\ln \frac{M_0}{M(t)} = \frac{V(t)}{c} \qquad (3.3)$$

rocket mass ratio

when the initial speed is taken as zero, corresponding to launching from rest. Equation (3.3) shows that the rocket speed at any time is dependent both on the instantaneous value of the *mass ratio* $M_0 \div M(t)$ and on the quality of the fuel as determined by c. Inverting (3.3) gives the rocket mass as an explicit function of time in the form

fundamental equation of rocket performance — deduced from mass and momentum conservation

$$M(t) = M_0 \exp\left(-\frac{V(t)}{c}\right) \qquad (3.4)$$

Equation (3.3) is termed the fundamental equation of rocket performance and gives the final, or *burnout*, velocity V_f as a function of c and the final value of the mass ratio $M_0 \div M_f$, where M_f is the rocket mass at the end of the combustion period, related to the propellant mass M_p by the equation $M_f = M_0 - M_p$. Then (3.3) indicates that

$$V_f = c \ln \frac{M_0}{M_f} \qquad (3.5)$$

A given value of the *characteristic velocity* V_f can evidently be realized in a variety of ways, because different fuel choices (i.e., values of c) are accommodated by adjustment of the mass ratio as determined by the quantity of fuel carried. Although a completely rational procedure in rocket design, including fuel selection, has not been developed, the concept of *propulsive efficiency* is taken as one measure of the quality of rocket performance. One definition gives this quantity as the ratio of the rate at which rocket thrust force T performs work divided by the sum of the same expression plus the rate at which the rocket loses energy in its jet wake:

propulsive efficiency — the perils of arbitrary definitions

$$\eta_P \equiv \frac{TV}{TV + [(V-c)^2/2](-\dot{M})} \qquad (3.6)$$

The effective thrust force, taken as the product of rocket mass times its acceleration, $M\dot{V}$, is seen from (3.2) to be equal to $-c\dot{M}$, and, when this value is substituted for T, (3.6) becomes

$$\eta_P = 2\frac{V}{c}\frac{1}{1+(V/c)^2} \quad (3.7)$$

peak value: unity — "perfect" efficiency

indicating a peak efficiency of unity when the rocket speed is just equal to the jet exhaust speed.

Exercise 2 (a) Show that $T = -c\dot{M}$ by the direct application of Newton's second law to a control volume containing the rocket but not the exhaust jet. (b) Interpret the condition $\eta_P = 1$ in (3.7) in terms of energy lost by the rocket.

Exercise 3 An alternative (and unsatisfactory) definition of rocket propulsive efficiency might consider the rate of energy release, $\dot{E}$, resulting from combustion in the rocket chamber, so that

$$\eta'_P \equiv \frac{TV}{\dot{E}} \quad (3.8)$$

where $\dot{E}$ corresponds to the differential energy dE in the energy balance, in analogy with the momentum balance (3.1). Thus

(Total kinetic energy)$|_t + dE =$ (total kinetic energy)$|_{t+dt}$ \quad (3.9)

Show that (3.9) leads to

$$dE = -dM\frac{c^2}{2} \quad (3.9a)$$

and that

$$\eta'_P = 2\frac{V}{c} \quad (3.10)$$

peak efficiency: better than unity?

What is the fallacy that leads to the excessive values (3.10)? HINT: Compare denominators of (3.8) and (3.10) and interpret $\frac{1}{2}(-\dot{M})V^2$.

Exercise 4 Still another possible definition would replace the "useful work rate" TV by the rate of increase of rocket kinetic energy, $\frac{d}{dt}\left(\frac{MV^2}{2}\right)$ in (3.6):

$$\eta''_P \equiv \frac{\frac{d}{dt}\left(\frac{MV^2}{2}\right)}{\frac{d}{dt}\left(\frac{MV^2}{2}\right) + \frac{(V-c)^2}{2}\left(-\dot{M}\right)} \quad (3.11)$$

Evaluate (3.11) as a function of the ratio V/c to prove that $\eta''_P < \eta_P$, and explain the difference by showing that the rocket kinetic energy increases less rapidly than TV. Decide whether (3.6) is an overstatement.

Exercise 5 Both of the definitions (3.6) and (3.11) indicate that the propulsive efficiency of a rocket approaches zero when $V \to 0$ and when $V \to \infty$. What is the physical significance of this characteristic?

Exercise 6 Investigate the error introduced in (3.4) by the neglect of gravity for the case of motion vertical upward when (3.2) must be more accurately written as

$$M \frac{dV}{dt} = -c \frac{dM}{dt} - Mg$$

estimation of error committed in neglect of gravity

leading to the equation

$$V(t) = -gt + c \ln \frac{M_0}{M(t)} \qquad (3.3a)$$

in place of (3.3). Under what conditions is it justified to neglect the gravity force?

4 SIMPLIFIED MOMENTUM THEORY OF PROPELLERS: PROPULSIVE EFFICIENCY

In conventional propulsive systems operating within the Earth's air or ocean atmospheres, screw propellers are commonly employed that impart momentum to the fluid medium in which they operate. *propulsion within a fluid medium* The forward thrust on the propeller is balanced by the fluid momentum it generates in the fluid, and the operating characteristics of the propeller are thereby related to the external flow conditions.

A propeller imparts energy to the fluid by increasing the pressure of each fluid particle during its brief passage from just ahead to just behind the propeller. As the distance of the fluid particle from the propeller subsequently increases, the pressure returns to the ambient value while the flow speed increases in accordance with the condition of Bernoulli's integral. A simplified one-dimensional analysis of the flow considers the fluid confined within the stream tube passing through the propeller blade tips. When the area of the propeller plane is denoted S, the stream tube area at a downstream section where the pressure is atmospheric will be smaller than S, the flow velocity being larger. By denoting it B, and introducing the stream tube area A far ahead of the propeller where the fluid velocity relative to the propeller is v, as in Fig. III-5, the areas and velocities are related by the continuity equation (II.2.1a) which can now be written as

$$\rho v A = \rho(v + v_1)S = \rho(v + v_2)B \qquad (4.1)$$

stream tube analysis of mass and momentum flux balances

expressing the constancy of fluid mass flux in steady flow through the stream tube. The fluid is assumed incompressible, and the velocity increments v_1 and v_2 refer to the flow at the propeller and

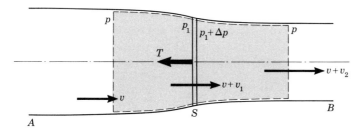

Fig. III-5 Idealized flow conditions in stream tube containing screw propeller.

far behind it. Atmospheric pressure is denoted p, and the pressures just ahead of and just behind the propeller are p_1 and $p_1 + \Delta p$, respectively, so that the Bernoulli integral (II.3.11) applied separately to the fluid in each of the two regions gives

$$\frac{p}{\rho} + \frac{v^2}{2} = \frac{p_1}{\rho} + \frac{(v+v_1)^2}{2} \qquad (4.2)$$

$$\frac{p}{\rho} + \frac{(v+v_2)^2}{2} = \frac{p_1 + \Delta p}{\rho} + \frac{(v+v_1)^2}{2} \qquad (4.3)$$

when variations of potential energy are neglected.

The propeller thrust force T can evidently be expressed in terms of the pressure increment Δp and the area S on which it acts. But T is also the force acting on the fluid and therefore given by the excess of the momentum flux at B as compared with A, the pressures being the same at both places (this is the condition that defines both A and B, in fact). The two relationships are therefore written as

$$T = \Delta p S = \rho(v+v_2)^2 B - \rho v^2 A \qquad (4.4)$$

thrust expressed in terms of momentum flux difference

Hence that the set of simultaneous equations (4.1)–(4.4) are six in number. It is appropriate to regard the thrust force T, the propeller area S, the atmospheric density and pressure ρ and p, and the forward speed v as known, or given, data for the flow problem, so that the stream tube areas A and B, the velocity increments v_1 and v_2, and the pressure p_1 and pressure increment Δp are treated as the "unknowns" to be determined from the solution of the set of equations. This is readily accomplished by judicious eliminations, since the equations are algebraic in nature, of degree not higher than quadratic in any of the unknowns, and not all of the unknown quantities appear in each of the separate equations.

six equations, six unknowns

A specific calculation that serves as a guide for the solution of the equations is the estimation of the *ideal propulsive efficiency* defined as the ratio of thrust-work Tv divided by the rate at which

propulsive efficiency defined ...

the propeller adds energy to the fluid on which it operates. The latter quantity, recognized as the power expended by the propeller, is evaluated at once by noting that the kinetic energy of the fluid is increased from the value $v^2/2$ per unit mass far ahead of the propeller to the value $(v + v_2)^2/2$ behind it, while the pressure is the same at both places. Since the total fluid mass affected in unit time is given by $\rho v A$, the power P is given by

$$P = \rho v A \left[\frac{(v + v_2)^2}{2} - \frac{v^2}{2} \right] \quad (4.5)$$

...and evaluated...

in the definition of the propulsive efficiency

$$\eta_P = \frac{Tv}{P} \quad (4.6)$$

The form of (4.5) suggests finding a similar expression for T, obtained by replacing B and v_2 in the last form of (4.1) by A and v in (4.4):

$$T = \rho v A v_2 \quad (4.4a)$$

When (4.4a) is substituted in (4.6), the power P is written as (4.5) and the expression for the propulsive efficiency is reduced to

...in terms of flow quantities

$$\eta_P = \frac{1}{1 + v_2/2v} \quad (4.6a)$$

showing that this quantity is fully determined by the incremental velocity v_2 in the region far behind the propeller. It is therefore of interest to solve the equations (4.1)–(4.4) in a manner that permits v_2 to be expressed in terms of the given data. The obvious advantage of low values of the slipstream velocity v_2 from the standpoint of efficiency is already apparent from (4.6a) without further calculation. Equation (4.6a) also demonstrates the corollary importance of maximizing the diameter of a propeller, since (4.4a) shows that a given value of the thrust is achieved with smaller speed v_2 when the area and hence the quantity of fluid affected are increased. In marine applications both the weight of the propeller and *cavitation* damage impose limitations in this regard which are not present for propellers operating in air, so that the generally higher propulsive efficiency of aircraft propeller operation can be ascribed to the relatively large diameters of airplane propellers.

propellers of large diameter achieve higher propulsive efficiencies

Exercise 7 Two independent expressions for the thrust are obtained by using the first of (4.1) in (4.4a) and by eliminating p_1 from the pair of equations (4.2) and (4.3) while replacing Δp by using (4.4). Compare the two expressions and thus show that

$$v_1 = \frac{v_2}{2} \quad (4.7)$$

solution indicates how propeller affects upstream flow

that is, that one-half of the slipstream speed increment v_2 is imparted to the fluid particles *before* they reach the propeller. What does this indicate concerning the pressure p_1 ahead of the propeller?

Exercise 8 Substitute (4.7) in the first of the two expressions for T obtained in the preceding exercise to obtain a quadratic equation for v_2 in terms of S, v, ρ and T. Solve this equation for the ratio v_2/v in terms of the dimensionless parameter termed *thrust coefficient*, defined as

$$C_T = \frac{2T}{\rho S v^2} \quad (4.8)$$

dimensionless coefficient

Select the one physically appropriate root and evaluate the efficiency in terms of the thrust coefficient C_T alone. Also express A and B, v_1, p_1, and Δp in terms of C_T and the given data.

Exercise 9 Show that

$$-1 + \sqrt{1 + x} \doteq \frac{x}{2} \quad \text{when} \quad |x| \ll 1$$

and hence that the ideal propulsive efficiency is given *approximately* by

$$\eta_P = \frac{1}{1 + C_T/4} \quad (4.9)$$

linearized approximation

when the propeller operation is characterized by small values of the parameter (4.8). It may be noted in passing that actual propulsive efficiency values greater than 90% are not uncommon in aeronautical practice.

Exercise 10 Evaluate the slipstream contraction ratios

$$\frac{A - S}{S} \quad \text{and} \quad \frac{S - B}{S}$$

in terms of C_T in approximation consistent with (4.9).

Exercise 11 Show that the *pressure drop* ahead of the propeller is given approximately by

$$\frac{p - p_1}{\rho v^2} = \frac{C_T}{4}$$

and explain how the propeller produces this effect on fluid with which it has not had direct contact.

5 MOTION IN A ROTATING REFERENCE FRAME — GEOSTROPHIC WINDS

The reader will note that the propulsive efficiency defined for screw propellers refers to instantaneous values in flow that can be regarded as steady, not fully in analogy with the usage in rocket propulsion analysis.

Exercise 12 Refer to Exercise 21 of Chapter I and evaluate the acceleration components $\ddot{X}, \ddot{Y}, \ddot{Z}$ of a particle moving with relative velocity components $u, v, 0$ in the rotating reference frame.

Exercise 13 Apply the coordinate transformation of Exercise 21 of Chapter I to the preceding result in order to evaluate the acceleration components in the *rotating* coordinate system x, y, z. Show that Euler's equations of plane fluid motion, referred to the rotating coordinates, become

$$\left.\begin{aligned}\frac{Du}{Dt} - 2\omega v - \omega^2 x + \frac{1}{\rho}\frac{\partial p}{\partial x} &= -\frac{\partial U}{\partial x} \\ \frac{Dv}{Dt} + 2\omega u - \omega^2 y + \frac{1}{\rho}\frac{\partial p}{\partial y} &= -\frac{\partial U}{\partial y}\end{aligned}\right\} \quad (5.1)$$

motion in a rotating coordinate system — Coriolis acceleration

in place of (I.8.3) for the motion of fluid *fixed* with respect to the rotating coordinates, and in place of (II.4.3) for moving fluid in an *inertial* coordinate system.

In addition to the already familiar centripetal acceleration terms in (5.1) that effectively combine with the real gravitational forces to form the apparent measurable gravity, fluid motion on the rotating Earth introduces the additional two ω-dependent terms in (5.1), known as the components of *Coriolis* acceleration. They represent the combined effect of rotation and relative motion and are of first importance in determining the forms of the large-scale wind systems in the Earth's atmosphere as well as currents in the oceans.

In order to apply equations (5.1) to the winds on the rotating Earth, it is recalled that at any place on the Earth where the latitude angle is denoted λ, the component of the Earth's axial rotation normal to the local horizon plane is given by the rotation speed times the sine of the latitude. Hence, if ω is interpreted as the Earth's rotation speed and equations (5.1) are applied to the nearly plane flow over the earth's surface at latitude λ, then ω is replaced by $\omega \sin \lambda$ in each of these equations. When x and y are taken parallel to the Earth's surface at the place in question (x measured eastward, y northward, for example), the effective gravity terms in these equations are very small and can be neglected (Fig. III-6).

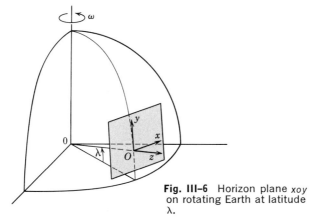

Fig. III-6 Horizon plane xoy on rotating Earth at latitude λ.

The equations then express the balance of acceleration terms in the rotating frame and the Coriolis and pressure gradient terms. It is not uncommon that the latter two terms are dominant, while the total derivative terms for each of the velocity components u and v are by comparison much smaller than these (this implies not simply a steady motion, but also motion on such a large scale that the velocity derivative terms like $\partial u/\partial x$ are small compared with ω).

horizontal motion — pressure balance by Coriolis term alone

In this case the pair of equations (5.1) is replaced by the pair of equations

$$\left.\begin{aligned} -2\omega \sin\lambda\, v + \frac{1}{\rho}\frac{\partial p}{\partial x} &= 0 \\ 2\omega \sin\lambda\, u + \frac{1}{\rho}\frac{\partial p}{\partial y} &= 0 \end{aligned}\right\} \quad (5.2)$$

from which the principal feature of the *geostrophic* winds can be directly inferred. Equations (5.2) indicate that the wind direction is *perpendicular* to the direction of the pressure gradient, i.e., the wind direction is parallel to the isobars (lines of constant pressure) when the pressure gradients are balanced only by the effective Coriolis force. This is seen by multiplying the first of (5.2) by u, the second by v, and adding:

motion is not directed from high pressure to low

$$\mathbf{q}\cdot\nabla p = u\frac{\partial p}{\partial x} + v\frac{\partial p}{\partial y} = 0$$

The vanishing of the scalar product of the two vectors $\mathbf{q}$ and ∇p indicates their perpendicularity (Fig. III-7).

Equations (5.2) tell also the wind directions at any location near the centers of meteorological activity. As these centers are identified with regions of low barometric pressure, it is seen from the first of (5.2), for example, that southerly winds ($v > 0$) are to be found to the east of the low-pressure centers, in the northern hemisphere, since both $\partial p/\partial x$ and $\sin\lambda$ are positive in this case. This finding conforms with common observation, as well as with the northerly winds that follow the passage of (eastward-moving) storm systems associated with low-pressure centers. Frictional effects in the lowest layers of the atmosphere near the surface of the Earth, neglected in the present calculation, modify somewhat the results just found, leading to wind directions with a component toward the low-pressure region. The characteristic anticlockwise direction (termed *cyclonic*) of wind motion around low-pressure storm centers and the converse sense around the typical high-pressure zones that accompany fair weather are familiar on published weather charts as well as in daily outdoor experience (Fig. III-7).

The phenomenon of *upwelling* in the oceans on the west coasts of large continents in the northern hemisphere furnishes an interesting example of the consequences of the motion just discussed, coupled with wind-sea surface interaction. Because of the high-pressure zones that become strongly established over the oceans, the winds form a clockwise sense of motion with northerly winds (i.e., winds coming from the north) at the eastern edges adjacent to the continental mass. The frictional stresses exerted by the winds on the surface layers of the ocean then represent a force directed southward, giving rise to a motion in the same direction,

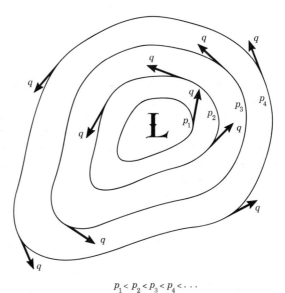

$p_1 < p_2 < p_3 < p_4 < \cdots$

Fig. III-7 Cyclonic wind flow: anticlockwise around low pressure center in northern hemisphere.

viz., southward so that $v < 0$. The first two terms of (5.1) show that a southward-moving particle is deflected to the right (i.e., to the west) by the Coriolis effect, and the surface waters are thus seen to be carried away from the shore. This water is replaced by water from greater depths, where the temperatures are reduced; the results are not only the comparatively colder bathing temperatures characteristic of west coast beaches, but also the more steeply sloped offshore beaches that contrast with conditions found on the east coast of the United States, for example.

*Exercise 14 Note that the total derivative of an arbitrary vector quantity **A** having components a_1, a_2, and a_3 in a reference frame rotating with uniform angular velocity ω is given by

$$\frac{D\mathbf{A}}{Dt} = \mathbf{i}\frac{Da_1}{Dt} + \mathbf{j}\frac{Da_2}{Dt} + \mathbf{k}\frac{Da_3}{Dt} + \boldsymbol{\omega} \times \mathbf{A} \qquad (5.3)$$

and show that, owing to the Earth's Coriolis effect, a particle in horizontal motion in the northern hemisphere appears to be deflected to the right.

6 IMPACT OF LIQUID JET STRIKING A SOLID SURFACE

When a jet of liquid impinges on a fixed surface, the momentum of each fluid particle is altered as its motion is deflected in the vicinity of the surface. The fluid pressure equals the atmospheric or ambient value at every point of the free surface between the liquid and the atmosphere in which it flows, and this is the pressure throughout the liquid wherever the flow is rectilinear and uniform. The integral statements of continuity and momentum conservation are applied directly to determine the total reaction of the wall surface and also the manner of spreading of the liquid deflected by the impact, when the motion is steady and two-dimensional (flat jet).

uniform pressure and gravity neglected: uniform flow speed

A jet of thickness a in the direction normal to its motion at uniform speed q is assumed to impinge on a fixed flat surface inclined at angle α to the jet. After a fluid particle has been deflected, it will be found either in one branch of the flow with a forward component of speed $q_1 \cos \alpha$ or in a second branch where a back-flow appears with component speeds $q_2 \cos \alpha$ (Fig. III-8) parallel to the incident jet and $q_2 \sin \alpha$ normal to it. When gravity and fluid friction are both neglected for simplicity and the deflected streams are considered to remain intact, i.e., not to break up into separate droplets, the thicknesses of the two branches can be denoted a_1 and a_2, respectively. If a control surface is taken as indicated by the dashed line to enclose the impact region completely, then the continuity

80 ELEMENTARY FRICTIONLESS FLOWS

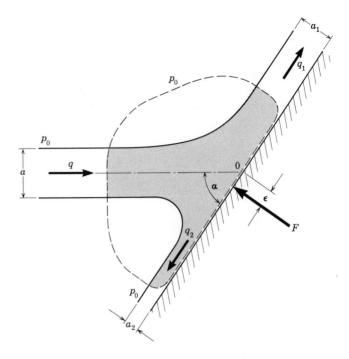

Fig. III-8 Oblique impingement of finite flat liquid jet, showing control surface employed in calculations.

principle is expressed as the vanishing sum of the outward mass flux terms for the incident jet and the two deflected branches:

continuity: vanishing total outward mass flux

$$0 = -\rho a q + \rho a_1 q_1 + \rho a_2 q_2 \tag{6.1}$$

Bernoulli's integral (II.3.11) may now be written as

gravity negligible: pressure depends on flow speed and vice versa

$$\frac{q^2}{2} + \frac{p}{\rho} = \text{const}$$

indicating that at all points where the pressure equals the atmospheric value p_0, the flow speed has the same value; thus $q_1 = q_2 = q$. It follows from (6.1) that the sum of the thicknesses of the two branches of deflected flow equals the thickness of the approaching jet:

continuity: total thickness of outflow branches equals a

$$a_1 + a_2 = a \tag{6.2}$$

Since the fixed wall surface is supposed to exert only normal stresses on the fluid, the total force is normal to the wall surface and is given by the balance of momentum components in that direction. Only the incident jet motion has a component normal to the wall as it crosses the control surface, the normal velocity component being $q \sin \alpha$. When this is multiplied by the mass flux $\rho q a$, the force F required to keep unit breadth of the wall stationary is obtained as

$$F = \rho a q^2 \sin \alpha \qquad (6.3)$$

force determined directly from momentum flux

This is independent of the manner in which the jet thickness a is divided into separate streams of thicknesses a_1 and a_2.

The latter quantities are readily determined, however, from the momentum condition that considers the flow components parallel to the wall. Summing the outward momentum flux terms for the jet and for the two deflected branches and equating to zero gives

$$-\rho a q^2 \cos \alpha + \rho a_1 q^2 - \rho a_2 q^2 = 0$$

When common factors are removed, this becomes

$$a_1 - a_2 = a \cos \alpha \qquad (6.4)$$

momentum consideration: another relationship for a_1 and a_2

Exercise 15 (a) Show that the thickness of the forward-flowing deflected stream is given by

$$a_1 = a \cos^2 \frac{\alpha}{2} \qquad (6.5)$$

and also find a_2. (b) Evaluate the displacement ϵ of the line of action of the resultant force F relative to the point 0 determined by the intersection of the median line of the jet with the plane wall. HINT: Consider the moment of momentum of each of the streams piercing the control surface.

Exercise 16 The jet shown in Fig. III-8 is supposed to be not flat but *axially symmetric* with cross section A. (a) Verify that the total reaction force of the wall (again denoted F) is given by an expression of the same form as (6.3). (b) Let the thickness of the deflected jet at any point of the plane wall be denoted $a(r, \beta)$ to indicate that a is a function of the distance r from the central impingement point 0 and of the polar (azimuth) angle β measured clockwise from the forward-directed half-line on the symmetry axis of the deflected flow plane. Prove that the continuity condition now takes the form of the integral relation

three-dimensional jet — one approach to an unsolved problem

$$\int_0^{2\pi} a(r, \beta) r \, d\beta = A \qquad (6.6)$$

instead of (6.2). Indicate what obstacles may be anticipated in attempting to extend the flat jet analysis to the case of genuine physical interest in which the jet is regarded as three-dimensional.

7 PRESSURE RECOVERY IN EXPANDING FLOW PASSAGE — GRADUAL VS. SUDDEN AREA CHANGE

The occurrence of constant pressure and constant velocity in *open jet* flows illustrated by the preceding example does not carry over when the flow is confined between solid boundary walls (*internal flow*) of variable geometry. The physical boundary surfaces are everywhere parallel to the flow and hence the stream surfaces also, while the average incompressible flow speed at any section varies inversely with the cross section of the passage [cf. stream tube form (II.2.1b) of the continuity equation]. When the flow is frictionless and Bernoulli's integral (II.3.11) can be applied, moreover, the pressure depends on the square of the flow speed; it follows that every change of flow cross section ordinarily entails changes in all the flow variables of internal flows. The dependence of flow pressure on the form of an expanding flow passage will now be examined more closely in order to demonstrate a fundamental difference between the case when Bernoulli's integral is useful and that when the form of the surface introduces frictional forces that must not be ignored.

internal flow — consequences when Bernoulli's integral is not applicable

Two cases are distinguished and analyzed separately: (a) gradual widening of flow passage, and (b) abrupt area change.

(a) Gradual widening

When the boundary surfaces are gently curved in the flow direction, the flow cross section S varies continuously; hence its value at another location near the first can be denoted $S + dS$. The continuity equation (II.2.1b) rewritten as

continuous flow, quasi-one-dimensional

$$\frac{dS}{S} + \frac{dq}{q} = 0 \qquad (7.1)$$

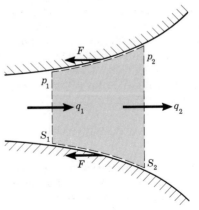

Fig. III–9 Gradual enlargement.

indicates that the average flow speed q also varies continuously; or the same equation can be written

$$q_2 S_2 = q_1 S_1 \qquad (7.2)$$

to relate conditions at two locations, 1 and 2. When 1 is taken for definiteness as an initial or reference flow condition regarded as given, the flow speed q_2 at location 2 is considered to be determined by the initial speed q_1 when the cross section areas S_1 and S_2 are specified, as shown in Fig. III-9, where $S_2 > S_1$. It follows that the flow is decelerated as it enters a widening section, i.e., $q_2 < q_1$ by continuity.

When gravity forces are again neglected, Bernoulli's integral (II.3.11) gives the total pressure increment from 1 to 2 in terms of the flow speeds.

$$\frac{p_2 - p_1}{\rho} = \frac{q_1^2 - q_2^2}{2} \qquad (7.3)$$

pressure recovery — ideal flow

expressible also as a function of the areas S_1 and S_2 by means of (7.2). The "pressure recovery" $p_2 - p_1$ has thereby been obtained on the basis of momentum principles without regard for the force F exerted by the solid boundary on the fluid in contact with it. The derivation of (II.5.5) shows that the same result is also obtained by an energy argument.

Exercise 17 (a) Apply force-momentum principles to all the fluid within the dashed control surface of Fig. III-9 in order to evaluate the thrust force F parallel to the mean flow direction exerted on the boundary walls:

$$F = \rho q_1 S_1 (q_2 - q_1) + p_2 S_2 - p_1 S_1 \qquad (7.4)$$

(b) Use (7.2) and (7.3) to express the dimensionless thrust coefficient C_F in terms of the pressure coefficient C_p and initial flow conditions p_1, q_1, S_1, where

$$C_F \equiv \frac{F}{\rho(q_1^2 S_1/2)} \qquad C_p \equiv \frac{p_2 - p_1}{\rho(q_1^2/2)}$$

force and pressure coefficients

Show that for small values of C_p, i.e., $|C_p| \ll 1$, the force is proportional to the pressure recovery; that is,

$$C_F \doteq \frac{C_p}{2} \frac{p_1}{\rho(q_1^2/2)} \qquad (7.5)$$

Exercise 18 Apply energy considerations to the fluid contained within the control surface to find the pressure recovery as (7.3), by balancing rate of work and energy flux terms. (Note that the fluid to the left of the element performs work on it at the rate given by the product of force $p_1 S_1$ and velocity q_1, while the flux of kinetic energy $\tfrac{1}{2}q_1^2$ at the same section is obtained by multiplying by the mass flux rate $\rho q_1 S_1$).

The more detailed differential form (II.2.4a) of the continuity equation can be used to show that the flow component v normal to the axis of the flow passage is given in order of magnitude by the axial component u and the rate of area change dS/dx as

$$v \sim u \frac{dS}{dx}$$

The reader can then easily verify that the flow speed $q = \sqrt{u^2 + v^2}$ differs from u only by a term of second order in dS/dx, so that the present approximate treatment as "quasi-one-dimensional" is justified for channel flow of slowly varying cross section for which $dS/dx \ll 1$.

(b) Abrupt change of section

It is an experimental fact that the pressure at the mouth of a jet that empties directly into an abruptly widened flow passage equals the value p_1 in the jet upstream of the sudden expansion. Although a fully detailed description of the flow as it spills into the enlarged section S_2 seems never to have been given, shearing action at the interface of the surrounding fluid appears to be *unstable*, giving rise to sizable friction forces and strong mixing. Pressure recovery to a peak value p'_2 occurs only at a certain distance from the mouth of the jet where the flow speed q_2 is nearly uniform across the section. A force-momentum balance may still be performed for a control surface that contains the mixing zone, and inspection of Fig. III-10 shows that this gives

sudden expansion — pressures and velocities not given by Bernoulli's integral

$$p_1 S_2 - p'_2 S_2 = \rho q_1 S_1 (q_1 - q_2)$$

The continuity equation (7.2) is again valid; hence the pressure recovery is found from the preceding equation by removing a common factor:

$$\frac{p'_2 - p_1}{\rho} = q_2(q_1 - q_2) \tag{7.6}$$

The pressure recovery (7.6) in a sudden flow expansion is accordingly less than the value (7.3); the loss of pressure recovery obtained as the difference of these two quantities is given by

diminished pressure recovery— "loss due to impact"

$$\frac{p_2 - p'_2}{\rho} = \frac{(q_1 - q_2)^2}{2} \tag{7.7}$$

and is termed "loss due to impact." The loss in recovered pressure is associated with the nonideal flow conditions that occur in the

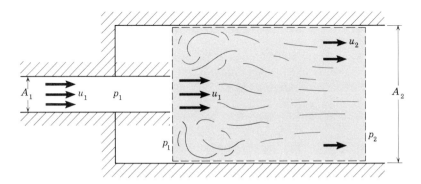

Fig. III–10 Sudden enlargement.

region of abrupt flow expansion. The terminology is based on the form of the expression (7.7) as the square of a velocity difference, since a similar form appears in the evaluation of kinetic energy lost in an inelastic collision of two solid bodies. The usage is confusing because the velocity difference in the latter case refers to the two speeds before (or after) impact, and not to one initial and one final speed as in (7.7).

By performing an energy balance in the present case it is shown that sudden expansion entails an *entropy* rise and heat absorption (per unit mass of fluid) in mechanical units, equal to the expression (7.7). Although the full physical interpretation involves concepts not yet introduced in our discussion, it may be mentioned that the entropy rise is associated with frictional heating in the entrance region of the suddenly expanded flow, where intensive flow mixing occurs. Precise relationships between fluid energy and frictional stresses are shown in section IX.8, for laminar flow.

thermodynamic consequences

Exercise 19 Consider two bodies of masses m and M having respective parallel velocities u_1 and u_2 before colliding, and use the principle of momentum conservation to show that the kinetic energy lost at an inelastic impact is given by

$$E = \frac{1 - e^2}{2} \frac{mM}{m + M} (u_1 - u_2)^2 \qquad (7.8)$$

where e denotes Newton's empirical coefficient of restitution defined as

$$e \equiv -\frac{(v_2 - v_1)}{(u_2 - u_1)} \qquad (7.9)$$

energy loss in inelastic collision

and v_1, v_2 denote the respective parallel speeds of the masses m and M after impact. Explain the terminology "loss due to impact" with reference to (7.7).

S1 OVERALL PROPULSIVE EFFICIENCY — EVALUATION BY NEWTON'S ITERATION METHOD

The risk of ambiguity when arbitrarily defined quantities like the propulsive efficiency are introduced has been demonstrated in Section 3. To show further that without careful interpretation the results may also be misleading, it is sufficient to consider that in both of the definitions (3.6) and (3.8) the energy expended in hauling fuel is counted as useful work, while the function of the fuel is to accelerate *payload* as well as remaining fuel. A more reasonable definition leads to more modest efficiency estimates with peak value substantially less than unity.

The objection cited above is removed by counting as useful work only the kinetic energy imparted to the final rocket mass M_f at the burnout velocity V_f. Attention is thus focused no longer on the instantaneous *rate* of energy expenditure, but on the total energy of the complete propulsive phase. It is therefore also appropriate to consider the total increment of kinetic energy E corresponding to (3.9), so that an *overall* propulsive efficiency is defined as

$$\eta_{P_0} \equiv \frac{M_f(V_f^2/2)}{E} \tag{S1.1}$$

From (3.9) it follows that

$$E = (M_0 - M_f)\frac{c^2}{2}$$

and substitution of (3.4) gives

$$E = (1 - e^{-x})\frac{c^2}{2} M_0 \quad \text{where} \quad x \equiv \frac{V_f}{c} \tag{S1.2}$$

(The latter symbol is introduced for conciseness of notation.) It follows at once that

$$\eta_{P_0} = \frac{x^2}{e^x - 1} \tag{S1.3}$$

which again vanishes both for very small values and for very large values of V_f (i.e., $x \to 0$ and $x \to \infty$). The efficiency (S1.3) is evidently positive for all $x > 0$, and it is of particular interest to determine both its maximum value and the condition (i.e., the value of x) at which the maximum is realized.

In spite of the clumsiness of transcendental expressions like (S1.3) from the standpoint of numerical evaluation, a technique of successive approximations based on the iteration procedure invented by Sir Isaac Newton leads to a quick and useful solution with the aid of nothing more than a slide rule. To find the maximum value of x given by (S1.3), the vanishing of the x-derivative of the expression on the right is investigated. As it is necessary only to consider the numerator in the derivative expression, the value of x is sought which is the solution of the equation

Newton's approximate solution technique

$$f(x) \equiv (2 - x)e^x - 2 = 0 \qquad (S1.4)$$

The principle of Newton's method is that an initial estimate x_0 for the root of an identity such as (S1.4) gives an improved estimate x_1 according to a tangent approximation to the function $f(x)$:

$$x_1 = x_0 - \frac{f(x_0)}{f'(x_0)} \qquad (S1.5)$$

where $f'(x_0)$ denotes the value of the derivative of $f(x)$ evaluated at x_0. Then the values of $x_1, f(x_1)$, and $f'(x_1)$ lead to a new value x_2 in the same manner, and the process is repeated until the incremental corrections become as small as desired. When reasonable caution is exercised in selecting the initial estimate x_0, the root is commonly found with good slide rule accuracy in no more than two or three iterations. In the present example it is found in this manner that $x \doteq 1.60$, implying peak efficiency in the present overall sense when the burnout velocity $V_f \doteq 1.60c$. From (S1.3) it is then seen that the value of the peak efficiency is closely given by 0.65, the results contrasting sharply with the more commonly perferred values of (3.6).

numerical example

The fact that, according to the present definition, the ideal efficiency occurs for burnout speed V_f *greater* than c can be interpeted in the following manner. During the very low-speed propulsion phase, and again at speeds much greater than c, considerable energy is imparted to the wake and "lost" to the rocket, but this effect is greatly reduced when $V \doteq c$ (cf. Exercise 5). Overall efficiency is maximized when the range of speeds $V \doteq c$ is included in the interval $(0, V_f)$, as shown by the present calculation.

physical interpretation

IV
Irrotational Flow—Velocity Potential

1 FLUID ROTATION: VORTICITY

Although such calculations as those shown in the preceding chapter are invaluable for qualitative purposes, more detailed information like velocity and pressure variations in the associated flow depends on the study of differential equations developed in Chapter II. The characteristic nonlinearity of the system of equations (II.5.6) and (II.5.7) and physically equivalent systems is a source of major difficulty whenever analytical solutions are sought that are the most useful from the standpoint of interpretation. A saving feature in the case of water, air, and other fluids possessing small internal friction (viscosity) is revealed by the absence of *fluid rotation* and the consequent reduction of the equations to *linear* form. Most of the classical hydrodynamics and aerodynamics treated in subsequent chapters follows easily by supplementing the Euler equations with the condition of *irrotationality* resulting from the circumstance just mentioned.

fluids of small viscosity — physical basis for an important simplification of Euler equations

absence of fluid rotation: irrotationality

A preliminary idea of the utility of a statement concerning rotation may be gained by considering the quantity of data required to specify a flow. In the general case a motion at any instant is specified by giving the velocity components for each of the (infinitely numerous) points of the region under consideration. When all the particles in the vicinity of a point move as if in solid-body rotation relative to one point, however, the entire motion is described by the linear velocity of that point and a single angular velocity. The reduction of complexity is enormous, even for actual fluid flows in which the rotational motion is expressed in different terms.

solid-body rotation

When a rigid body rotates at angular speed ω, the velocity components of a point located by its coordinates x, y are

$$u = -\omega y \qquad v = \omega x \qquad (1.1)$$

[(cf. equations (I.8.1)]. Different pairs of velocity components are given by the equations (1.1) for different points x, y. One statement equivalent to all of these is found by elimination of both of the variables x and y from the pair of equations. This is accomplished by differentiations; the resulting equation

$$2\omega = \left\{ \frac{\partial v}{\partial x} - \frac{\partial u}{\partial y} \right\} \tag{1.2}$$

essential feature: same value of rotation for every point

contains derivatives of the velocity components but no *explicit* dependence on the coordinates x and y. The particular combination of derivatives on the right side of (1.2) is therefore independent of location of an individual particle; hence the relationship is valid for all particles. The bracketed quantity that leads to such great economy of statement is termed the *vorticity*, and it is equal to twice the angular rotation speed ω.

vorticity equals twice the angular velocity

The vorticity of fluid motion in the x, y plane is likewise defined by (1.2) and serves as a measure of the fluid rotation, varying in general from one point to another according to the values of the velocity derivatives it contains.

Exercise 1 Sketch the streamlines and evaluate the vorticity for each of the following flows:

(a) $u = U = \text{const}$ $v = 0$ uniform flow parallel to x-axis
(b) $u = Ay$ $v = 0$ uniform shear flow ($A = \text{const}$)
(c) $u = 0$ $v = Ay$ uniform *dilatation* in y-direction
(d) $u_r = V = \text{const}$ $u_\theta = 0$ radial flow   [(cf. (II.1.10))]
(e) $u_r = 0$ $u_\theta = Ar$ pure rotation

Exercise 2 (a) Use the result of Exercise 3, Chapter I, and the fact that an arbitrary polygon can be decomposed into mutually continguous triangles to show that *uniform* pressure acting on fluid bounded by any closed curve creates a vanishing force moment. (b) Use Green's theorem (A.5) to express the moment about an origin 0 of pressure forces on the boundary C of an arbitrary closed curve as a double integral over the region S contained within C:

$$M_0 = \oint_C p(x, y)(x\, dx + y\, dy) = \iint_S \left(y \frac{\partial p}{\partial x} - x \frac{\partial p}{\partial y} \right) dx\, dy$$

and evaluate for flow when the pressure is given by $p(x, y) = x^2 + y^2$. Can it be correctly concluded that the moment vanishes for arbitrary $p(x, y)$?

The relationship between vorticity and fluid rotation in the general case of plane flow is now established directly from the velocity components u and v. For this purpose the motions of two points A

90 IRROTATIONAL FLOW — VELOCITY POTENTIAL

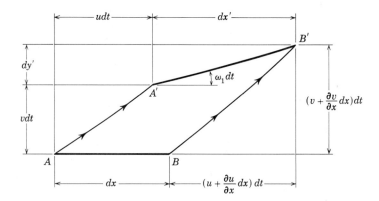

Fig. IV-1 Rotation of fluid line element.

evaluation of vorticity in terms of fluid velocity components

and B at the extremities of a fluid line element dx are compared during a short time interval dt (Fig. IV-1). The rate of change of the slope of the fluid line, ω_1 for example, gives its rotation because

$$\omega_1\, dt = \frac{dy'}{dx'}$$

where dx' and dy' are the Cartesian projections of the line element after the interval dt.

When the velocity components of the point A are taken as u and v, those of B are $u + (\partial u/\partial x)\, dx$ and $v + (\partial v/\partial x)\, dx$, and it is seen that

$$dy' + v\, dt = \left\{ v + \frac{\partial v}{\partial x} dx \right\} dt$$

$$dx' + u\, dt = dx + \left\{ u + \frac{\partial u}{\partial x} dx \right\} dt$$

so that

$$dy' = \frac{\partial v}{\partial x} dx\, dt$$

and

$$dx' = \left\{ 1 + \frac{\partial u}{\partial x} dt \right\} dx \doteq dx$$

Using these values gives

$$\omega_1 = \frac{\partial v}{\partial x} \tag{1.3}$$

to the first order in small quantities dx, dy, and dt.

Exercise 3 Calculate the rotation ω_2 of the line element AC extending in the y-direction from A, to show that

$$\omega_2 = -\frac{\partial u}{\partial y} \tag{1.3a}$$

The average value of the rotation rates ω_1 and ω_2 of the mutually orthogonal line elements dx and dy, taken as half the sum of (1.3) and (1.3a), again denoted ω, is thus

$$\omega = \frac{\omega_1 + \omega_2}{2} = \frac{1}{2}\left\{\frac{\partial v}{\partial x} - \frac{\partial u}{\partial y}\right\} \tag{1.4}$$

vorticity related to fluid rotation

Thus the vorticity of a general fluid motion is seen to be twice the *average* rotation of the fluid in the sense indicated.

Exercise 4 Verify that the *vorticity*

$$\zeta \equiv \frac{\partial v}{\partial x} - \frac{\partial u}{\partial y} \tag{1.5}$$

of flow in the x, y plane is given by the component, normal to that plane, of the curl of the velocity vector field, i.e., by

$$\nabla \times \mathbf{q} = \begin{vmatrix} \mathbf{i} & \mathbf{j} & \mathbf{k} \\ \dfrac{\partial}{\partial x} & \dfrac{\partial}{\partial y} & \dfrac{\partial}{\partial z} \\ u & v & w \end{vmatrix} \tag{1.6}$$

vorticity is a vector quantity

where the right side of (1.6) is expanded according to the ordinary rules for evaluating determinants, the differentiation elements of the second row applying to the velocity components u, v, w regarded as functions of the Cartesian coordinates.

2 FLOW CIRCULATION AND IRROTATIONALITY — KELVIN'S THEOREM

The vorticity ζ gives an indication of the rotational character of fluid motion at a given point in plane flow, and the form of the expression (1.5) of this quantity as the sum of a pair of derivatives suggests that its integral over a region S of the flow plane can be simplified with the aid of Green's theorem. If it is noted specifically that the vorticity expressed as the curl of the velocity, (1.6) conforms exactly with Stokes' transformation (A.7) it is seen that the surface integral

$$\iint_S \nabla \times \mathbf{q}\, dx\, dy$$

integral of vorticity...

is evaluated as the line integral of the tangential component of the velocity on the boundary C of the region S, i.e., as

$$\mathbf{k} \cdot \iint_S \nabla \times \mathbf{q} \, dx \, dy = \Gamma \equiv \oint_C \mathbf{q} \cdot \boldsymbol{\tau} \, dl$$

...expressible as flow circulation

$$= \oint_C \{u \, dx + v \, dy\} \qquad (2.1)$$

Because of the particular importance of the integral (2.1), it is termed the *circulation* and commonly denoted by the special symbol Γ. The circulation then expresses the total rotation of all the fluid particles within a region S as an integral around the boundary C of the region. The line integral (2.1) evaluated on an *open* curve is also termed the circulation.

Exercise 5 Use Gauss' transformation (A.6) to show that the surface integral of the divergence of the velocity is given by the normal component of the flow velocity evaluated around the boundary contour of the surface; this is analogous to the relationship between vorticity and circulation. From the fact that the divergence of the velocity of an incompressible fluid is everywhere zero, establish the continuity equation (II.2.4a).

Exercise 6 Verify the Cartesian form (1.5) of the expression for the vorticity ζ by selecting as area S the rectangular element having sides dx and dy at the point denoted A where the velocity components are u and v, and applying Stokes' transformation (2.1) directly. Specifically, note that

infinitesimal fluid volume element, Cartesian form..

$$\mathbf{k} \cdot \iint \nabla \times \mathbf{q} \, dx \, dy \doteq \mathbf{k} \cdot (\nabla \times \mathbf{q}) \, dx \, dy = \zeta \, dx \, dy$$

to the first order, while the line integral is evaluated on each of the four segments AB, BC, CD, and DA on which only one of the velocity components is required (Fig. IV-2), so that

$$\int_{AB} \mathbf{q} \cdot \boldsymbol{\tau} \, dl = u \, dx$$

and so on.

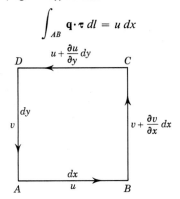

Fig. IV-2 Infinitesimal contour for vorticity evaluation; Cartesian form.

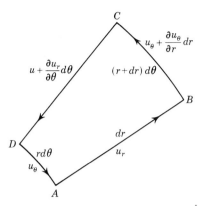

Fig. IV-3 Infinitesimal contour for vorticity evaluation; polar form.

Exercise 7 Obtain the expression for the vorticity ζ in polar coordinates by applying the transformation (2.1) to the quasi-rectangular element having sides dr, $r\,d\theta$, $(r + dr)\,d\theta$ and velocities as shown in Fig. IV-3.

... and polar

The time derivative of the circulation measures the rate at which the rotation of a fluid-filled region increases as it moves in the flow. This is given by the total derivative of the line integral in (2.1)

$$\frac{D\Gamma}{Dt} = \frac{D}{Dt} \oint \mathbf{q} \cdot \boldsymbol{\tau}\, dl$$

$$= \oint_C \frac{Du}{Dt}\, dx + \oint_C \frac{Dv}{Dt}\, dy + \oint_C u\, \frac{D(dx)}{Dt}$$

$$+ \oint_C v\, \frac{D(dy)}{Dt} \qquad (2.2)$$

rate of change of circulation with time

where the derivatives of the line elements have the meanings already indicated in Fig. IV-1. Specifically, the fundamental definition of a derivative as the limit of a difference quotient gives

$$\frac{D(dx)}{Dt} = \lim_{dt \to 0} \left\{ \frac{dx' - dx}{dt} \right\}$$

analysis of total derivative of a fluid line element

in the notation of the figure, and

$$dx' - dx = \left(u + \frac{\partial u}{\partial x}\, dx\right) dt - u\, dt = \frac{\partial u}{\partial x}\, dx\, dt$$

Hence

$$\frac{D(dx)}{Dt} = \frac{\partial u}{\partial x}\, dx$$

and a similar relationship is obtained for the derivative of the moving line element dy.

The last two integrals of (2.2) are therefore of the form

$$\oint_C u \frac{\partial u}{\partial x} dx = \oint_C d\left(\frac{u^2}{2}\right) = 0$$

line integral of a perfect differential equals zero

vanishing because the velocity components have the same value at the identical upper and lower limits of integration on the closed curve C.

The remaining terms on the right side of (2.2) are also seen to vanish by noting from (II.5.7) that both Du/Dt and Dv/Dt are expressible as partial derivatives of the quantity $p/\rho + U$. Hence they are

$$\oint_C \frac{\partial}{\partial x}\left(\frac{p}{\rho} + U\right) dx + \oint_C \frac{\partial}{\partial y}\left(\frac{p}{\rho} + U\right) dy \qquad (2.2')$$

representing the integral around a closed curve of the total differential, vanishing as before. It is therefore seen that the circulation is constant

circulation remains constant — Kelvin's theorem

$$\frac{D\Gamma}{Dt} = 0 \qquad (2.3)$$

the result (2.3) being known as Kelvin's theorem. Because of the arbitrary form of the curve C, hence also of the enclosed area S, (2.1) shows that the vorticity is constant at each point in the flow: fluid rotation remains unchanged in an incompressible frictionless fluid when the only forces acting are those which, like gravity, are derived from a potential. Thus, if a flow is uniform in one region, and therefore without rotation, this property is conserved as the fluid moves into a region of space where the motion is not uniform, and the flow is termed *irrotational*. Irrotational fluid motion is of great importance in flows of air or water that are only very slightly affected by fluid friction or compressibility, external forces being either negligible or conservative in character.

flow irrotational at one place remains irrotational

Exercise 8 The velocity field of a plane flow is given by

$$u = \frac{x^2 - y^2}{(x^2 + y^2)^2} \qquad v = \frac{2xy}{(x^2 + y^2)^2} \qquad (2.4)$$

(a) Evaluate the circulation Γ_{AB} along the straight line joining points $A(1, 0)$ and $B(1, 1)$. Repeat for Γ_{BC} where the coordinates of C are $(0, 1)$.
(b) Evaluate the circulation Γ_{AC} along the circular arc joining points A and C, and compare with the sum of Γ_{AB} and Γ_{BC}. HINT: Use polar coordinates and velocity components for the calculation of Γ_{AC}.

circulation value does not depend on path of line integral

3 VORTICITY IN NONBAROTROPIC FLUID — SEA BREEZE

In addition to friction and attendant tangential stresses as a source of vorticity in real fluids, there are also other conditions of Kelvin's theorem which, when not met, lead to commonly observed flows that are *not* irrotational. Two examples considered in this and the subsequent sections, respectively, relate to fluid of nonuniform density and to nonconservative forces.

In the first instance the atmosphere is regarded as composed of horizontally stratified constant-pressure layers, as in aerostatic equilibrium. When the Sun's rays heat one portion of the sea-level surface more strongly than an adjacent region, thermal expansion leads to reduced densities in the preferentially heated region. It follows that the *isopycnic* (constant-density) surfaces slope upward in passing from land to an ocean or lake, since water is more transparent to the Sun's radiation and is therefore the less strongly heated. In such cases the density of the air is not a single-valued function of the pressure, as in the aerostatics of Section I.5, and the fluid is termed *nonbarotropic:* the isopycnic surfaces and the *isobaric* surfaces are not parallel to each other. The demonstration of Kelvin's theorem is hindered by the fact that an integral of dp/ρ cannot be written as p/ρ as in (2.2') and the growth of circulation depends on the precise relationship of the isopycnics and the isobarics to each other.*

<small>non-constant vorticity: *nonbarotropic* fluid</small>

<small>*isopycnic* surfaces not parallel to *isobaric* surfaces</small>

The configuration of isopycnics and isobarics is taken as shown in Fig. IV-4, each family of surfaces being regarded as plane for simplicity. A pair of isopycnics and a pair of isobarics then determine a parallelogram $ABCD$ as an elementary cell or "solenoid" in which the growth of circulation and fluid rotation can be directly calculated. The circulation around $ABCD$ is again denoted by Γ_{ABCD} and evaluated from the definition (2.1) as

$$\Gamma_{ABCD} = \left[\int_{AB} + \int_{BC} + \int_{CD} + \int_{DA}\right] \mathbf{q} \cdot \boldsymbol{\tau} \, dl \quad (3.1)$$

It is sufficient to consider that $ABCD$ is fixed in space in order to obtain the ordinary time derivative of the circulation there as

$$\frac{d\Gamma_{ABCD}}{dt} = \int \frac{D\mathbf{q}}{Dt} \cdot \boldsymbol{\tau} \, dl \quad (3.2)$$

<small>circulation evaluated on typical vorticity "solenoid"</small>

without requiring terms like the last two in (2.2). In the present

---

*It may be pointed out here that the validity of Kelvin's theorem is not limited to incompressible fluids, but also follows when a single-valued *barotropic* relationship exists so that $\rho = \rho(p)$ and dp/ρ is an exact differential.

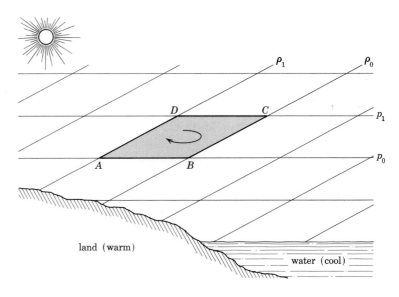

Fig. IV-4 Vorticity solenoids formed by nonparallel isobaric and isopycnic surfaces.

case the introduction of the total derivative symbol in (3.2) instead of the partial derivative $\partial/\partial t$ is justified by supposing the motion to be so small that the nonlinear convective terms in (II.1.11) are negligible.

Ordinary gravity forces contribute nothing to the growth of circulation, and it follows that only the pressure gradient term of (II.5.7a) is required in the evaluation of (3.2); it gives for the line integral around $ABCD$

$$-\int \frac{1}{\rho} \nabla p \cdot \boldsymbol{\tau} \, dl \tag{3.3}$$

The pressure is constant on AB and also on CD; therefore the gradient terms vanish for the corresponding integrals. When the constant density on BC is denoted by ρ_0, that on DA by ρ_1, while pressures on AB and CD are denoted, respectively, p_0 and p_1, the remaining integrals of (3.3) are written at once as

$$\frac{d\Gamma}{dt} = -\frac{1}{\rho_0}[p_1 - p_0] - \frac{1}{\rho_1}[p_0 - p_1]$$

$$= -\left(\frac{1}{\rho_1} - \frac{1}{\rho_0}\right)(p_0 - p_1) < 0 \tag{3.4}$$

circulation intensified steadily

sea-breeze and lake-breeze at day . . .

According to (3.4) circulation increases in the negative sense of rotation (viz., clockwise) when the more intense heating occurs to the left, as shown in Fig. IV-4. Thus motion develops which is directed toward the land at the lower levels, upward in the region above the land, and outward and downward toward the water at upper levels and directly over the water. This is the familiar sea breeze effect noticeable after a few hours of direct sunshine in the vicinity of ocean and large lake shores. It is also interesting to note that the motion built up by the action of adjacent solenoids over a large region takes the opposite sense during evening hours when clear skies are conducive to greater radiative cooling of the land surface and temperature differentials are generated in the reverse sense to those of the present calculation. In both cases the rotation imparted to the individual solenoids is associated with a shift in the location of the center of mass away from the center of the figure, the pressure forces passing through the latter point as before.

...reversed at night

Exercise 9 The density variation within a square fluid element of horizontal and vertical dimensions h on each of its sides is given by

$$\rho(x, y) = \rho_0 \left\{ 1 - \frac{\epsilon}{h}(y - \alpha x) \right\} \qquad 0 \leq x, y \leq h$$

fundamental interpretation: forces are balanced but not aligned

The maximum fractional density variation from top to bottom of the element is therefore

$$\frac{\delta \rho}{\rho_0} = \epsilon$$

and α gives the slope of the isopycnics relative to the horizontal isobars. Show that the horizontal displacement of the mass centroid relative to the center of the square is expressed in dimensionless form as

$$\frac{x_c - h/2}{h} = \frac{\delta \rho}{\rho_0} \frac{\alpha}{12}$$

so that pressure and gravity forces on the element form a couple given quantitatively as

detailed evaluation

$$\frac{h^3 g \alpha}{12} \delta \rho$$

for unit depth of element. Compare with (3.4) and discuss.

4 NONCONSERVATIVE FORCES — VORTICITY AT EDGE OF LIFTING SURFACE

When a submerged body like an airplane wing experiences a lift force normal to the direction of its motion, the reaction on the surrounding fluid can be represented as a fictitious body force acting in a limited region of fluid near the wing-fluid interface. Because the force in this case is not derivable from a potential as

98 IRROTATIONAL FLOW — VELOCITY POTENTIAL

nonconservative forces

gravity forces are, the conditions of Kelvin's theorem are again violated. The nonvanishing circulation associated with the applied force is calculated now and later seen to play an essential role in the analysis of flow past aerofoils and hydrofoils.

The Euler equations (II.5.7) are modified by designating $X(x, y)$ and $Y(x, y)$ as force components per unit volume so that

Euler equations, modified

$$\left. \begin{array}{l} \dfrac{Du}{Dt} + \dfrac{1}{\rho}\dfrac{\partial p}{\partial x} = \dfrac{1}{\rho}X \\[2ex] \dfrac{Dv}{Dt} + \dfrac{1}{\rho}\dfrac{\partial p}{\partial y} = \dfrac{1}{\rho}Y \end{array} \right\} \quad (4.1)$$

Exercise 10 Expand each of the total derivatives as in (II.1.11), differentiate the first of (4.1) with respect to y, the second with respect to x, and combine to show that the growth of vorticity in an incompressible fluid is given by

vorticity growth in terms of the curl of the forces acting

$$\frac{D\zeta}{Dt} = \frac{1}{\rho}\left\{\frac{\partial Y}{\partial x} - \frac{\partial X}{\partial y}\right\} \quad (4.2)$$

when fluid friction is neglected; $D\zeta/Dt$ has the usual meaning:

$$\frac{\partial \zeta}{\partial t} + u\frac{\partial \zeta}{\partial x} + v\frac{\partial \zeta}{\partial y} \quad (4.3)$$

It is clear from (4.2) that vorticity is not produced in regions where the forces X and Y are either zero or constant in value, but, even if $X = 0$ and $Y = \text{const} \neq 0$ in a limited region, the first derivative of Y, appearing on the right side of (4.2), must be nonzero at the edge of that region; increasing vorticity results. This is the case when a solid boundary of the flow, like a wing surface, sustains a force exerted by the fluid and provides a reaction force on the fluid. The growth of circulation is calculated as follows.

The force $Y(x)$ is taken as downward in direction, of constant magnitude f in the region $x > 0$, except for a narrow zone of width ϵ where it drops off to the zero value in the region $x < 0$:

analysis of force near edge of loaded boundary surface

$$Y(x) = \begin{cases} 0 & x < 0 \\ -f\dfrac{x}{\epsilon} & 0 < x < \epsilon \\ -f & x > \epsilon \end{cases} \quad (4.4)$$

as indicated in Fig. IV-5. Then (4.2) becomes

$$\frac{D\zeta}{Dt} = \frac{1}{\rho}\left(\frac{-f}{\epsilon}\right) \quad 0 < x < \epsilon \quad (4.5)$$

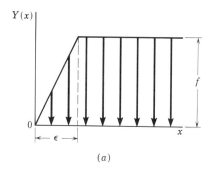

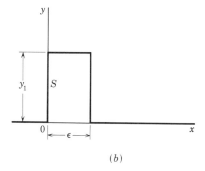

Fig. IV-5 Nonconservative force representation; (a) force variation; (b) zone of vorticity generation.

and, if the force (4.4) acts in a region thickness y_1, the integral of (4.5) over the rectangular region S of area ϵy_1 gives for the rate of increase of circulation around the boundary C of the region:

vorticity growth, and circulation...

$$\frac{D\Gamma}{Dt} = -\frac{f}{\rho} y_1 \qquad (4.6)$$

The left side of (4.6) is related to the surface integral of the vorticity by the transformation (2.1). Finally, as the thickness y_1 of the layer in which the fictitious force is presumed to act is reduced, while the magnitude of the force f is increased in such a manner that their product gives the pressure p when $y_1 > 0$, (4.6) becomes

$$\frac{D\Gamma}{Dt} = -\frac{p}{\rho} \qquad (4.7)$$

...in terms of loading

If the force drops off to zero at the edge $x = 0$, (4.7) indicates that circulation is generated by reason of an intensive production of vorticity in the immediate vicinity of the edge. This is understood as the consequence of fluid at pressure p "escaping" around the leading edge of a load-carrying surface, and (Fig. IV-6) the sense of motion conforms with the sign of (4.7).

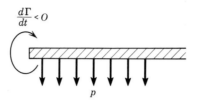

Fig. IV-6 Circulation growth at edge of load-supporting boundary surface.

5 BERNOULLI'S INTEGRAL IN IRROTATIONAL FLOW: VELOCITY POTENTIAL

When all the conditions of Kelvin's theorem are satisfied and the flow is irrotational, the vanishing of the vorticity ζ introduces an additional relationship between velocity components:

irrotational flow — another restriction on the velocity components

$$\frac{\partial v}{\partial x} - \frac{\partial u}{\partial y} = 0 \quad (5.1)$$

according to (1.5). Although it might appear at first sight that (5.1) in conjunction with equations (II.5.6) and (II.5.7) forms a mathematically redundant system (four equations to determine three flow quantities, u, v, and p), a basic feature of irrotational flows is revealed by demonstrating that the two equations (II.5.7) now are actually not independent of each other but simply two different expressions of a single statement.

If steady flow is considered for simplicity, the Euler equations (II.4.3) are

$$\left. \begin{array}{l} u\dfrac{\partial u}{\partial x} + v\dfrac{\partial u}{\partial y} + \dfrac{1}{\rho}\dfrac{\partial p}{\partial x} = -\dfrac{\partial U}{\partial x} \\[6pt] u\dfrac{\partial v}{\partial x} + v\dfrac{\partial v}{\partial y} + \dfrac{1}{\rho}\dfrac{\partial p}{\partial y} = -\dfrac{\partial U}{\partial y} \end{array} \right\} \quad (5.2)$$

and substitution of (5.1) shows that the nonlinear terms can be written as exact differentials. Thus

acceleration components identified as derivatives of kinetic energy

$$u\frac{\partial u}{\partial x} + v\frac{\partial u}{\partial y} = \frac{\partial}{\partial x}\left(\frac{u^2 + v^2}{2}\right) = \frac{\partial}{\partial x}\left(\frac{q^2}{2}\right)$$

and the first two terms of the second of (5.2) are likewise seen to be the y-derivative of the same quantity. The two equations (5.2) are therefore recognized as the x- and y-derivatives, respectively, of the single equation

Bernoulli's theorem — the same const for all streamlines

$$\frac{q^2}{2} + \frac{p}{\rho} + U = \text{const} \quad (5.3)$$

when $\rho = \text{const}$. This equation is identical in form with (II.5.5). Both equations are therefore referred to as Bernoulli's theorem or integral, but an essential distinction is noted by observing that (5.3) is a relationship between flow speed, pressure, and potential valid at all points of the flow, i.e., the *same* value of the constant energy sum is found on *all* the streamlines in the region of irrotational flow where (5.2) is valid. Both of the equations (5.2) are obtained by differentiation of (5.3) and the latter is therefore seen to contain both of the former; it is now the basic relationship that connects the pressures with the velocities.

The continuity equation (II.5.6) of incompressible flow

$$\frac{\partial u}{\partial x} + \frac{\partial v}{\partial y} = 0 \qquad (5.4)$$

serves as the third of the set of equations which, together with (5.1) and (5.3), determine the two velocity components and the pressure.

The fact that the pair of equations (5.1) and (5.4) contains only the two velocity components means that the pressure may be disregarded in the calculation of these quantities. Once the velocities are found, Bernoulli's integral (5.3) gives the pressure directly in terms of them. It will be seen directly that the application of this solution procedure is further simplified by reason of the particular form of the equation (5.1) of irrotationality.

calculate velocities separately from (5.1) and (5.4), evaluate pressure from (5.3)

The two terms on the left side of (5.1) are first derivatives, one with respect to one of the independent variables, the other with respect to the second. When it is recalled that the elementary property of mixed second partial derivatives according to which the order of the different operations is immaterial, viz.,

$$\frac{\partial^2 \phi}{\partial x \, \partial y} = \frac{\partial^2 \phi}{\partial y \, \partial x}$$

for any function $\phi(x, y)$ that possesses these derivatives, this property can easily be capitalized on. Specifically, it is supposed that the velocity components u and v appearing in (5.1) are of precisely the form

velocity potential defined...

$$u = -\frac{\partial \phi}{\partial x} \qquad v = -\frac{\partial \phi}{\partial y} \qquad (5.5)$$

Hence the two terms in (5.1) add to zero. Then (5.4) indicates that the *velocity potential* $\phi(x, y)$ satisfies the equation obtained by direct substitution of the expressions (5.5):

$$\frac{\partial^2 \phi}{\partial x^2} + \frac{\partial^2 \phi}{\partial y^2} = 0 \qquad (5.6)$$

...satisfies Laplace's equation

Therefore the introduction of an arbitrary (differentiable) function $\phi(x, y)$ according to (5.5) assures that the condition of irrotationality

is satisfied, while the continuity equation that takes the form (5.6) indicates that only those functions $\phi(x, y)$ that are solutions of (5.6) are suitable to represent the plane irrotational flow of an incompressible fluid. Equation (5.6) is known by the name of Laplace, and it is regarded as the most important single partial differential equation of mathematics, physics, and engineering. Its solutions describe physical processes in a great many other fields of science in addition to fluid mechanics.

The reader will not fail to note the similarity between the gravitational potential U that gives the forces as the negative partial derivatives in the respective coordinate directions (cf. Section I.7) and the velocity potential that gives the velocity components according to (5.5) in the same manner. A solution of (5.6) affords the convenience of furnishing both velocity components at a single stroke.

{gravitational velocity} potential
{U: force, ϕ: velocity} components by partial differentiation

Exercise 11 (a) Find the Cartesian velocity components of the flow specified by the potential

$$\phi(x, y) = \frac{x}{x^2 + y^2}$$

and sketch the flow field. Compare with the flow in Exercise 8. (b) Determine the equations of the streamlines from the property (II.1.6). HINT: Integration is facilitated by replacing y by a new variable η by means of the transformation $y(x) = x\eta(x)$. (c) Note that the x-axis is a streamline of the flow, and observe that it may be regarded as a physical boundary of flow in the half-plane $y > 0$. Take the pressure at $(1, 0)$ to be zero and calculate the pressure force on unit thickness of boundary in the interval $1 < x < 2$ by integration of (5.3) neglecting gravity. (d) Differentiate the velocity $u(x, y)$ to show that it is a solution of Laplace's equation (5.6). Establish the same result directly from (5.1) and (5.4).

flow defined by arbitrarily chosen function is irrotational if ϕ is a solution of (5.6)

6 MINIMAL ENERGY PROPERTY OF IRROTATIONAL FLOW

A basic property of flow with zero vorticity is demonstrated by evaluating the kinetic energy of the motion and comparing it with the corresponding quantity in the analogous flow *not* subject to the restrictive assumption of irrotationality. In the first case the velocity components are given by (5.5), and the total kinetic energy of the flow in a region S of the x, y plane is

total kinetic energy of irrotational flow in region S

$$T = \frac{\rho}{2} \iint_S \left\{ \left(\frac{\partial \phi}{\partial x}\right)^2 + \left(\frac{\partial \phi}{\partial y}\right)^2 \right\} dx\, dy \qquad (6.1)$$

When the velocity components in the "comparison flow" are denoted by u' and v', and the associated energy by T',

$$T' = \frac{\rho}{2}\iint_S \{u'^2 + v'^2\}\, dx\, dy \qquad (6.2)$$

kinetic energy of comparison flow

and the difference $T' - T$ is formed. This is expressible as the integral of a pair of differences, one of which is

$$u'^2 - \left(\frac{\partial \phi}{\partial x}\right)^2 = \left(u' + \frac{\partial \phi}{\partial x}\right)^2 - 2\left(\frac{\partial \phi}{\partial x}\right)^2 - 2u'\frac{\partial \phi}{\partial x}$$

and the other is obtained by replacing u' by v' and writing derivatives with respect to y instead of x. Then the energy difference becomes

$$T' - T = \frac{\rho}{2}\iint_S \left\{\left(u' + \frac{\partial \phi}{\partial x}\right)^2 + \left(v' + \frac{\partial \phi}{\partial y}\right)^2\right\} dx\, dy$$

$$- \rho \iint_S \left\{\frac{\partial \phi}{\partial x}\left(\frac{\partial \phi}{\partial x} + u'\right) + \frac{\partial \phi}{\partial y}\left(\frac{\partial \phi}{\partial y} + v'\right)\right\} dx\, dy \qquad (6.3)$$

form the difference of energies

In an incompressible fluid the continuity equation is satisfied both by the irrotational flow and by the flow with component speeds u' and v'; therefore the second integral of (6.3) is equal to

$$\iint_S \left(\frac{\partial}{\partial x}\left\{\phi\left[\frac{\partial \phi}{\partial x} + u'\right]\right\} + \frac{\partial}{\partial y}\left\{\phi\left[\frac{\partial \phi}{\partial y} + v'\right]\right\}\right) dx\, dy \qquad (6.4)$$

reduction based on continuity equation

Each of the terms in braces in (6.4) may now be regarded as one of the components of a vector **A**, as in Appendix A and (A.6), which is Gauss' transformation indicates that (6.4) is expressed by the line integral around the boundary C of the region S as

$$\int_C \phi\left\{\left[\frac{\partial \phi}{\partial x} + u'\right](-dx\mathbf{j}) + \left[\frac{\partial \phi}{\partial y} + v'\right](dy\mathbf{i})\right\}$$

velocity differences at boundary surface C, necessarily zero for comparable flows

The terms in each of the square brackets are now recognized as the difference between the velocity components of the comparison flow and the irrotational flow at points of the boundary C: each of the differences must be zero if the flows compared are analogous except with regard to irrotationality. Hence (6.4) vanishes and the energy difference depends only on the first term on the right side of (6.3). The integrand in this term consists of a sum of squares of real quantities; hence it is positive, or at least zero but not negative. For the energy difference to be zero it is necessary for each of the squared terms to vanish separately at every point of the flow, and this implies that

$$u' = -\frac{\partial \phi}{\partial x} \quad \text{and} \quad v' = -\frac{\partial \phi}{\partial y} \qquad (6.5)$$

104 IRROTATIONAL FLOW—VELOCITY POTENTIAL

conclusion: irrotational flow energy is minimal

The conclusion is that the kinetic energy of an irrotational flow is *less* than that of any other flow subject to the same boundary conditions. The interpretation depends on the fact that, when the action of fluid friction is accounted for, a continual reduction of flow energy attributable to this effect is verified in all cases *except* irrotational flow, which may therefore be regarded as the natural ultimate state of motion of any real fluid (see Chapter IX).

7 EXAMPLES OF FLOW DEFINED BY ELEMENTARY VELOCITY POTENTIAL FUNCTIONS

A variety of simple but basic flows is obtained by choosing functions $\phi(x, y)$ and applying (5.5) to define Cartesian velocity components $u(x, y)$ and $v(x, y)$. In each of the examples that follows a different subscript is employed to distinguish the corresponding flows.

Example 1

uniform flow

$$\phi_1 = -Ux \text{ where } U = \text{const} \qquad (7.1)$$

From (5.5) it is seen that

$$u = u_1 = -\frac{\partial \phi_1}{\partial x} = U = \text{const}$$
$$v = v_1 = -\frac{\partial \phi_1}{\partial y} = 0 \qquad (7.1\text{a})$$

circular streamlines, concentric

so that (7.1) represents uniform flow in the x-direction at speed U. The streamlines are therefore all parallel to the x-axis and the pressure is uniform everywhere.

Example 2

$$\phi_2 = -A \tan^{-1}\frac{y}{x} \text{ where } A = \text{const} \qquad (7.2)$$

Evaluation of velocity components as before gives

$$u_2(x, y) = -\frac{A\,y}{x^2 + y^2} \quad v_2(x, y) = \frac{A\,x}{x^2 + y^2} \qquad (7.2\text{a})$$

The nature of the flow is seen by determining the form of the streamlines according to (II.1.6). Thus

$$\left.\frac{dy}{dx}\right|_{\text{S.L.}} = -\frac{x}{y} \quad \text{and} \quad x^2 + y^2 = \text{const}$$

so that the streamlines are all *circles* concentric with the origin of coordinates, each streamline corresponding to a distinct value of the constant of integration.

Exercise 12 (a) Evaluate the circulation of the flow about the closed path $ABCD$ as in (3.1), where $A = (1, -1), B = (1, 1), C = (-1, -1)$ and $D = (-1, -1)$. Thus find that

as in Exercise II

$$\Gamma_{AB} = \int_{-1}^{-1} v_2(1, y)\, dy = A\frac{\pi}{4} \quad \text{etc.}$$

and obtain Γ as the sum of four such integrals.
(b) Calculate the circulation around the circle of unit radius inscribed in $ABCD$ as the integral

$$\Gamma = \int_0^{2\pi} u_\theta\, d\theta$$

where u_θ is obtained from (7.2a) by using transformations (II.1.10) and $r = 1$. Does the result agree with part (a)?

Example 3

$$-\phi_3(x, y) = \frac{\partial \phi_2}{\partial y} = -\frac{A\,x}{x^2 + y^2} \quad A = \text{const} \quad (7.3)$$

streamlines are hyperbolae

The flow is identical in form to the case examined in Exercise 11, while the Cartesian velocity components are given by (2.4).

Example 4

$$-\phi_4(x, y) = A(x^2 - y^2) \quad A = \text{const} \quad (7.4)$$

It is readily seen that

$$u_4(x, y) = 2Ax \quad v_4(x, y) = -2Ay \quad (7.4a)$$

and that the streamlines are rectangular hyperbolae asymptotic to the Cartesian coordinate axes.

Example 5

$$-\phi_5(x, y) = e^{-ay} \sin bx \quad a \text{ and } b \text{ are constants} \quad (7.5)$$

In this case,

$$u_5 = b\,e^{-ay} \cos bx \quad \text{and} \quad v_5 = -a\,e^{-ay} \sin bx \quad (7.5a)$$

Exercise 13 (i) Under what restrictions on the constants a and b can ϕ represent the plane flow of an incompressible fluid? (ii) If the fluid is *not* incompressible, can ϕ_5 represent a plane flow regardless of the values of a and b? Indicate why Bernoulli's theoreme (5.3) is not valid when a and b are arbitrarily chosen.

8 IRROTATIONALITY CONDITION USING STREAMLINE COORDINATES AND HODOGRAPH VARIABLES

The vanishing of the flow vorticity has already been expressed in two particular coordinate systems in Section 2 by equating to zero the expressions for circulation around appropriate infinitesimal flow regions. When the streamline coordinates depicted in Fig. II-9 is adopted, this condition assumes an especially simple form as

streamline coordinates, primitive expression of irrotationality

$$\frac{\partial}{\partial n}(q\,\Delta s) = 0 \qquad (8.1)$$

which is analogous to the continuity expression (II.8.1). The elementary length Δs is eliminated by noting that its variation in the direction normal to flow is given by the streamwise variation of flow direction angle; thus

$$\Delta n \frac{\partial \Delta s}{\partial n} = -\frac{\partial \theta}{\partial s} \Delta s\, \Delta n$$

The expanded form of (8.1) then becomes

final form, companion equation to (II.8.2)

$$\frac{\partial q}{\partial n} - q\frac{\partial \theta}{\partial s} = 0 \qquad (8.2)$$

as the irrotationality condition in streamline coordinates. Equation (8.2) is the counterpart of the continuity equation (II.8.2) for plane incompressible flow, written in streamline coordinates.

The result (8.2) may of course also be obtained by transformation of the Cartesian expression (5.1) with the aid of the Jacobian determinant technique employed in Section II.7. For the typical derivative $\partial v/\partial x$, for example,

verification by direct transformation

$$\frac{\partial v}{\partial x} = \frac{\partial [q \sin \theta, y]}{\partial (x, y)} = \frac{\partial [q \sin \theta, y)}{\partial (s, n)} J$$

where J denotes the Jacobian transformation determinant

$$J \equiv \frac{\partial (s, n)}{\partial (x, y)}$$

relating the streamline coordinates s, n and the Cartesian coordinates x, y.

Exercise 14 Evaluate the derivative $\partial u/\partial y$ in terms of the streamline coordinates in order to recover (8.2) from (5.1) when $J \neq 0$.

The pair of equations (II.8.2) and (8.2) are sufficient, in principle, for the determination of the flow velocity at each point of an ir-

IV. 8 / IRROTATIONALITY CONDITION: STREAMLINE AND HODOGRAPH SYSTEMS

rotational flow, because they contain the two dependent variables q and θ but not the pressure. The flow inclination angle θ is eliminated from these equations by differentiations and addition to give the single equation for the flow speed,

$$\frac{\partial}{\partial s}\left(\frac{1}{q}\frac{\partial q}{\partial s}\right) + \frac{\partial}{\partial n}\left(\frac{1}{q}\frac{\partial q}{\partial n}\right) = 0 \quad (8.3)$$

single equation for flow speed, nonlinear but...

Exercise 15 Determine the transformation $\tau = \tau(q)$ of the dependent variable q in such a manner that (8.3) becomes a linear equation for τ as a function of s and n,

$$\frac{\partial^2 \tau}{\partial s^2} + \frac{\partial^2 \tau}{\partial n^2} = 0 \quad (8.4)$$

...readily transformed to linear equation already studied

[Note that (8.4) is identical in form with (5.6) and may therefore also be termed Laplace's equation, satisfied by a function of the *flow speed* when the streamline coordinates are treated like ordinary Cartesian coordinates in the role of independent variables.] *Ans.* $\tau = \ln q$.

Equation (8.4) is subject to the practical inconvenience that the coordinates s and n can be related to common axes only after the flow has been determined [cf., for example, equations (II.8.3)]. If the flow geometry is known in advance, the velocity variation is readily found from (8.4) or from the equivalent pair of equations from which it was obtained. In terms of the variable τ, these equations are seen to be

$$\frac{\partial \theta}{\partial n} + \frac{\partial \tau}{\partial s} = 0 \quad \text{and} \quad \frac{\partial \tau}{\partial n} - \frac{\partial \theta}{\partial s} = 0 \quad (8.5)$$

Two elementary flow examples are now considered.

In one example, flow is parallel to a fixed direction, $\theta = $ const. Both of the derivatives of θ appearing in (8.5) vanish, and the same is true of the derivatives of τ. But $\tau = $ const implies that the speed $q = $ const, according to Exercise 15, and it is shown that a strictly parallel flow of incompressible fluid is irrotational only if the speed is constant.

a basic flow identified from the equations (8.5): uniform flow

In the other example the flow is radial outward, $\partial \theta / \partial s = 0$. Denoting distance from center of flow by r as usual and the plane polar angle by θ means that $ds = dr$, $dn = r\, d\theta$. Then the first term in the first of (8.5) gives

$$\frac{\partial \theta}{\partial n} = \frac{1}{r}\frac{\partial \theta}{\partial \theta} = \frac{1}{r} = -\frac{\partial \tau}{\partial r} \quad (8.6a)$$

and the second of (8.5) indicates that

$$\frac{1}{r}\frac{\partial \tau}{\partial \theta} = 0 \tag{8.6b}$$

From (8.6a) and (8.6b) it follows that $\tau + \ln r = \text{const}$, while the meaning of τ already established implies that $\ln(rq) = \text{const}$, or

source flow
$$q = \frac{\text{const}}{r} \tag{8.7}$$

It is thus shown that a radial plane flow of incompressible fluid is irrotational if the speed varies inversely with the distance from the center (termed plane source flow).

Interchanging the roles of dependent variables and independent variables in (8.2) in the same manner as in Exercise 22 of Chapter II gives the equation

irrotationality, polar hodograph variables
$$\frac{\partial s}{\partial \theta} - q\frac{\partial n}{\partial q} = 0 \tag{8.8}$$

which expresses irrotationality by using the velocity coordinates q, θ as independent variables, so that (8.8) is again in linear form, the *polar* hodograph coordinates appearing as in Section II.8.

Exercise 16 Use the Jacobian determinant transformation technique to show that

$$\frac{\partial \theta}{\partial s} = -\frac{\partial n}{\partial q}\frac{\partial(q, \theta)}{\partial(s, n)} \quad \text{and} \quad \frac{\partial q}{\partial n} = -\frac{\partial s}{\partial \theta}\frac{\partial(q, \theta)}{\partial(s, n)}$$

and prove (8.8).

some curious-looking equations displaying the inverse relationships of velocities and coordinates in irrotational flow

Exercise 17 Use the Jacobian determinant transformation technique to show that the derivatives appearing in (5.1) are related to those in the hodograph variables u, v by the equation

$$\frac{\partial v}{\partial x} : \frac{\partial u}{\partial y} = \frac{\partial y}{\partial u} : \frac{\partial x}{\partial v}$$

Show that the hodograph form of the irrotationality condition (5.1) is

$$\frac{\partial y}{\partial u} - \frac{\partial x}{\partial v} = 0 \tag{8.9}$$

Exercise 18 Combine (8.9) with the corresponding hodograph form of the continuity equation in order to show that the Cartesian coordinates are solutions of Laplace's equation when the independent variables are taken as the hodograph Cartesian components u, v:

$$\frac{\partial^2 x}{\partial u^2} + \frac{\partial^2 x}{\partial v^2} = 0 \quad \text{and} \quad \frac{\partial^2 y}{\partial u^2} + \frac{\partial^2 y}{\partial v^2} = 0 \tag{8.10}$$

V
Two-Dimensional Perfect Fluid Motions

1 QUANTITY OF FLOW — THE STREAM FUNCTION

The formal similarity of the continuity and irrotationality conditions (II.2.4a) and (IV.5.1) for plane incompressible flow, viz.,

$$\frac{\partial u}{\partial x} + \frac{\partial v}{\partial y} = 0 \quad \text{and} \quad \frac{\partial v}{\partial x} - \frac{\partial u}{\partial y} = 0$$

suggests introducing another velocity-related function of the same nature as the velocity potential ϕ, which is even more useful than the latter in certain respects. A physical motivation will be employed; it is based on extension of the condition (II.6.3) of zero normal flow across an element of solid boundary surface. In particular, the volume flux across an arbitrary line element dl is determined by the angle formed by the velocity vector and the unit normal vector, as

$$\mathbf{q} \cdot \mathbf{n} \, dl = u \, dy - v \, dx \qquad (1.1)$$

volume flux in two-dimensional flow...

for unit breadth of flow.

The right side of (1.1) is integrated between two points, A and B, to give the quantity of fluid crossing an arbitrarily drawn path C that connects the two points. We denote this fluid quantity as $-\psi_{AB}$ so that

$$-\psi_{AB} = \int_C (u \, dy - v \, dx) \qquad (1.2)$$

analogous to the elementary circulation expression Γ_{AB} appearing in Exercise IV.8. When the initial point A is regarded as fixed while $B(x, y)$ is variable, the *stream function* ψ is defined by the integral (1.2) regarded as a function of its upper limit:

...defines another potential function called the stream function ψ

$$-\psi(x, y) = \int^{B(x,y)} (u \, dy - v \, dx) \qquad (1.3)$$

The equivalent differential form is therefore

$$-d\psi(x, y) = u\, dy - v\, dx \tag{1.3a}$$

so that the analogy with the velocity potential follows by writing

velocity components by differentiation

$$u(x, y) = -\frac{\partial \psi}{\partial y} \qquad v(x, y) = \frac{\partial \psi}{\partial x} \tag{1.4}$$

quantity of flow evaluated...

The expressions (1.4) for the velocity components in terms of the stream function might evidently also have been obtained, without the volume flux interpretation, directly from the continuity equation (II.2.4a): the equality of the two mixed partial derivatives and the introduction of different signs in (1.4) assures that the continuity equation is satisfied, no matter what choice is made for the function $\psi(x, y)$.

...independent of integration path

Exercise 1 For the flow represented by (IV.2.4), and the points A, B, and C defined in Exercise IV.8, calculate $-\psi_{AB}$, $-\psi_{BC}$, and $-\psi_{AC}$ along the paths employed earlier for the corresponding circulation calculations, and compare as before.

Exercise 2 Show that the volume flux $-\psi_{AB}$ of an arbitrary flow $u(x, y)$, $v(x, y)$ does not depend on the path between the points A and B. Justify the omission from (1.3) of path specification, and establish the uniqueness of the relationships (1.4).

Exercise 3 Use (1.4) and (IV.5.1) to show that the stream function $\psi(x, y)$ of a two-dimensional incompressible irrotational flow is a solution of Laplace's equation (IV. 5.6).

Exercise 4 (a) Verify that the velocity field given by (IV.2.4) is represented by the stream function

function $\psi(x,y)$ determines flow field

$$\psi(x, y) = -\frac{y}{x^2 + y^2} + \text{const} \tag{1.5}$$

and check the value found in Exercise 1 for ψ_{AC} by noting that

$$\psi_{AC} = \psi(0, 1) - \psi(1, 0)$$

(b) Determine whether the flow (1.5) is irrotational.

Exercise 5 From the definition (1.3a) of the stream function, show that the value of ψ is the same at all points lying on the same streamline, i.e., that the stream function is constant on each streamline. Find the slope of the streamline at any point x, y as

streamlines are identical with family of lines $\psi(x,y)$ = const

$$\left.\frac{dy}{dx}\right|_\psi = -\frac{(\partial \psi / \partial x)}{(\partial \psi / \partial y)} = \frac{v}{u}$$

and prove that the streamlines $\psi(x, y) = $ const are everywhere orthogonal to the potential lines $\phi(x, y) = $ const.

From the fact that the volume flux $-\psi_{AB}$ passing between two points A and B of a plane flow of unit breadth does not depend on the choice of coordinates, it follows that velocity components in different coordinate systems can also be expressed in terms of derivatives of the same nature as (1.4). In the case of plane polar coordinates r, θ and the corresponding velocity components u_r and u_θ, (II.1.10) and (1.4) show that

$$u_r = -\left\{\frac{\partial \psi}{\partial y}\cos\theta - \frac{\partial \psi}{\partial x}\sin\theta\right\} \qquad u_\theta = \left\{\frac{\partial \psi}{\partial y}\sin\theta + \frac{\partial \psi}{\partial x}\cos\theta\right\}$$

and

$$\cos\theta = \frac{1}{r}\frac{\partial y}{\partial \theta} \qquad \sin\theta = -\frac{1}{r}\frac{\partial x}{\partial \theta}$$

so that

$$u_r = -\left\{\frac{\partial \psi}{\partial y}\frac{\partial y}{r\,\partial\theta} + \frac{\partial \psi}{\partial x}\frac{\partial x}{r\,\partial\theta}\right\} = -\frac{1}{r}\frac{\partial \psi}{\partial \theta} \qquad u_\theta = \frac{\partial \psi}{\partial r} \quad (1.6)$$

stream function likewise determines *polar* velocity components

Exercise 6 Show that the continuity equation of plane incompressible flow written in polar coordinates by neglecting the last term of (II.2.5) is identically satisfied when the velocity components u_r and u_θ are defined as in (1.6).

Exercise 7 Obtain Laplace's equation in plane polar coordinates by setting the vorticity $\zeta = 0$ in Exercise IV.7 and substituting for u_r and u_θ from (1.6).   Ans.
$$\frac{1}{r}\left[\frac{\partial}{\partial r}\left(r\frac{\partial \psi}{\partial r}\right) + \frac{\partial}{\partial \theta}\left(\frac{1}{r}\frac{\partial \psi}{\partial \theta}\right)\right] = 0 \quad (1.7)$$

$\psi(r,\theta)$ also satisfies Laplace's equation

Exercise 8 Indicate why it is not possible to define a stream function that will satisfy the continuity equation of three-dimensional flow (II.2.5) in general, i.e., when all three terms may differ from zero. Determine how u_r and w may be defined in terms of a stream function in the case of axially symmetric flow ($\partial/\partial\theta \equiv 0$).

Exercise 9 Resolve the unit normal vector $\mathbf{n}$ of Fig. II-7 into radial and transverse components to show that

$$\mathbf{n}\,dl = -\mathbf{e}_\theta\,dr + \mathbf{e}_r r\,d\theta$$

independent deduction of (1.6)

and write the velocity as $\mathbf{q} = u_r\mathbf{e}_r + u_\theta\mathbf{e}_\theta$ instead of (1.1) to recover (1.6).

2 SIMPLE FLOWS REPRESENTED BY STREAM FUNCTION

Every function of x and y that has derivatives can be regarded either as a potential function or as a stream function, the corresponding velocity components being determined either by (IV.5.5) or by (1.4). If the function is also a solution of Laplace's equation (IV.5.6), then the conditions of continuity and irrotationality are both satisfied for an incompressible fluid. Several examples are now considered.

arbitrary functions $\psi(x,y)$ or $\psi(r,\theta)$ determine plane flows

Example 1. Uniform Flow Parallel to x-Axis.

$$\psi_1 = -Uy \qquad (2.1)$$

The velocity components, according to (1.4), are

uniform flow

$$u_1 = -\frac{\partial \psi_1}{\partial y} = U \qquad v_1 = \frac{\partial \psi_1}{\partial x} = 0 \qquad (2.1a)$$

when U is constant, and the flow is uniform, everywhere parallel to the x-axis (Fig. V-1). It is clear that a linear function of x and y gives a flow inclined to the x-axis, while a linear function of x alone corresponds to motion parallel to the axis of y. Cf. (IV.7.1)

Example 2. Vortex at the Origin.

$$\psi_2 = A \ln \sqrt{x^2 + y^2} \qquad (2.2)$$

The (Cartesian) velocity components are obtained now as

vortex

$$u_2 = -\frac{Ay}{x^2 + y^2} \qquad v_2 = \frac{Ax}{x^2 + y^2} \qquad (2.2a)$$

or more conveniently in polar form by writing the right side of (2.2) as $A \ln r$ and using (1.6) to give

$$u_{r_2} = 0 \qquad u_{\theta_2} = \frac{A}{r} \qquad (2.2b)$$

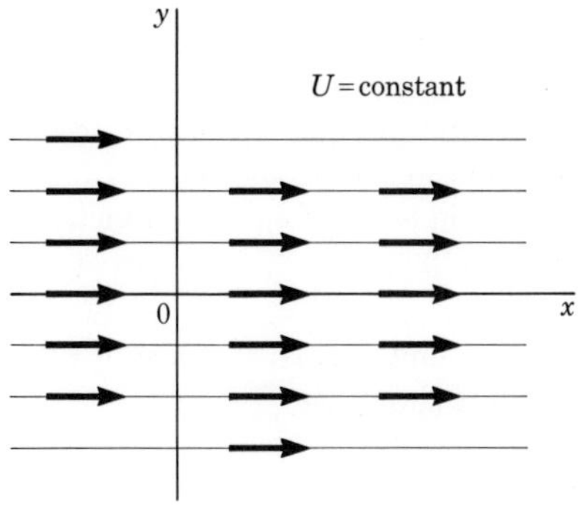

Fig. V-1 Uniform flow.

V. 2 / SIMPLE FLOWS REPRESENTED BY STREAM FUNCTION

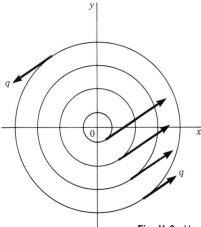

Fig. V-2   Vortex flow.

The result of Exercise 5 shows that the streamlines are given by $\psi_2(x, y) = $ const, from (2.2), or in polar form $\psi_2(r, \theta) = $ const, as the family of circles $r = $ const concentric with the origin of coordinates. There is thus no radial motion, and the transverse velocities vary inversely with the distance from the origin, according to (2.2b). This circulating motion centered at the origin is termed a two-dimensional *vortex* at the origin (Fig. V-2). Cf. (IV.7.2a)

Exercise 10 Use the result of Exercise IV.7 to evaluate the vorticity of the vortex flow (2.2) from the form of the velocities in polar coordinates (2.2b). Verify the calculation by using the values (2.2a) in the vorticity expression (IV.1.2).

Exercise 11 Determine directly whether $\psi_2(x, y)$ is a solution of Laplace's equation (IV.5.6). Is this finding consistent with the results of the preceding exercise?

Example 3. Two-Dimensional Source at the Origin.

$$\psi_3 = -A \tan^{-1} \frac{y}{x} \qquad (2.3) \qquad \text{source}$$

The right side of (2.3) is $-A\theta$ in polar form, and the polar velocity components are found from (1.6) as

$$u_{r_3} = \frac{A}{r} \qquad u_{\theta_3} = 0 \qquad (2.3a)$$

These components indicate a radial flow away from the origin, with speed diminishing as the first power of the distance from the

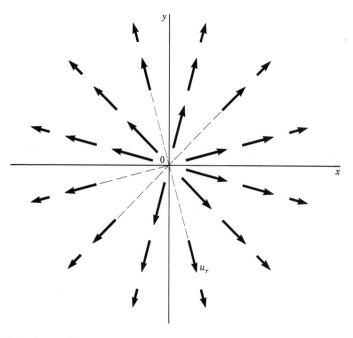

Fig. V-3 Source flow.

origin, the streamlines all passing through the origin. When $A > 0$, the motion is directed outward, away from this point, and it is natural to refer to the flow as a two-dimensional source at the origin (Fig. V-3). *Sink* flow corresponds to $A < 0$. Contrast with (IV.7.2).

Exercise 12 Establish that the flow (2.3) is irrotational by determining its velocity potential.

Example 4.

$$\psi_4 = \text{Re}\{z^3\} \quad \text{where} \quad z \equiv x + iy \quad \text{and} \quad i = \sqrt{-1} \quad (2.4)$$

complex functions

By expanding and isolating the real part of z^3, it is seen that

$$\psi_4(x, y) = x^3 - 3y^2 x$$

so that

$$u_4(x, y) = 6yx \quad v_4(x, y) = 3x^2 - 3y^2 \quad (2.4a)$$

derivatives of ψ are likewise stream functions, but see Exercise 14

Example 5.

$$\psi_5 = u_4(x, y) = 6yx \quad (2.5)$$

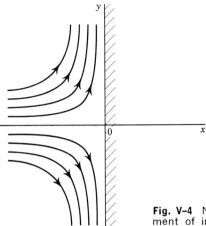

Fig. V-4 Normal impingement of infinite jet, represented by Eq. (2.5a).

The velocity components are

$$u_5(x, y) = -6x \qquad v_5(x, y) = 6y \qquad (2.5a)$$

and (2.5) shows that the streamlines $\psi_5 = $ const are rectangular hyperbolae asymptotic to the Cartesian coordinate axes (Fig. V-4). Since the x-axis is itself one of the streamlines, the flow may be interpreted as interior to a jet flow in the half-plane $x < 0$ impinging at right angles on the boundary wall at $x = 0$. CF. (IV.7.4)

It has already been noted that each of the flows (2.1)–(2.5) is shown to be irrotational by verifying that each of the stream functions ψ_i ($i = 1, 2, 3, 4, 5$) are solutions of Laplace's equation (IV.5.6),

$$\frac{\partial^2 \psi_i}{\partial x^2} + \frac{\partial^2 \psi_i}{\partial y^2} = 0 \qquad (2.6)$$

and it is clear that the sum or difference of any two of these functions is also a solution of (2.6). This most important property of linear differential equations (i.e., those in which the dependent variable, ψ_i in this case, and its derivatives appear to the first power only in each term of the equation) is termed the principle of superposition of solutions. Starting with a few solutions, therefore, a large number of different solutions is obtained, and the technique is useful in the study of physically important flows. When it is recalled, moreover, that a derivative is formed as the limit of a difference ratio, it follows that *differentiation* of a stream function or potential function also yields new and more complex flows starting such elementary flows as (2.1)–(2.5). Cf. Equation IV. 7.3.

method of *generating* solutions representing plane irrotational flow

Exercise 13 Prove that the x- or y-derivative of a solution of (2.6) is also a solution of the same equation. HINT: Differentiate (2.6) and recall that the order of differentiations is immaterial, as in Section IV.5.

Exercise 14 Laplace's equation (1.7) in plane polar coordinates determined in Exercise 7 can be written

$$\frac{1}{r}\frac{\partial}{\partial r}\left(r\frac{\partial \psi}{\partial r}\right) + \frac{1}{r^2}\frac{\partial^2 \psi}{\partial \theta^2} = 0 \qquad (1.7)$$

Show that the r-derivative of a solution of (1.7) is *not* a solution of the same equation, but that the θ-derivative *is* a solution. Explain the difference in the two cases in terms of the coefficients in (1.7).

Exercise 15 Using the same argument that led to (1.6), show that the radial and transverse velocity components are represented by the velocity potential as a function of polar coordinates r, θ by the formulae

$$u_r = -\frac{\partial \phi}{\partial r} \qquad u_\theta = -\frac{\partial \phi}{r\,\partial \theta} \qquad (2.7)$$

polar velocity components likewise given by velocity potential

Determine the stream function of the flow considered in Exercise IV.11 and show that

$$\frac{\cos \theta}{r} \qquad (2.8)$$

is the stream function of an irrotational flow. Demonstrate that the stream function (2.8) can be regarded as the superposition of a source and a sink [i.e., source of negative "strength" corresponding to sign reversal in (2.3)] at the origin, by recovering (2.8) as a derivative of (2.3). The flow represented by (2.8) is termed a two-dimensional *doublet* (or *dipole*) at the origin.

3 SUPERPOSITION OF ELEMENTARY FLOWS — CIRCULAR CYLINDER IN UNIFORM MOTION

The stream function obtained as the θ-derivative of (2.8) also represents a doublet, its "axis" parallel to the x-direction. When this is added to the stream function (2.1) of uniform flow in the same direction, the new stream function

circular cylinder flow: superposition of uniform flow and doublet

$$\psi = -Uy + \frac{Ua^2}{r}\sin\theta = -Ur\left(1 - \frac{a^2}{r^2}\right)\sin\theta \qquad (3.1)$$

is obtained; it is most easily interpreted from the polar coordinate representation in (3.1). All points of the circle $r = a$ lie on the streamline $\psi = 0$; hence the circle can be regarded as a physical boundary of flow exterior to it. The vanishing of $\sin\theta$ on the positive and negative x-axes indicates that they are parts of the

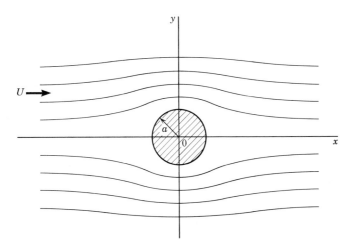

Fig. V-5 Streamline pattern, uniform flow past circular cylinder without circulation.

same streamline. Thus the external flow is parallel to the axis of x, and it is also clear that the velocity approaches the constant value U with increasing distance from the circle centered at the coordinate origin. The motion is indicated in Fig. V-5, showing the streamline pattern of a uniform flow which is divided as it passes around the cylinder having the circle as projection, becoming uniform again downstream of the cylinder.

Exercise 16 When gravity is neglected, the Bernoulli integral (IV.5.3) is evaluated in terms of the pressure p_0 at points distant from the cylinder where the velocity is U, so that

$$\frac{q^2}{2} + \frac{p}{\rho} = \frac{U^2}{2} + \frac{p_0}{\rho} \tag{3.2}$$

Denote the pressure p_0 by zero, i.e., measure pressures relative to the free-stream value, and show that the pressure $p(a, \theta)$ at an arbitrary point of the cylinder is given by

$$p(a, \theta) = \frac{U^2}{2}(1 - 4\sin^2\theta) \tag{3.3}$$

Sum the x-components of the pressure forces at all points of the first quadrant AB of the cylinder (i.e., $0 < \theta < \pi/2$), and repeat for the y-components. Ans. $X_{AB} = \rho(aU^2/6)$.

Exercise 17 Prove that the resultant force on the cylinder is zero.

In Exercise 10 the vortex flow (2.2) is shown to be irrotational at all points away from the origin. It is also clear from (2.2b) that the circulation about any circular path enclosing the origin is $2\pi A$, from which it is inferred that the flow (2.2) represents a concentration of vorticity at the origin, in a motion having zero rotation at all other points. Hence the irrotational flow (3.1) past a circular cylinder *without* circulation is extended to flow with finite circulation $-\Gamma$ by adding the vortex stream function to obtain

preceding flow, with circulation

$$\psi(r, \theta) = -Ur\left(1 - \frac{a^2}{r^2}\right) \sin \theta - \frac{\Gamma}{2\pi} \ln \frac{r}{a} \qquad (3.4)$$

Because of the vanishing of the last term of (3.4) when $r = a$, the circle $r = a$ remains part of the zero streamline.

4 IRROTATIONAL FLOW WITH CIRCULATION: AERODYNAMIC LIFT

The symmetry of the streamline pattern in Fig. V-5 obtained as the superposition of a doublet and a uniform flow shows that velocities are equal at corresponding points on the upper and lower surfaces of the circular cylinder. It follows that the pressures (3.3) are also equal and hence that a net force can be accounted for only by a more complete flow representation than (3.1). On the basis of the calculation of Section IV.4 it may be expected that inclusion of a circulation term is appropriate for this purpose. Aerofoil contours are in fact so determined as to assure a finite circulation in the sense indicated in (3.4), and a direct calculation now permits evaluation of the resultant force in terms of Γ.

Exercise 18 Show that the radial and transverse velocity components on the surface of the circular cylinder $r = a$ are obtained from (3.4) as

$$u_r(a, \theta) = 0 \qquad u_\theta(a, \theta) = -2U \sin \theta - \frac{\Gamma}{2\pi a} \qquad (4.1)$$

Evaluate the force components X and Y per unit breadth of cylinder according to (3.2) to show that

lift force proportional to flow circulation

$$X = -\int_0^{2\pi} p(a, \theta) \cos \theta \, a d\theta = 0$$

$$Y = -\int_0^{2\pi} p(a, \theta) \sin \theta \, a d\theta = \rho U \Gamma \qquad (4.2)$$

The second equation of (4.2), showing that the force *normal* to the direction of flow is given by the product of the circulation and the mass flux ρU, is known as the Kutta-Joukowski theorem.

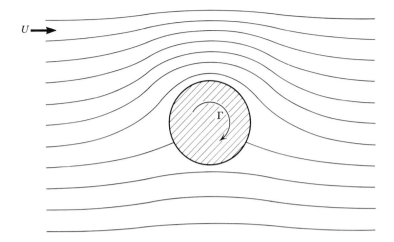

Fig. V–6 Cylinder flow with circulation.

The practical interest of the Kutta-Joukowski theorem is that it gives a good approximation to aerodynamic lift force not only on circular cylinders but also on common airplane wing sections. The total absence of a force component parallel to the flow direction, friction being neglected, is also noted and is seen to correspond to the symmetry of the streamline pattern in Fig. V-6: both upstream and downstream of the cylinder the velocities are equal at corresponding points. Flows above and below the cylinder are different in this case, and the closer streamline spacing near its upper surface is recognized as an indication of higher speeds and hence lower pressures than at the under surface. The same general characteristics are well confirmed in flow past actual airfoil sections, the stream functions for which are found with the aid of calculations based on (3.4).

An indication that the lift force expression in (4.2) is also valid for noncircular sections may be obtained directly from the Bernoulli integral (3.2) when the circulation is accounted for by incremental velocities u' small compared with U. With an airfoil section represented as a thin rectangle, thickness in direction normal to the flow small compared with the *chord l*, on the upper surface of which the resultant flow speed is $U + u'$, on the lower surface $U - u'$, the circulation is evaluated directly from the definition (IV.2.1). Counting the circulation as positive in the clockwise sense, as in (3.4) leads to

verification of (4.2) for slender wing section

$$\Gamma = \int_{AB} \mathbf{q} \cdot \boldsymbol{\tau} \, dl + \int_{B'A'} \mathbf{q} \cdot \boldsymbol{\tau} \, dl$$
$$= (U - u')(-l) + (U + u')l = 2u'l \qquad (4.3)$$

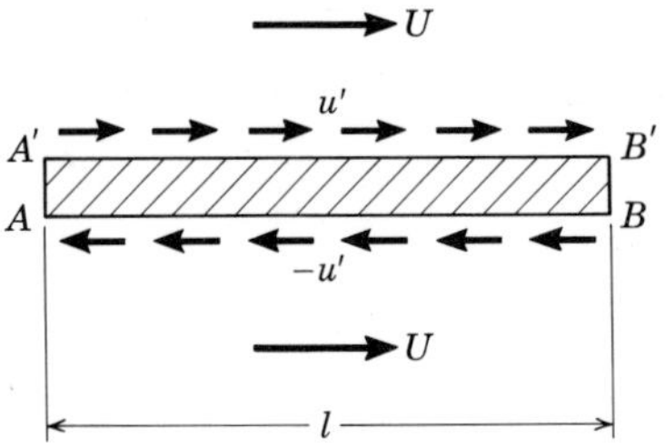

Fig. V-7 Illustrating lift force dependence on circulation.

Exercise 19 Show that the pressures p_{AB} and $p_{A'B'}$ on the lower and upper surfaces, respectively, in Fig. 5-7 are given by $\rho U u'$ and $-\rho U u'$ when p_0 is again taken as zero, and hence that the resultant upward force Y is expressed as

$$Y = p_{AB}l - p_{A'B'}l = \rho U \Gamma$$

in accord with (4.2).

5 METHODS FOR DETERMINING IRROTATIONAL FLOWS

The calculation of irrotational flows starting with Laplace's equation (2.6) is greatly facilitated by the availability of a variety of general methods of constructing solutions of this equation which serve as stream functions and velocity potentials. Five distinct procedures that either have been or will be employed are reviewed here.

Method I.

superposition

Superposition of Known Solutions. The stream function (3.4) is the sum of three elementary flow solutions, and it is clear that any other combination of *harmonic functions*, i.e., solutions of (2.6) or of any of its equivalent forms like (1.7), may likewise be added to represent different two-dimensional incompressible flows.

Method II.

fundamental solutions

Fundamental Solutions. Laplace's equation in two dimensions contains the two independent variables x and y in Cartesian form, r and θ in polar form. The stream function (2.1) depends on only one of the independent variables in (2.6), while (2.2) and (2.3) each contain only one of the polar coordinates appearing in (1.7). Solutions depending on one of the variables alone are easy to obtain because the equation simplifies considerably in each case, becoming an ordinary differential equation of the second order the solutions of which are usually obtainable by inspection.

Method III.

Differentiation of a Known Solution. Although differentiation is a limiting case of addition (superposition), a flow obtained in this manner is basically different from the initial flow. The stream function (2.5) and the general conclusions of Exercises 13–15 are applicable.

differentiation

Method IV.

Real and Imaginary Parts of Complex Functions. When i again denotes "the imaginary unit" treated algebraically according to the rule $i^2 = -1$, every ordinary function of $x + iy$ can be expressed as the sum of a real part plus i times an imaginary part. Thus, if $x + iy$ is denoted by z,

complex functions

$$z^2 = (x + iy)^2 = (x^2 - y^2) + i2xy$$

so that the real part of z^2 is $x^2 - y^2$, the imaginary part $2xy$. An arbitrary function $W(z)$ having real and imaginary parts $f(x, y)$ and $g(x, y)$ respectively, that is,

$$W(z) = f(x, y) + ig(x, y)$$

is shown in the theory of complex variables to possess the usual properties of differentiability [viz., the function $W(z)$ has a derivative dW/dz] whenever the real and imaginary parts are interrelated by the two equations

$$\frac{\partial f}{\partial x} = \frac{\partial g}{\partial y} \qquad \frac{\partial f}{\partial y} = -\frac{\partial g}{\partial x} \qquad (5.1)$$

When the two equations (5.1) are seen to imply that both $f(x, y)$ and $g(x, y)$ are harmonic functions, it is clear that either one of them can be taken as a stream function, or as a potential function, of a plane flow. The required properties of f and g are established at once, however, by appropriate differentiations and addition of the two equations. The importance of complex variables for the study of plane flows is therefore clear, particulary as (5.1) are recognized to be identical in form with the equations of continuity and irrotationality.

Method V.

Product-Type Solutions. When the function ψ is assumed to be the product of one function depending on one of the independent variables multiplied by another function that depends on the second of these, a *separation* into two ordinary differential equations is achieved. Each of these is solved, giving ψ as the product

product-type solutions

of two solutions of equations that are usually familiar and therefore easy to solve. The method is illustrated by the calculation of the next section and applied repeatedly in subsequent chapters.

Exercise 20 Specifically identify the functions f and g with the potential and stream functions, respectively, and define the *complex potential* $W(z)$ as

complex potential $\phi(x,y) + i\psi(x,y)$

$$W(z) = \phi(x, y) + i\psi(x, y) \tag{5.2}$$

Determine the velocity components and streamlines of the flow corresponding to setting

$$W(z) = \frac{1}{z}$$

Relate doublet flow (2.8) to vortex flow $W(z) = A \ln z$ and show that $-(dW/dz) = u - iv$, in general.

Exercise 21 Laplace's equation in spherical polar coordinates is

Laplace's equation in spherical polar coordinates

$$\frac{1}{r^2}\left[\frac{\partial}{\partial r}\left(r^2 \frac{\partial \phi}{\partial r}\right) + \frac{1}{\sin \theta}\frac{\partial}{\partial \theta}\left(\sin \theta \frac{\partial \phi}{\partial \theta}\right) + \frac{1}{\sin^2 \theta}\frac{\partial^2 \phi}{\partial \omega^2}\right] = 0 \tag{5.3}$$

Find the fundamental solution corresponding to $\phi = \phi(r)$, and show that it leads to the three-dimensional counterpart of the source flow (2.3) when the radial velocity component is again taken as $u_r = -(\partial \phi/\partial r)$.

Exercise 22 Demonstrate that the product-type solution of (IV.5.6), viz., $\phi(x,y) = X(x)Y(y)$ leads to the pair of ordinary differential equations

an important example

$$\begin{aligned} X''(x) + k^2 X(x) &= 0 \\ Y''(y) - k^2 Y(y) &= 0 \end{aligned} \tag{5.4}$$

having trigonometric and exponential functions as solutions, respectively, k^2 representing an undetermined positive constant.

6 LINEARIZED ANALYSIS OF UNIFORM FLOW PAST A WAVY WALL

If one of the streamlines of the uniform flow (2.1) is modified so as to be everywhere only nearly parallel to the x-axis, it is to be expected that the velocities elsewhere will differ but slightly from the value U and that associated pressure variations will be correspondingly small. Important general properties of incompressible and irrotational flow are demonstrated by considering the flow non-uniformities related to the introduction of a slight waviness of one boundary streamline. Specifically, the flow of a uniform stream

nearly uniform flow — velocity perturbation

adjacent to a slightly curved boundary wall of sinusoidal form is analyzed; it is defined by the equation

$$y = h \sin \lambda x \qquad (6.1)$$

where the maximum displacement of the boundary from the *x*-axis is given by h. Since there is no "characteristic length" or scale in a completely uniform flow like (2.1) the smallness of h in (6.1) must be taken with reference to the length associated with the parameter λ in (6.1) itself. Since the length of a whole cycle of the sine wave is given by $L = 2\pi/\lambda$, the requirement that h be small is given definite meaning by the inequality

$$h \ll L \qquad (6.2) \qquad \text{slight waviness}$$

The possibility of directly determining the flow in the half-plane $y > 0$ past the boundary (6.1) depends on the simplifying approximations related to the assumption (6.2).

The flow velocity components are assumed to differ only slightly from the values $U, 0$ and can therefore be written

$$u = U + u' \qquad v = v' \qquad (6.3)$$

where the *velocity perturbations* u' and v' evidently vanish with h and are therefore small compared with U:

$$u' \ll U \qquad v' \ll U \qquad (6.4)$$

Since the irrotationality condition (5.1) retains its original form when the velocities (6.3) are introduced, that is,

$$\frac{\partial v'}{\partial x} - \frac{\partial u'}{\partial y} = 0 \qquad (6.5)$$

the velocity potential $\phi(x, y)$ may be taken to refer only to the perturbations

$$u' = -\frac{\partial \phi}{\partial x} \qquad v' = -\frac{\partial \phi}{\partial y} \qquad (6.6) \qquad \begin{array}{l}\phi \text{ is the velocity potential} \\ \text{for the flow perturbation}\end{array}$$

so that (6.3) becomes

$$u = U - \frac{\partial \phi}{\partial x} \qquad v = -\frac{\partial \phi}{\partial y} \qquad (6.3a) \qquad \text{total flow}$$

In the form (6.3a) it appears that small deviations from flow uniformity are entirely related to the potential ϕ, and we suppose that all products of ϕ-derivatives can be regarded as small to a higher order than the velocities (6.6) themselves. The pressure perturbations are accordingly given, to the first order of small quantities, by evaluating the pressure perturbation from (3.2) as

$$p(x, y) - p_0 = \rho U \frac{\partial \phi}{\partial x} \qquad (6.7) \qquad \begin{array}{l}\text{simplified pressure} \\ \text{expression}\end{array}$$

Here p_0 again denotes the pressure in those parts of the flow where the velocity reaches its uniform value U, and forces are found as in (4.2).

The continuity equation and (6.3a) again show that the velocity potential is a harmonic function:

differential equation...

$$\frac{\partial^2 \phi}{\partial x^2} + \frac{\partial^2 \phi}{\partial y^2} = 0 \tag{6.8}$$

It is necessary to determine the particular one of the many solutions of Laplace's equation (6.8) that corresponds to the present flow, which is characterized by the form of the boundary (6.1) and by the condition that the disturbance velocities (6.4) vanish at very large distances from the wall, i.e., as $y \to \infty$. This is assured if the potential ϕ vanishes in the limit represented by writing

...and one boundary condition...

$$\phi(x, \infty) = 0 \tag{6.9}$$

The requirement of vanishing normal component of flow at the boundary (6.1) is expressed by the statement that the slope of the velocity vector there equals the boundary slope:

$$\frac{v}{u} = h\lambda \cos \lambda x \tag{6.10}$$

while the left side of (6.10) is written in terms of the potential as

$$-\frac{\partial \phi / \partial y}{U - \partial \phi / \partial x} \doteq -\frac{1}{U} \frac{\partial \phi}{\partial y} \tag{6.11}$$

when terms of second and higher order in the small quantities are neglected compared with the first-order term retained on the right side of (6.11). The condition (6.11) is strictly applicable at the boundary surface (6.1), but, when it is considered that this differs by only a small amount from $y = 0$ and that the velocity perturbations are also small, a consistent approximation is obtained by again neglecting second-order terms and writing for the *approximate* boundary condition

...and another

$$-\frac{\partial \phi}{\partial y}(x, 0) = Uh\lambda \cos \lambda x \tag{6.12}$$

The meaning of (6.12) is, then, that the tangential flow condition that it represents is applied not at the actual surface (6.1), but at the mean line that is taken to be everywhere close to it. Clearly the limit $h=0$ corresponds to the solution of (6.8) of flow with straight streamlines, vanishing velocity perturbations, and uniform pressure as in (2.1). For small but nonzero values of the wall thickness h, the differential equation (6.8) and the boundary conditions (6.9) and (6.12), as well as the pressure relationship (6.7) are all seen to be *linear* in the ϕ-derivatives, according to what is usually termed the *linearized approximation* or the small-disturbance approximation.

The solution of (6.8) is given by the pair of functions X and Y determined in Exercise 22. They can be written

$$X(x) = A \cos k x + B \sin kx$$
$$Y(y) = Ce^{-ky} + De^{ky} \qquad (6.13)$$

tentative solution

so that

$$\phi(x, y) = \{A \cos kx + B \sin kx\}\{Ce^{-ky} + De^{ky}\} \qquad (6.14)$$

is the velocity potential of the flow. For the condition (6.9) to be satisfied, it is necessary for the positive exponential term in (6.14) to be absent; this implies $D = 0$. Then there remain constants AC and BC to determine, and it is clear that we can put $C = 1$ without loss of generality. Finally (6.12) indicates that

$$Uh\lambda \cos \lambda x = k\{A \cos kx + B \sin kx\}$$

which requires that A, B, and k satisfy the relationships

$$Uh\lambda = kA \qquad B = 0 \qquad k = \lambda \qquad (6.15)$$

evaluation of constants

With these values the potential function is expressed in terms of the flow and boundary parameters as

$$\phi(x, y) = Uhe^{-\lambda y} \cos \lambda x \qquad (6.16)$$

solution completely determined

while the pressure is given by (6.7) as

$$p(x, y) = p_0 - \rho U^2 h\lambda \sin \lambda x \, e^{-\lambda y} \qquad (6.17)$$

pressure in the flow...

at any point in the flow. The linear approximation to the pressure on the boundary wall $y = 0$ follows at once as

$$p(x, 0) = p_0 - \rho U^2 h\lambda \sin \lambda x \qquad (6.17a)$$

... and approximated at the boundary surface

Equation (6.17a) shows that the pressure is a minimum at the peaks of the wavy wall, i.e., where $\lambda x = \pi/2$ and at integral multiples of the wavelength L in both directions, while the pressure maxima occur at the wave troughs midway between the successive crests or peaks. The reduction of pressure at points of convex boundary curvature as in the present instance is a common feature of incompressible flows. Because the wall pressures (6.17a) are "in phase" with the profile (6.1), the resultant pressure force on each complete cycle of the wavy wall is zero. The pressures are indicated in Fig. V-8, where the streamwise symmetry of the streamlines is also noted, as in the flows considered in Sections 3 and 4.

zero resultant force

126 TWO-DIMENSIONAL PERFECT FLUID MOTIONS

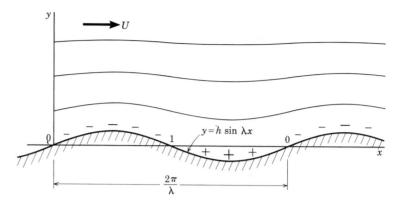

Fig. V-8 Streamline pattern and pressures on wavy wall in uniform flow.

influence of a plane boundary at distance H

Exercise 23 Instead of flow of unlimited extent in the y-direction represented by (6.16) consider the case in which a plane boundary is located at $y = H$ while $h \ll H$ and all other conditions of flow are as before. Evaluate the coefficients in the second of (6.13) by requiring that $v(x, H) = 0$ instead of the condition (6.9), showing that the y-dependence is now given by

$$\frac{\cosh \lambda(H - y)}{\sinh \lambda H} \qquad (6.18)$$

hyperbolic functions

instead of the exponential $e^{-\lambda y}$ in (6.16). Recall that the *hyperbolic cosine* and the *hyperbolic sine* functions appearing in (6.18) are the sums of exponentials defined, respectively, by the equations

$$\cosh x = \frac{e^x + e^{-x}}{2} \qquad \sinh x = \frac{e^x - e^{-x}}{2} \qquad (6.19)$$

Exercise 24 Show that the pressure on the wavy wall of the preceding exercise are modified by the factor $\coth \lambda H$ as compared with (6.17a), where the hyperbolic cotangent function is given by the quotient of the first of (6.19) divided by the second.

Exercise 25 Evaluate the "lift" and "drag" force components on one cycle of the wall for the flow considered in Exercise 23 by noting that they are approximated as

$$F_y = \int_0^L -p(x, 0)\, dx \quad \text{and} \quad F_x = \int_0^L p(x, 0)\frac{dy}{dx}(x, 0)\, dx \qquad (6.20)$$

respectively, where the derivative in the second of (6.20) gives the slope of the wall defined by (6.1). Discuss and interpret the results.

7 KINETIC ENERGY OF FLOW DETERMINED AT BOUNDARY POINTS: APPARENT MASS

The concept of energy, which is of most fundamental importance in all branches of physical theory, is introduced in the dynamics of plane flows in particularly simple form with the aid of either the stream function or the velocity potential. The fluid kinetic energy associated with the passage of a solid body through a fluid medium otherwise at rest is considered for definiteness. It is preferable to regard the flow as *unsteady*, i.e., to view it from a reference frame that is stationary with respect to the undisturbed fluid far from the body. In this manner the kinetic energy of an individual particle approaches zero at large distances from the body, the total value of the kinetic energy for all particles being

body moving through fluid at rest

$$T = \rho \iint_S \frac{u^2 + v^2}{2} \, dx \, dy \qquad (7.1)$$

as in (IV.6.1), where the region of integration, S, is now of unlimited extent. Because of the smallness of the flow velocity at distant points, the energy (7.1) is determined mainly by the motion in the vicinity of the moving boundary C, as shown in Fig. V-9.

When the velocity components are written in terms of the stream function according to (1.4), the flow being irrotational, the integrand in (7.1) becomes

$$\left(\frac{\partial \psi}{\partial x}\right)^2 + \left(\frac{\partial \psi}{\partial y}\right)^2 = \frac{\partial}{\partial x}\left(\psi \frac{\partial \psi}{\partial x}\right) + \frac{\partial}{\partial y}\left(\psi \frac{\partial \psi}{\partial y}\right)$$
$$= \nabla \cdot (\psi \nabla \psi) \qquad (7.2)$$

kinetic energy expressed in terms of stream function, per unit volume

on account of (2.6). The final vector form of (7.2) shows that the energy expression (7.1) is the divergence of a vector, integrated over a surface. Gauss' transformation (A.6) is therefore applicable; it permits the kinetic energy (7.1) to be written in terms of the

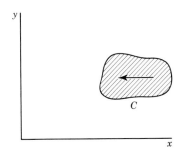

Fig. V-9 "Unsteady" flow representation — submerged body advancing into fluid at rest.

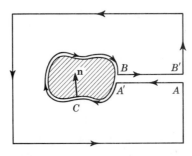

Fig. V-10 Integration path, Eq. (7.3).

normal component of the vector $\psi \nabla \psi$ integrated over the bounding contour C:

energy of entire flow field

$$T = \frac{\rho}{2} \int_C \psi \frac{\partial \psi}{\partial n} \, dl \qquad (7.3)$$

where the n-component of the gradient is simply the derivative in the direction of **n**. According to the convention adopted in Appendix A, the vector **n** is directed to the right, as the complete fluid boundary is described in the clockwise sense with the region of integration on the left. In the present case n is therefore directed *inward* at boundary points C, while integration over remaining portions of the complete contour shown in Fig. V-10 are omitted either because of mutual cancellation (AA' and BB') or because velocities are very small at large distances from the body.

As an application of the formula (7.3), the flow past a circular cylinder with circulation is obtained in "unsteady" form by superposing a uniform velocity U on the motion given by (3.4), so that

$$\psi(r, \theta) = U \frac{a^2}{r} \sin\theta - \frac{\Gamma}{2\pi} \ln\frac{r}{a} \qquad (7.4)$$

and the normal derivative $\partial\psi/\partial n = -\partial\psi/\partial r$, giving these values at the boundary C, where $r = a$:

$$\psi(a, \theta) = Ua \sin\theta \qquad (7.4a)$$

$$\frac{\partial\psi(a, \theta)}{\partial n} = U \sin\theta + \frac{\Gamma}{2\pi}\frac{1}{a} \qquad (7.4b)$$

Substitution of (7.4a) and (7.4b) in (7.3) leads to

flow energy evaluated ...

$$T = \frac{\rho}{2} U a^2 \int_0^{2\pi} d\theta \, \sin\theta \left[U \sin\theta + \frac{\Gamma}{2\pi}\frac{1}{a} \right] = \frac{U^2}{2} \rho\pi a^2 \qquad (7.5)$$

... and interpreted

The final expression (7.5) gives the kinetic energy of the unsteady flow as the product of one-half of the square of the velocity U of the cylinder multiplied by the mass $\rho\pi a^2$. The latter quantity is evidently equal to the mass of fluid in a circular cylinder of radius a

and unit thickness. It therefore appears that the fluid velocities of the innumerable particles moving at various speeds together account for a total energy the same as if an *additional apparent volume* πa^2 of fluid of density ρ were set in motion at the speed U. The product of these two quantities, termed the *apparent mass*, provides a measure of the force required to establish the flow: in addition to the kinetic energy of the solid cylinder itself, force is required to impart energy (7.5) to the fluid which it agitates by its motion in the fluid.

apparent mass of a steady flow

Exercise 26 Exhibit the unsteady character of the flow (7.4) by: (a) expressing the stream function in terms of Cartesian coordinates x and y according to the transformation (II.1.10) and observing that the velocity approaches zero far from the cylinder (i.e., $u \to 0$, $v \to 0$ as $x^2 + y^2 \to \infty$) while the flow normal to the cylinder at an arbitrary point specified by x and $y = \sqrt{a^2 - x^2}$ is nonzero; (b) obtaining ψ as an explicit function of time t and coordinates x_0, y_0 attached to the fluid distant from the cylinder, related to x and y by the transformation equations

$$x_0 = x - Ut$$
$$y_0 = y$$
(7.6)

Show that

$$\psi(x_0, y_0, t) = Ua^2 \frac{y_0}{(x_0 + Ut)^2 + y_0^2} - \frac{\Gamma}{4\pi} \ln \frac{(x_0 + Ut)^2 + y_0^2}{a^2} \quad (7.7)$$

explicit form of stream function for unsteady cylinder flow

Exercise 27 Determine the explicit form of the equation of the moving boundary $f(x_0, y_0, t) = 0$ in order to prove that the boundary condition (II.6.3a) is actually satisfied by the flow (7.7) and evaluate u_0, v_0, on the cylinder.
Ans. $f(x_0, y_0, t) = (x_0 - Ut)^2 + y_0^2 - a^2$

boundary surface in motion

*8 REDUCTION TO STEADY FLOW — GALILEAN INVARIANCE

Although the energy and apparent mass are found as in the preceding section only for flows viewed as unsteady, it will be recalled that force calculations and the determination of the flows themselves are ordinarily simplified by adopting the viewpoint of an observer for whom the flow appears to be steady. Even though it is quite apparent that the two viewpoints are merely different descriptions of a single flow condition, only a careful inspection of the underlying equations reveals the precise elements of equivalence in the two cases.

equivalence of "steady" and "unsteady" representations

130 TWO-DIMENSIONAL PERFECT FLUID MOTIONS

A useful starting point is provided by the "steady" flow representations of Sections 3 and 4; it is recognized of course that the practical application of such calculations is usually in the study of the motion of a body passing at uniform speed through a fluid otherwise at rest. The more naturally appropriate frame of reference employing a system of axes x_0, y_0, having origin O_0 fixed with respect to the fluid at rest far from the body, is adopted, distinct from coordinates x, y with origin O at a definite point attached to the body moving at constant speed V. The two sets of coordinates and the velocities then transform according to the equations

axes x_0, y_0 are "fixed"; x and y motion at relative speed V

two sets of independent variables: x_0, y_0 and t_0; and x, y, and t

$$x = x_0 - Vt_0 \qquad u = u_0 - V$$
$$y = y_0 \qquad v = v_0 \qquad (8.1)$$
$$t = t_0$$

as seen from Fig. V-11. For purposes of easy identification of the different sets of variables, time is denoted t_0 in the "fixed" frame of reference and by t in the "moving" frame. It is now shown that the *general* form of the equations of incompressible fluid motion in two dimensions (II.5.6) and (II.5.7) remains unchanged when the axes are transformed according to (8.1).

It is convenient to write partial derivatives in Jacobian determinant notation, proceeding as in Section II.7 so that

transformations with aid of Jacobian determinants

$$\frac{\partial u}{\partial x} = \frac{\partial(u, y, t)}{\partial(x, y, t)} = \frac{\partial(u_0 - V, y_0, t_0)}{\partial(x_0, y_0, t_0)} J$$

$$= \begin{vmatrix} \frac{\partial u_0}{\partial x_0} & 0 & 0 \\ 0 & 1 & 0 \\ 0 & 0 & 1 \end{vmatrix} J = \frac{\partial u_0}{\partial x_0} J \qquad (8.2)$$

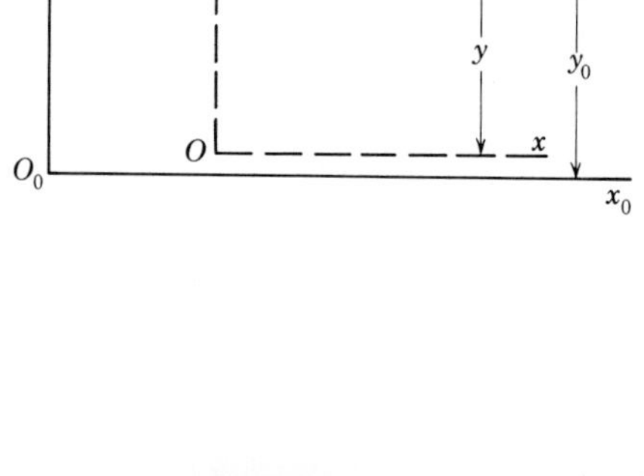

Fig. V-11 Coordinates of reference frames illustrating Galilean invariance.

where J denotes the transformation determinant

$$J \equiv \frac{\partial(x_0, y_0, t_0)}{\partial(x, y, t)}$$

assumed finite and nonzero.

Exercise 28 Show that only three of the elements of the corresponding determinant expression for $\partial v/\partial y$ are zero, and that

$$\frac{\partial v}{\partial y} = \frac{\partial v_0}{\partial y_0} J$$

so that the continuity equation (II.5.6) becomes

$$\frac{\partial u_0}{\partial x_0} + \frac{\partial v_0}{\partial y_0} = 0 \qquad (8.3)$$

continuity equation is *invariant* with respect to uniform translation

Exercise 29 (a) Prove that

$$\frac{\partial u}{\partial t} = \left\{ \frac{\partial u_0}{\partial t_0} + V \frac{\partial u_0}{\partial x_0} \right\} J$$

(b) Evaluate $\partial u/\partial y$ in terms of the quantities on the right sides of (8.1),
(c) Combine the results with (8.2) to show that

$$\frac{Du}{Dt} \equiv \frac{\partial u}{\partial t} + u \frac{\partial u}{\partial x} + v \frac{\partial u}{\partial y} = \frac{\partial u_0}{\partial t_0} + u_0 \frac{\partial u_0}{\partial x_0} + v_0 \frac{\partial u_0}{\partial y_0} \equiv \frac{Du_0}{Dt_0} \qquad (8.4)$$

(d) Transform the derivatives of pressure and gravitational potential in (II.5.7) in order to prove that they become

$$\left. \begin{array}{l} \dfrac{Du_0}{Dt_0} + \dfrac{1}{\rho} \dfrac{\partial p}{\partial x_0} = -\dfrac{\partial U}{\partial x_0} \\[2ex] \dfrac{Dv_0}{Dt_0} + \dfrac{1}{\rho} \dfrac{\partial p}{\partial y_0} = -\dfrac{\partial U}{\partial y_0} \end{array} \right\} \qquad (8.5)$$

Euler equations likewise invariant in the absence of relative acceleration of the coordinate systems

Thus the form of the equations of fluid motion, written in the x, y, t coordinate system of "steady" flow and also in the x_0, y_0, t_0 system for "unsteady" flow, is identical when *all* terms are included (the same result does not follow if the vanishing terms $\partial u/\partial t$, $\partial v/\partial t$, are neglected at the outset). The uniform relative velocity V of the two sets of coordinate axes does not appear in (8.3) or in (8.5). The term *Galilean invariance* is used to designate the transformation properties just demonstrated, valid whenever the relative motion of the reference frames is limited to uniform translation without acceleration [cf. the case represented by (III.5.1) for rotating coordinates].

counter-example

Exercise 30 Let the relative motion of two coordinate systems be specified by the transformation

$$x = x_0 - At_0^2 \tag{8.6}$$

in place of the first equation of (8.1), where A is a constant. Investigate the expression obtained by transforming the streamline derivative Du/Dt in the manner of (8.4). Are the Euler equations invariant in this case?

The importance of Galilean invariance, from the logical point of view, can hardly be exaggerated, since the conditions of "absolute rest" (which would otherwise be required for the reference frame in which Newton's laws are presumed to be valid) are physically meaningless and there is no method of determining the "absolute" speed of any reference frames in the universe.

VI
Irrotational Flow in Three Dimensions

1 ELEMENTARY AND DERIVATIVE FLOWS IN SPHERICAL COORDINATES

The vorticity of three-dimensional flow is a vector quantity having x- and y-components, denoted ξ and η, respectively, that are inferred from the form of ζ by simple cyclic permutation of Cartesian coordinates and velocity components, thus:

$$\xi = \frac{\partial w}{\partial y} - \frac{\partial v}{\partial z} \qquad \eta = \frac{\partial u}{\partial z} - \frac{\partial w}{\partial x} \qquad \zeta = \frac{\partial v}{\partial x} - \frac{\partial u}{\partial y} \qquad (1.1)$$

in accord with (IV.1.6). All three vorticity components vanish whenever the velocity is derived from a potential according to the rules

vorticity components and velocity potential

$$u = -\frac{\partial \phi}{\partial x} \qquad v = -\frac{\partial \phi}{\partial y} \qquad w = -\frac{\partial \phi}{\partial z} \qquad (1.2)$$

extending (IV.5.5). Hence an arbitrary function $\phi(x, y, z)$ determines an irrotational flow in three dimensions in exactly the same manner as was shown in Chapter IV for plane flow.

When spherical polar coordinates are introduced as in (II.2.6), i.e., related to the Cartesian coordinates by the transformation equations

$$x = r \sin \theta \cos \omega \qquad y = r \sin \theta \sin \omega \qquad z = r \cos \theta \qquad (1.3)$$

the radial derivatives of the Cartesian coordinates are

$$\left. \begin{array}{l} \dfrac{\partial x}{\partial r} = \sin \theta \cos \omega = \dfrac{x}{r} \\[6pt] \dfrac{\partial y}{\partial r} = \sin \theta \sin \omega = \dfrac{y}{r} \qquad \dfrac{\partial z}{\partial r} = \cos \theta = \dfrac{z}{r} \end{array} \right\} \qquad (1.3\text{a})$$

Since the last term in each of the equations (1.3a) is recognized as one of the direction cosines of the radial direction at x, y, z, the radial velocity component u_r at any point is obtained from $u, v,$ and w as

$$u_r = u\frac{\partial x}{\partial r} + v\frac{\partial y}{\partial r} + w\frac{\partial z}{\partial r} = -\frac{\partial \phi}{\partial x}\frac{\partial x}{\partial r} - \frac{\partial \phi}{\partial y}\frac{\partial y}{\partial r} - \frac{\partial \phi}{\partial z}\frac{\partial z}{\partial r}$$
$$= -\frac{\partial \phi}{\partial r} \quad (1.4)$$

spherical polar coordinates: velocity components

Proceeding in the same manner for the velocity components u_θ and u_ω, viz., by summing the θ-components of the Cartesian velocities, and likewise for the ω-components, leads to the formulae

$$u_\theta = -\frac{\partial \phi}{r\, \partial \theta} \qquad u_\omega = -\frac{1}{r \sin \theta}\frac{\partial \phi}{\partial \omega} \quad (1.4a)$$

according to which the continuity equation in spherical polar coordinates, (II.2.6), becomes

velocity potential satisfies Laplace's equation — polar coordinates...

$$\frac{1}{r^2}\left[\frac{\partial}{\partial r}\left(r^2 \frac{\partial \phi}{\partial r}\right) + \frac{1}{\sin\theta}\frac{\partial}{\partial \theta}\left(\sin\theta \frac{\partial \phi}{\partial \theta}\right) + \frac{1}{\sin^2 \theta}\frac{\partial^2 \phi}{\partial \omega^2}\right] = 0 \quad (1.5)$$

as already indicated without proof in Section V.5. In the same manner, substituting the expressions (1.2) in (II.2.4) gives

...Cartesian coordinates

$$\frac{\partial^2 \phi}{\partial x^2} + \frac{\partial^2 \phi}{\partial y^2} + \frac{\partial^2 \phi}{\partial z^2} = 0 \quad (1.5a)$$

Equations (1.5) and (1.5a) are known as Laplace's equation in three dimensions, in polar and Cartesian forms, respectively. It is noteworthy that solutions of these equations are again found by the various methods enumerated in section V.5, a slight modification being required only in the case of Method IV. Thus a fundamental solution of (1.5) is found by considering a velocity potential that depends on the distance r alone, i.e., $\phi = \phi(r)$. The first term of (1.5) then indicates that

$$r^2 \frac{\partial \phi}{\partial r} = \text{const}$$

from which (cf. Exercise V.21) a fundamental solution is found as

fundamental solution; three-dimensional source flow

$$\phi(r) = \frac{A}{r} \quad (1.6)$$

corresponding to flow with velocity in the radial direction only:

$$u_r = \frac{A}{r^2} \qquad u_\theta = 0 \qquad u_\omega = 0 \quad (1.7)$$

The volume flux across a spherical surface of radius r centered at the origin, the surface area being $4\pi r^2$, is $4\pi A$ and is independent

VI. 1 / ELEMENTARY AND DERIVATIVE FLOWS IN SPHERICAL COORDINATES

of the radius of the sphere. It is therefore natural to refer to the flow (1.6) as a three-dimensional source at the origin, of *strength* $4A$. According to the argument presented earlier and confirmed by the form of (1.5a), the x-derivative of a solution of Laplace's equation is also a solution. Hence inverting (1.3) and noting that $r = [x^2 + y^2 + z^2]^{1/2}$, so that

$$\frac{\partial r}{\partial x} = \frac{x}{r}$$

gives the x-derivative of (1.6) as

$$-\frac{A \sin \theta \cos \omega}{r^2} \qquad (1.8) \qquad \text{doublet}$$

and also represents an irrotational flow in three dimensions. As in the differentiation of plane source flow, (1.8) is referred to as a three-dimensional doublet or dipole.

The fundamental solution (1.6) may also be regarded as a product-type solution in which both the θ-function and the ω-function are taken as constants. When only one of these functions is taken as constant, a product-type solution depending on two variables is obtained for the Laplace equation in three dimensions. A case of importance is found when the potential depends on the radial distance r from a center of coordinates and on a polar angle θ but is independent of the longitude angle ω. When the function $\phi(r, \theta)$ is assumed in the form of a power of r multiplied by an unspecified function $P(\theta)$ of the angle θ

$$\phi(r, \theta) = r^n P(\theta) \qquad (1.9) \qquad \text{product-type solution — extension of fundamental solution}$$

the two r-derivatives in (1.5) and the factor r^2 recover the n-th power of r, so that r^n is a common factor. When it is removed, the equation for $P(\theta)$ becomes

$$P''(\theta) + \cot\theta\, P'(\theta) + n(n + 1)P(\theta) = 0 \qquad (1.10) \qquad \text{Legendre's equation}$$

where primes refer to differentiations of the function P with respect to θ. When $n = -1$, (1.10) shows that $P_{-1}(\theta) = \text{const}$ is a solution, and this evidently corresponds to the source flow (1.6). When $n = 2$, the *Legendre function* of order two, denoted $P_2(\theta)$, is given by

$$P_2(\theta) = \frac{3\cos^2\theta - 1}{2} \qquad (1.11) \qquad \text{Legendre function } P_2(\theta)$$

as verified by substitution in (1.10). The explicit form of the solution (1.9) is then

$$\phi(r, \theta) = r^2 \frac{3\cos^2\theta - 1}{2} \qquad (1.12)$$

It is easy to see that a large number of solutions may be obtained by proceeding in the manner just indicated. In succeeding sections physically important flows are studied with the aid of the functions found above.

Exercise 1 Denote by N the product $n(n+1)$ appearing in (1.10) and solve the quadratic equation for n in terms of N. Thus show that, if one root is denoted n_1, the other is given by $n_2 = -n_1 - 1$, so that $r^{-3}P_2(\theta)$ is also a solution of (1.10).

2 OCEAN TIDE HEIGHT, ELEMENTARY CALCULATION

The spherical form of a large liquid mass in hydrostatic equilibrium is modified by the attraction of a distant mass such as the Moon or the Sun, and the solutions just found play a central role in the theory of the resulting ocean tides. When the attracting mass is denoted M and referred to as the Moon for the sake of definiteness, the gravitational potential of the Moon at a point of the Earth's surface, $r = a$, is conventionally taken as

tide-raising potential caused by the Moon's gravitational attraction

$$U_M(a, \theta) = \frac{-k^2 M}{D}\left(\frac{a}{D}\right)^2 P_2(\theta) \tag{2.1}$$

where D is the distance between the Earth-Moon centers. The polar angle θ in (2.1) is measured from the line of centers, and (2.1) differs from (1.12) only by a multiplicative constant. Bernoulli's integral (IV.5.3) is assumed to be applicable, and, in the absence of motion, $q = 0$ so that the height of the ocean surface (at constant pressure) is determined by the total gravitational potential consisting of the potential of the Earth, U_E, plus the Moon's potential (2.1). For very small displacements $\eta(\theta)$ of the water surface from the sphere, $r = a$, the approximation

potential of nearly spherical Earth

$$U_E(a + \eta, \theta) = \frac{-k^2 E}{a + \eta} \doteq \frac{-k^2 E}{a}\left(1 - \frac{\eta}{a}\right) \tag{2.2}$$

is made. The sum of (2.1) and (2.2) is constant if the θ-dependent terms add to zero, and this gives the height of the "equilibrium tide" as

classical tide theory — "equilibrium" tide height

$$\eta(\theta) = \frac{M}{E}\frac{a^4}{D^3}P_2(\theta) \tag{2.3}$$

As it is seen from (1.11) that $P_2(\theta)$ has its maximum value (unity) at $\theta = 0$ and at $\theta = \pi$, the figure for the oceans given by (2.3) is elongated toward the Moon, indicating high tide at any place when the Moon is directly overhead. It follows that low tide should be expected roughly six hours earlier and later, since this is the time required for the Earth to make one-quarter of a revolution, and (1.11) is minimized when $\theta = \pi/2$. The same conclusion holds when various refinements of the theory are made, and it is curious to note that, in the words of Sir George Darwin, the "theory is nearly as much wrong as possible, in respect to the time of high water." There are few instances in which the efforts of generations of the world's most able mathematicians have met with so little success as in this aspect of the hydrodynamic theory of tides.

Exercise 2 Verify that $f(x, y, z)$ is a solution of Laplace's equation (1.5a) when: (a) $f(x, y, z) = [(x \cos \sigma + y \sin \sigma) + iz]^n \equiv \lambda^n$ where σ and n are constants and $i = \sqrt{-1}$, so that

$$\frac{\partial f}{\partial x} = n\lambda^{n-1}\cos \sigma \qquad \frac{\partial^2 f}{\partial x^2} = n(n-1)\lambda^{n-2}\cos^2 \sigma, \text{ etc.}$$

three-dimensional analog of complex variable technique in plane flows

(b) f is an arbitrary function of λ, so that

$$\frac{\partial f}{\partial x} = f'\cos \sigma \qquad \frac{\partial^2 f}{\partial x^2} = f'' \cos^2 \sigma, \text{ etc.}$$

(c)
$$f(x, y, z) = \int_a^b F(\lambda)g(\sigma) \, d\sigma \qquad (1.13)$$

where the function $g(\sigma)$ and the integration limits a and b, as well as the function $F(\lambda)$, are arbitrary. Solutions represented by (1.13) yield three-dimensional irrotational flows in the same manner that Method IV of the preceding chapter is applied in the two-dimensional case.

Exercise 3 Set $F(\lambda) = A\lambda^2, g(\sigma) = 1, a = 0, b = 2\pi$ in (1.13) and find the value of the constant A for which the resulting solution $f(x, y, z)$ is identical to (1.12).

Exercise 4 The "universal coefficient" of lunar tide height

$$\frac{3}{2} \frac{M}{E}\left(\frac{a}{D}\right)^3 a \qquad (2.4)$$

"universal coefficient" — lunar tide

is obtained as the difference between maximum and minimum values of (2.3) at $\theta = 0$ and $\theta = \pi/2$. Thus (2.4) gives the total lunar tidal "variation" from high tide to low tide. Compare its value with the universal coefficient of *solar* tide height obtained by replacing the constants M and D in (2.4) by the solar mass S and distance L, where

$$\frac{M}{E} = \frac{1}{81.3} \qquad \frac{a}{D} = \frac{1}{60.3} \qquad a = 6.38 \times 10^8 \text{ cm}$$

$$\frac{S}{E} = 332{,}484 \qquad \frac{a}{L} = \frac{1}{389 \times 60.3}$$

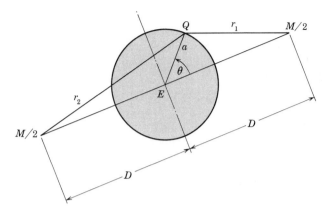

Fig. VI-1 Positions of Moon and anti-Moon for tide calculations.

The sum of lunar and solar tidal variations thus obtained can be compared with typical observed tide amplitudes of the order of a few meters to explain why accurate tidal predictions are largely based on empirical methods.

Exercise 5 The tide height values given by (2.3) indicate an increment of ocean volume, relative to the undisturbed height $r = a$, represented by

(2,3) satisfies continuity requirement

$$2\pi a \, \sin\theta \, \eta(\theta) \, a \, d\theta \qquad (2.5)$$

for places located by the zenith angle θ. Substitute (2.3) in (2.5) and integrate over the complete surface to show that the total incremental volume is zero, i.e., that the continuity condition is satisfied by the tide height formula (2.3).

potential of Moon and anti-Moon

Exercise 6 Show that the potential (2.1) is obtained by considering one-half of the Moon's mass to be located at distance D from the center of the Earth (distance r_1 from points Q located by the angle θ as shown in Fig. VI-1), and also one-half of the Moon's mass located at the "anti-Moon" position at distance D from the Earth's center in the opposite direction from the Moon. The expression for the resulting potential is thus

$$\frac{-k^2 M}{2}\left\{\frac{1}{r_1} + \frac{1}{r_2}\right\}$$

where r_2 is the distance of the anti-Moon from Q and reduces to (2.1), when first-order terms in a/D are retained as well as those of second order, while third and higher powers of this ratio are neglected.

3 EULER EQUATIONS IN POLAR FORM

Although the Cartesian equations (II.5.7) are readily extended to three-dimensional form, the introduction of spherical polar coordinates by means of transformations like (1.3) and (1.4) becomes tedious. A direct and more instructive procedure uses the vector expression of fluid momentum conservation

$$\mathbf{a} + \frac{1}{\rho} \nabla p = -\nabla U \qquad (3.1)$$

and expresses the fluid acceleration $\mathbf{a}$ in terms of the spherical polar velocity components given by

$$\mathbf{q} = u_r \mathbf{e}_r + u_\theta \mathbf{e}_\theta + u_\omega \mathbf{e}_\omega \qquad (3.2)$$

vector velocity referred to spherical base vectors

In the evaluation of the acceleration $\mathbf{a}$ as the total time derivative of the velocity, account must now be taken of the fact that the *base* (unit) *vectors* $\mathbf{e}_r$, $\mathbf{e}_\theta$, and $\mathbf{e}_w$, unlike the unit vectors $\mathbf{i}, \mathbf{j}, \mathbf{k}$ of the Cartesian system in (II.1.3), vary in direction from one point in the flow to another. Hence (3.2) gives the acceleration as

$$\mathbf{a} = \frac{D\mathbf{q}}{Dt} = \frac{Du_r}{Dt} \mathbf{e}_r + \frac{Du_\theta}{Dt} \mathbf{e}_\theta + \frac{Du_\omega}{Dt} \mathbf{e}_\omega$$
$$+ u_r \frac{D\mathbf{e}_r}{Dt} + u_\theta \frac{D\mathbf{e}_\theta}{Dt} + u_\omega \frac{D\mathbf{e}_\omega}{Dt} \qquad (3.3)$$

acceleration equals total derivative of velocity vector

where the last three terms involve the rotation of the unit vectors associated with nonradial components of flow velocity. These terms, like the terms in (III.5.1), can again be written as the vector product of the rotational velocity of the base vectors and the particle velocity.

The base vectors are related to the Cartesian unit vectors, and their total time derivatives are evaluated by writing the position vector $\mathbf{r}$ in terms of the coordinates and base vectors of each system:

$$\mathbf{r} = \mathbf{e}_r r = x\mathbf{i} + y\mathbf{j} + z\mathbf{k}$$

so that (1.3) gives the first of the relationships

$$\left.\begin{aligned}
\mathbf{e}_r &= \sin\theta \cos\omega\, \mathbf{i} + \sin\theta \sin\omega\, \mathbf{j} + \cos\theta\, \mathbf{k} \\
\mathbf{e}_\theta &= \cos\omega \cos\theta\, \mathbf{i} + \sin\omega \cos\theta\, \mathbf{j} - \sin\theta\, \mathbf{k} \\
\mathbf{e}_\omega &= -\sin\omega\, \mathbf{i} + \cos\omega\, \mathbf{j}
\end{aligned}\right\} \qquad (3.4)$$

spherical and Cartesian base vectors interrelated

Differentiating each of equation (3.4) successively leads to the formulae

$$\begin{aligned}
\frac{D\mathbf{e}_r}{Dt} &= +\dot{\theta}\mathbf{e}_\theta + \dot{\omega}\sin\theta\,\mathbf{e}_\omega = \frac{u_\theta}{r}\mathbf{e}_\theta + \frac{u_\omega}{r}\mathbf{e}_\omega \\
\frac{D\mathbf{e}_\theta}{Dt} &= \qquad\qquad\qquad\;\; = -\frac{u_\theta}{r}\mathbf{e}_r + \frac{u_\omega}{r}\cot\theta\,\mathbf{e}_\omega \\
\frac{D\mathbf{e}_\omega}{Dt} &= \qquad\qquad\qquad\;\; = -\frac{u_\omega}{r}\mathbf{e}_r - \frac{u_\omega}{r}\cot\theta\,\mathbf{e}_\theta
\end{aligned} \quad (3.5)$$

where velocity components replace the coordinate derivatives on the right side of (3.5) according to the relationships

velocity components expressed as coordinate derivatives

$$u_r = \dot{r} \qquad u_\theta = r\dot{\theta} \qquad u_\omega = r\sin\theta\,\dot{\omega} \qquad (3.6)$$

Exercise 7 Verify the third equation of (3.4) by examining the x- and y-Cartesian projections of the unit vector $\mathbf{e}_\omega$ parallel to the ω-coordinate direction. Also find $\mathbf{e}_\theta$ as the vector product $\mathbf{e}_\omega \times \mathbf{e}_r$.

Exercise 8 Solve the system of equations (3.4) to give $\mathbf{k}$ in terms of the unit vectors $\mathbf{e}_r$, $\mathbf{e}_\theta$, and $\mathbf{e}_\omega$ and refer the total vector rotation $\boldsymbol{\Omega}$ of this triad of base vectors

$$\boldsymbol{\Omega} = \dot{\omega}\mathbf{k} + \dot{\theta}\mathbf{e}_\omega$$

to its polar coordinate form. Show that $\boldsymbol{\Omega} \times \mathbf{q}$ equals the last three terms in (3.3).

Substituting (3.5) and (3.3) in (3.1) and resolving into components parallel to the coordinate directions gives Euler's equations in spherical polar form as

polar form of Euler equations

$$\begin{aligned}
\frac{Du_r}{Dt} - \frac{u_\theta^2 + u_\omega^2}{r} + \frac{1}{\rho}\frac{\partial p}{\partial r} &= -\frac{\partial U}{\partial r} \\
\frac{Du_\theta}{Dt} + \frac{u_r u_\theta}{r} - \frac{\cot\theta\, u_\omega^2}{r} + \frac{1}{\rho}\frac{1}{r}\frac{\partial p}{\partial \theta} &= -\frac{1}{r}\frac{\partial U}{\partial \theta} \\
\frac{Du_\omega}{Dt} + \frac{u_r u_\omega}{r} + \frac{\cot\theta\, u_\theta u_\omega}{r} + \frac{1}{\rho}\frac{1}{r\sin\theta}\frac{\partial p}{\partial \omega} &= -\frac{1}{r\sin\theta}\frac{\partial U}{\partial \omega}
\end{aligned} \quad (3.7)$$

In addition to the usual acceleration components of steady flow present in the Cartesian counterparts of the equations (3.7), the polar equations also exhibit acceleration components *normal* to the flow, in each equation of (3.7). Thus, even in the absence of meridional motion ($u_\theta = 0$), velocities parallel to the ω-direction at any point contribute to the meridional component of the pressure gradient, as seen from the second of (3.7). This is readily understood in terms of the curvature of the coordinate lines and acceleration toward the center of the curvature of motion.

Exercise 9 Obtain (3.7) by consideration of momentum flux, as in Exercise IV.10.

Exercise 10 Given the general expression for the vorticity in an arbitrary orthogonal curvilinear coordinate system in determinant form

$$\nabla \times \mathbf{q} = \frac{1}{h_1 h_2 h_3} \begin{vmatrix} h_1 \mathbf{e}_1 & h_2 \mathbf{e}_2 & h_3 \mathbf{e}_3 \\ \frac{\partial}{\partial s_1} & \frac{\partial}{\partial s_2} & \frac{\partial}{\partial s_3} \\ h_1 u_1 & h_2 u_2 & h_3 u_3 \end{vmatrix} \quad (3.8)$$

— vorticity vector, general orthogonal system of coordinates

find the vorticity components in spherical polar coordinates by taking the coordinates s_1, s_2, s_3, as $r, \theta,$ and ω, respectively, so that the *metric elements* h_1, h_2, h_3 and velocity components $u_1, u_2,$ and u_3 are as tabulated:

$$\left. \begin{array}{lll} s_1 = r & h_1 = 1 & u_1 = u_r \\ s_2 = \theta & h_2 = r & u_2 = u_\theta \\ s_3 = \omega & h_3 = r \sin \theta & u_3 = u_\omega \end{array} \right\} \quad (3.9)$$

— specialized for spherical polar coordinates

Show that, if the radial velocity is defined as in (1.4) as the derivative of a potential, then the remaining velocity coefficients must be of the form (1.4a) if the flow is irrotational.

Exercise 11 Use the results of Exercise 10 to show that, for steady irrotational flow, Euler's equations (3.7) yield the Bernoulli integral in the form (IV.5.3).

When *cylindrical polar coordinates* r, θ, z are adopted, the acceleration components are most easily found in the same manner as employed above for spherical polars. Thus, from the expression for the position of a fluid particle, written as

$$\mathbf{R} = r\mathbf{e}_r + z\mathbf{e}_z = x\mathbf{i} + y\mathbf{j} + z\mathbf{k} \quad (3.10)$$

in both curvilinear and Cartesian forms, the fluid velocity and accelerations appear as successive derivatives of (3.10). The rates of change of the base vectors $\mathbf{e}_r, \mathbf{e}_\theta,$ and $\mathbf{e}_z$ associated with a moving fluid particle are again obtained from the transformation equations relating the curvilinear and the (fixed) Cartesian coordinates:

$$x = r \cos \theta \qquad y = r \sin \theta \qquad z = z \quad (3.11)$$

Exercise 12 Use (3.10) and (3.11) to determine the position dependence of the unit vectors; i.e., show that

$$\mathbf{e}_r = \mathbf{i}\cos\theta + \mathbf{j}\sin\theta \qquad \mathbf{e}_z = \mathbf{k}$$

$$\mathbf{e}_\theta = \mathbf{e}_z \times \mathbf{e}_r = -\mathbf{i}\sin\theta + \mathbf{j}\cos\theta \qquad (3.12)$$

so that

$$\frac{D\mathbf{e}_r}{Dt} = \dot\theta\mathbf{e}_\theta \equiv \frac{u_\theta}{r}\mathbf{e}_\theta \qquad \frac{D\mathbf{e}_\theta}{Dt} = -\frac{u_\theta}{r}\mathbf{e}_r \qquad (3.13)$$

Exercise 13 Differentiate (3.10) to obtain the velocity and acceleration as

$$\mathbf{q} = \frac{Dr}{Dt}\mathbf{e}_r + r\frac{D\mathbf{e}_r}{Dt} + \frac{Dz}{Dt}\mathbf{e}_z + z\frac{D\mathbf{e}_z}{Dt} \equiv u_r\mathbf{e}_r + u_\theta\mathbf{e}_\theta + u_z\mathbf{e}_z \qquad (3.14)$$

and

$$\mathbf{a} = \left(\frac{Du_r}{Dt} - \frac{u_\theta^2}{r}\right)\mathbf{e}_r + \left(\frac{Du_\theta}{Dt} + \frac{u_r u_\theta}{r}\right)\mathbf{e}_\theta + \frac{Du_z}{Dt}\mathbf{e}_z \qquad (3.15)$$

The first two terms on the right side of (3.15) are recognized as components of fluid momentum flux in the coordinate directions, in agreement with (II.4.4). It is clear that (3.15) may be calculated in this manner by considering an appropriate fluid volume element.

A third manner of obtaining (3.15) should also be noted; it is based on the observation that the vector acceleration is expressible in terms of a gradient and the curl of the velocity (i.e., the vorticity $\boldsymbol{\omega}$), as

$$\mathbf{a} = \frac{D\mathbf{q}}{Dt} = \frac{\partial \mathbf{q}}{\partial t} + \nabla\left(\frac{q^2}{2}\right) + \boldsymbol{\omega} \times \mathbf{q} \qquad (3.16)$$

Verification of (3.16) is immediate, in the case of Cartesian coordinates, by substitution of (1.1) in (3.16) and comparison with (II.4.3) or (II.5.7) in their fully written out forms. In the present case it is only necessary to observe that the gradient takes the form

$$\nabla f = \frac{1}{h_1}\frac{\partial f}{\partial s_1}\mathbf{e}_1 + \frac{1}{h_2}\frac{\partial f}{\partial s_2}\mathbf{e}_2 + \frac{1}{h_3}\frac{\partial f}{\partial s_3}\mathbf{e}_3 \qquad (3.17)$$

in the notation introduced in (3.8). For cylindrical coordinates,

$$\begin{aligned} s_1 &= r & h_1 &= 1 & u_1 &= u_r \\ s_2 &= \theta & h_2 &= r & u_2 &= u_\theta \\ s_3 &= z & h_3 &= 1 & u_3 &= u_z \end{aligned} \qquad (3.18)$$

and it follows that

$$\nabla\left(\frac{q^2}{2}\right) = \mathbf{e}_r\left\{u_r\frac{\partial u_r}{\partial r} + u_\theta\frac{\partial u_\theta}{\partial r} + u_z\frac{\partial u_z}{\partial r}\right\} + \cdots \qquad (3.19)$$

with corresponding components in the directions of $\mathbf{e}_\theta$ and $\mathbf{e}_z$ containing terms involving differentiations in the respective directions.

The components of vorticity are found from (3.8) with the values given by (3.18); their vector sum is

$$\boldsymbol{\omega} = \frac{\mathbf{e}_r}{r}\left\{\frac{\partial u_z}{\partial \theta} - \frac{\partial(ru_\theta)}{\partial z}\right\} + \mathbf{e}_\theta\left\{\frac{\partial u_r}{\partial z} - \frac{\partial u_z}{\partial r}\right\}$$
$$+ \frac{\mathbf{e}_z}{r}\left\{\frac{\partial(ru_\theta)}{\partial r} - \frac{\partial u_r}{\partial \theta}\right\} \quad (3.20)$$

... and vorticity

Exercise 14 (a) Calculate the vector product $\boldsymbol{\omega} \times \mathbf{q}$ using (3.20). (b) Add the separate components of (3.19) and the expression obtained in (a) to show that (3.16) takes exactly the form shown as (3.15).

The equations of fluid motion, in cylindrical polar coordinates, are finally written by equating the separate components of (3.15) to the sum of corresponding pressure and gravity terms in the usual manner:

$$\left.\begin{array}{l} \dfrac{Du_r}{Dt} - \dfrac{u_\theta^2}{r} + \dfrac{1}{\rho}\dfrac{\partial p}{\partial r} = -\dfrac{\partial U}{\partial r} \\[6pt] \dfrac{Du_\theta}{Dt} + \dfrac{u_r u_\theta}{r} + \dfrac{1}{\rho r}\dfrac{\partial p}{\partial \theta} = -\dfrac{\partial U}{r\,\partial \theta} \\[6pt] \dfrac{Du_z}{Dt} + \dfrac{1}{\rho}\dfrac{\partial p}{\partial z} = -\dfrac{\partial U}{\partial z} \end{array}\right\} \quad (3.21)$$

Euler equations, cylindrical coordinates

4 AXIAL SYMMETRY — SPHERE FLOW

Since the continuity equation of three-dimensional flow contains three derivative terms [cf. (II.2.4), (II.2.5), (II.2.6)], their sum cannot be made to vanish by the same expedient (V.1.4) employed in plane flow for which the same equation contains only two terms. In any special case for which one of the terms is zero, however, as in plane flow, the derivative form of the remaining terms permits the introduction of a stream function that defines velocities by means of its partial derivatives. This is the case when velocities are parallel to the planes containing the polar axis $\theta = 0$ in (1.5), so that $u_\omega = 0$ (the equation assumes the same form even when this component is nonzero, provided that the flow is axially symmetric $\partial/\partial\omega = 0$, of course). A stream function for axially symmetric flow is shown to have essentially the same physical interpretation as in plane

stream function for axially symmetric flow

motion. The flow components are related to this function as before then the equation it satisfies when the flow is irrotational is found.

Proceeding in the same manner as in Section V.1 gives the quantity of flow in an axially symmetric motion by evaluating the normal component of velocity across an element dl as

differential volume flux expression

$$\mathbf{q}\cdot\mathbf{n}\,dl = u_r r\,d\theta - u_\theta\,dr \tag{4.1}$$

and by summing along a path AB from point A on the symmetry axis to an arbitrary point located by coordinates r and θ. The line elements now lie on a circle of radius $r \sin \theta$ at all points of which the velocity components are the same. Since the circle perimeter is $2\pi r \sin \theta$, the counterpart of (V.1.3a) giving the volume flux is now written

defines Stokes stream function

$$-2\pi\,d\psi = 2\pi r \sin \theta\,(u_r r\,d\theta - u_\theta\,dr) \tag{4.2}$$

where the slightly different meaning of ψ associated with the introduction of the factor 2π is justified by the advantage that the velocity components are now related to the *Stokes stream function* $\psi(r, \theta)$ by the equations

velocity components

$$u_r(r, \theta) = \frac{-1}{r^2 \sin \theta}\frac{\partial \psi}{\partial \theta} \qquad u_\theta(r, \theta) = \frac{1}{r \sin \theta}\frac{\partial \psi}{\partial r} \tag{4.3}$$

that no longer contain this factor.

Exercise 15 Show that an arbitrary function $\psi(r, \theta)$ defines a velocity field according to (4.3) such that the continuity equation of axially symmetric flow

$$\frac{\partial}{\partial r}(r^2 u_r) + \frac{1}{\sin \theta}\frac{\partial}{\partial \theta}(r \sin \theta\,u_\theta) = 0 \tag{4.4}$$

is identically satisfied.

Exercise 16 From the condition of irrotationality obtained from (3.8) as

$$\frac{\partial(r u_\theta)}{\partial r} - \frac{\partial u_r}{\partial \theta} = 0 \tag{4.5}$$

show that the Stokes stream function of irrotational flow is a solution of a differential equation *not* reducible to Laplace's equation, namely,

equation satisfied by Stokes stream function

$$\frac{\partial^2 \psi}{\partial r^2} + \sin \theta\,\frac{\partial}{\partial \theta}\left\{\frac{1}{r^2 \sin \theta}\frac{\partial \psi}{\partial \theta}\right\} = 0 \tag{4.6}$$

Equation (4.6), like Laplace's equation (1.5), is a linear partial differential equation of the second order, but general methods of solution have not been so extensively developed as for (1.5), and as a result meaningful flows are not easily found from it. Specifically, attempting to find product-type solu-

Exercise 17 Uniform flow parallel to the polar axis is given by the velocity potential

$$-\phi_1 = Uz = Ur\cos\theta \tag{4.7}$$

Using (4.3) and either (1.4) or (1.4a), show that the corresponding Stokes stream function is

equivalence of representations by velocity potential and stream function

$$\psi_1 = -\frac{Ur^2}{2}\sin^2\theta \tag{4.7a}$$

Exercise 18 For the three-dimensional doublet flow with axis parallel to z, analogous to (1.8), obtain the Stokes stream function $\psi_2(r, \theta)$.

Because of the linearity of (4.6) it follows that the sum $\psi_1 + \psi_2$ of any two solutions is also a solution and therefore represents an irrotational axially symmetric flow. By adding the result of Exercise 18 to (4.7a), a flow is thus determined as

$$\psi(r, \theta) = -U\frac{r^2}{2}\sin^2\theta \left\{1 - \frac{a^3}{r^3}\right\} \tag{4.8}$$

sphere flow given by stream functions of two elementary flows

According to the defining property of the function ψ in (4.2) it follows that $\psi = $ const is the equation of a stream surface in the flow, and (4.8) shows that the sphere $r = a$ has this property. $\psi = 0$ also on the polar symmetry axis; hence the superposition of uniform flow and doublet in (4.8) yields the *sphere flow* in exactly the same manner as the same two-dimensional flows gave the stream function of uniform flow past a circular cylinder in (V.3.1). The potential function corresponding to (4.8) is less easily interpreted by inspection.

5 VORTEX DYNAMICS — RECTILINEAR VORTICES

When rotational motion is represented by an angular velocity vector in three-dimensional flow, the number of its components is three, instead of the one that appears in plane motion as in (IV.1.2). The corresponding vorticity components have been given in (IV.1.6) as

$$\xi = \frac{\partial w}{\partial y} - \frac{\partial v}{\partial z} \quad \eta = \frac{\partial u}{\partial z} - \frac{\partial w}{\partial x} \quad \zeta = \frac{\partial v}{\partial x} - \frac{\partial u}{\partial y} \tag{5.1}$$

Because of the interchangeability of the order of differentiations in a second derivative and the particular form of the terms on the right sides of (5.1), it is seen that the vorticity components are interrelated:

vorticity "continuity"

$$\frac{\partial \xi}{\partial x} + \frac{\partial \eta}{\partial y} + \frac{\partial \zeta}{\partial z} = 0 \qquad (5.2)$$

Comparison with the continuity equation (II.2.4) shows that the velocity components of an incompressible flow satisfy the same equation as the vorticity components; both are recognized as the vanishing of the divergence of a vector. The main properties of fluid vorticity are demonstrated by developing this analogy.

Exercise 19 Show that the divergence of the curl of a vector is equal to the curl of its divergence in general. By identifying the vector with the velocity field of an incompressible fluid motion in three dimensions, justify (5.2).

The instantaneous rotation directions of a flow are portrayed by a family of lines in space in the same manner that the streamlines indicate directions of particle velocity. From (5.1) the differential equations of the *vortex lines* thus defined are

differential equations of vortex lines

$$\frac{dx}{\xi} = \frac{dy}{\eta} = \frac{dz}{\zeta} \qquad (5.3)$$

just as the streamlines are given by the equations

$$\frac{dx}{u} = \frac{dy}{v} = \frac{dz}{w}$$

that extend (II.1.6) to three-dimensional motion. All the vortex lines passing through the points of an arbitrary curved line in the flow then define a *vortex surface*. The vortex lines passing through a closed curve also determine a *vortex tube;* when it is of very small cross section, it is called a *vortex filament*, or simply a vortex. They have properties reminiscent of the stream quantities shown in Fig. II-1, so that vortex lines do not intersect each other nor terminate abruptly at interior points of flow, etc. The particular elaboration of the geometrical characteristics of vortices is due to Helmholtz, and their remarkable properties are described in theorems known by his name.

vortex lines
vortex surfaces
vortex tubes (and filaments)

Of greatest importance is the "indestructibility" of vorticity, seen by examining the flux of its components across the coordinate surfaces of an elementary volume in the manner of Fig. II-4, in the

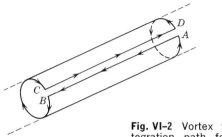

Fig. VI-2 Vortex filament; integration path for Eq. (5.4).

same way as the continuity equation (II.2.4) refers to flux of fluid volume. Specifically, by replacing the velocity components u, v, w by the vorticity components ξ, η, ζ, (5.2) shows that the net outward flux of vorticity is zero: each "entering" vortex line at one point of the boundary surface of the fluid element is also to be found "leaving" at another point. The mass flow through a stream filament passing through the volume remains constant, and the *vortex strength* or circulation (product of vorticity and cross-section area) of a vortex filament is similarly preserved.

properties of vortex lines

Exercise 20 Stokes' transformation (A.7) is extended to curved surfaces by introducing the normal vector **n** at each point of the surface, so that

$$\int_S (\nabla \times \mathbf{A}) \cdot \mathbf{n} \, dS = \int_C \mathbf{A} \cdot \boldsymbol{\tau} \, dl \qquad (5.4)$$

Show that the strength of a vortex filament is constant along its length by setting $\mathbf{A} = \mathbf{q}$, so that the integrand on the left side of (5.4) is the (zero) normal component of vorticity at a vortex line. The bounding curve C is shown in Fig. VI-2, consisting of two straight-line adjacent segments AB and CD running the length of the filament section and two circuitous paths BC and DA surrounding the filament.

conservation of vorticity

When vorticity is confined to a limited region of flow, as it frequently is in practical problems, a useful consequence of the result just established is the idealization it suggests in introducing *concentrated vortices*. Thus (V.2.2) represents a finite vortex strength, by means of an equivalent "infinite" vorticity of a filament of vanishing cross section, in a flow that is irrotational everywhere except at the origin $r = 0$.

In addition to the kinematical properties of vortex lines exemplified by regarding (5.2) as a kind of continuity equation for vorticity, the vector form (II.5.7a) of the Euler equations provides a convenient basis for showing how the vorticity moves under the combined actions of inertia, pressure gradient, and gravity force. Writing out the complete expressions for the acceleration terms in the scalar equations (II.5.7) readily reveals that their vector sum can be written as

$$\frac{\partial \mathbf{q}}{\partial t} + \nabla\left(\frac{q^2}{2}\right) - \mathbf{q} \times (\nabla \times \mathbf{q}) + \frac{1}{\rho}\nabla p = -\nabla U \qquad (5.5)$$

where the velocity vector $\mathbf{q}$ is regarded as three-dimensional. The *vector curl* of the velocity, $\nabla \times \mathbf{q}$ is recognized as the vorticity in (5.5) and is denoted ω:

$$\omega = \nabla \times \mathbf{q} \qquad (5.6)$$

It is observed that the curl of the gradient of any quantity contains only mixed second derivative terms with opposite signs, and it follows that the curl of the equation (5.5), i.e., the equality obtained for the curl of the separate terms added in the manner indicated, becomes

$$\frac{\partial \omega}{\partial t} + \nabla \times (\omega \times \mathbf{q}) = 0 \qquad (5.7)$$

Exercise 21 Prove the vector identity

$$\nabla \times (\mathbf{A} \times \mathbf{B}) = (\mathbf{B}\cdot\nabla)\mathbf{A} - (\mathbf{A}\cdot\nabla)\mathbf{B} + \mathbf{A}(\nabla\cdot\mathbf{B}) - \mathbf{B}(\nabla\cdot\mathbf{A}) \qquad (5.8)$$

and thus show that (6.5.7) becomes

$$\frac{\partial \omega}{\partial t} + (\mathbf{q}\cdot\nabla)\omega - (\omega\cdot\nabla)\mathbf{q} = 0 \qquad (5.7a)$$

by setting $\mathbf{A} = \omega$ and $\mathbf{B} = \mathbf{q}$ in (5.8) and recalling the vector forms of (5.2) and of (II.2.4).

In the total derivative notation the preceding result is expressed as

vorticity begets vorticity

$$\frac{D\omega}{Dt} = (\omega\cdot\nabla)\mathbf{q} \qquad (5.9)$$

which shows that, in regions of flow where there is no vorticity, $\omega = 0$, no vorticity is created [in agreement with Kelvin's theorem (IV.2.3), of course]. It also follows from (5.9) that a fluid particle remains on the *same* vortex line throughout its motion, even if the motion is unsteady (it will be recalled that a fluid particle remains permanently on the same streamline only if the flow is steady).

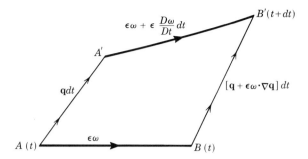

Fig. VI-3 Vortex lines associated with definite fluid particles.

This is shown by comparing the motions of two fluid particles situated at time t at two points A and B of the same vortex line so that the vector extending from A to B can be written as $\epsilon\omega$, where ϵ is a small number depending on the closeness of the two points. When the velocity of the particle at A is denoted $\mathbf{q}$, its displacement to new location A' at time $t + dt$ is given vectorially as $\mathbf{q}\,dt$. The velocity of the particle at B is $\mathbf{q} + \epsilon\omega\cdot\nabla\mathbf{q}$ (as seen by adding the scalar components), so that the vector displacement BB' is $\{\mathbf{q} + \epsilon\omega\cdot\nabla\mathbf{q}\}\,dt$. The vector $\mathbf{A'B'}$ can now be expressed in terms of $\mathbf{AB}$, $\mathbf{AA'}$, and $\mathbf{BB'}$ to give (Fig. VI-3)

$$\begin{aligned}\mathbf{A'B'} &= \mathbf{AB} + \mathbf{BB'} - \mathbf{AA'} \\ &= \epsilon\omega + (\mathbf{q} + \epsilon\omega\cdot\nabla\mathbf{q})\,dt - \mathbf{q}\,dt \qquad (5.10) \\ &= \epsilon\omega + \epsilon\omega\cdot\nabla\mathbf{q}\,dt\end{aligned}$$

When (5.9) is substituted in (5.10), it is seen that $\mathbf{A'B'}$, obtained as the subsequent relative position vector of particles originally on the same vortex line, is also part of the same vortex line, since

$$\mathbf{A'B'} = \epsilon\omega + \epsilon\frac{D\omega}{Dt}dt$$

simply indicates the variation of the location of the vortex line as it moves in the flow. It is now possible to speak of the motion of vorticity in much the same manner as for material fluid particles. A further development of this idea, which will not be considered in detail here, is the introduction of the "stream function giving the motion of vortices" in analogy with the stream function of Section V.1 that gives the motion of the fluid particles. The theory, elaborated by Kirchhoff and Routh nearly a century ago, has not found extensive practical applications.

6 EXTENSION TO CURVED VORTEX LINES — RING VORTICES

Helmholtz' vortex theorems described in Section 5 relate to various properties of vorticity in three-dimensional flow but give no indication of the velocity field associated with an arbitrary distribution of vorticity. When a flow is irrotational except for limited regions of vorticity, these regions can be conceived of as forming vortex tubes that take on the appearance of vortex filaments at great enough distances from the tubes. The further idealization as vortex lines reduces the question of determining the velocities to a generalization of the two-dimensional line vortex flow (V.2.2b).

vorticity specified, to determine the flow

A difficulty arises at this point which is analogous to the calculation of the magnetic field "induced" by a steady electric current in an infinitesimal element of conductor at points external to the conductor. The continuity requirement that implies the physical impossibility of finding an isolated element of electrical conductor carrying a steady current (which would entail the creation of electrons at one of its ends and their annihilation at the other) corresponds to the indestructibility of vorticity in a fluid motion, as expressed by the vanishing of its divergence (5.2). For present purposes it is sufficient to indicate an extension of (V.2.2b) which both recovers that velocity field as a special limiting case and also corresponds to the appropriate velocity potential for flow everywhere irrotational except on the vortex line itself. It is usual to speak of the flow "induced" by a vortex element in the same manner as in the electrical case. In both cases the formula developed below as (6.2b), known by the names of Biot and Savart and not subject to experimental verification for the reasons already stated, is accepted as the basis for further calculations.

"induction" of vorticity — Law of Biot-Savart

With a slight change of notation the velocity field of a line vortex coincident with the z-axis can be written in terms of the vortex strength as

$$q = \frac{\Gamma}{2\pi r} \tag{6.1}$$

The two-dimensional flow field (6.1) is now replaced by the flow induced by the finite segment AB of vortex line extending along the z-axis from $z = A$ to $z = B$ by writing for the velocity in the plane $z = 0$

generalization of (6.1)

$$q_{AB}(r, \theta, 0) = \frac{\Gamma}{4\pi r}(\cos \sigma_B - \cos \sigma_A) \tag{6.2}$$

where the angles σ_A and σ_B are formed by the vectors $\mathbf{R}_A$ and $\mathbf{R}_B$

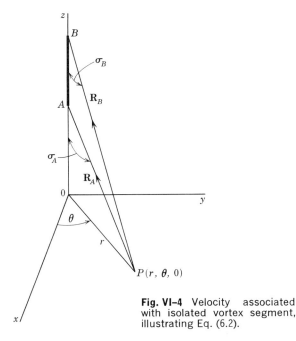

Fig. VI-4 Velocity associated with isolated vortex segment, illustrating Eq. (6.2).

extending from the *field point* $(r, \theta, 0)$ at which the velocity is given by (6.2) to the extremities of the vortex line (Fig. VI-4).

Exercise 22 Show that the limit of (6.2) when AB is the whole z-axis gives the velocity (6.1). Find the velocity induced by a semi-infinite vortex.

Exercise 23 For an infinitesimal element AB put $\sigma_A = \sigma$ and $\sigma_B = d\sigma + \sigma$ and prove that

$$(\cos \sigma_B - \cos \sigma_A) = -\sin \sigma \, d\sigma$$

and that (6.2) becomes

$$\delta q = \frac{-\Gamma}{4\pi r} \sin \sigma \, d\sigma \qquad (6.2a) \qquad \text{differential expression}$$

Exercise 24 By taking the vortex element length AB as dl and denoting its vector by $d\mathbf{l} = dl\,\mathbf{k}$ referred to the unit vector in the z-direction, obtain the vector expression of (6.2a) as

$$\delta \mathbf{q} = \frac{\Gamma}{4\pi} \frac{d\mathbf{l} \times \mathbf{R}}{R^3} \qquad (6.2b) \qquad \text{vector form}$$

where the slant distance R (magnitude of $\mathbf{R}$) replaces the normal distance r of the point P from the vortex axis. Equation (6.2b) is known as the Biot-Savart law of vortex flow induction. HINT: $|d\mathbf{l} \times \mathbf{R}| = dl\, R\, \sin\sigma$,

$R \sin \sigma = r$, and dz is related to $d\sigma$ by noting that $z \tan \sigma = r = $ const, while dz gives the vector line element $d\mathbf{l}$).

Exercise 25 Show that the velocity potential $\phi(r, \theta, z)$ in cylindrical coordinates for the line vortex flow (6.1)

$$\phi(r, \theta, z) = \frac{-\Gamma}{2\pi} \theta \tag{6.3}$$

is obtained from relationships (V.2.7) and (1.2) when $q = u_\theta$.

The velocity potential corresponding to the flow (6.2b) is obtained by noting that $\mathbf{R} = (0 - x)\mathbf{i} + (0 - y)\mathbf{j} + (\zeta - 0)\mathbf{k}$, where ζ locates the vortex element dl on the z-axis, so that $d\mathbf{l} \times \mathbf{R}$ is found as $\mathbf{e}_\theta r\, d\zeta$ and

$$-\frac{\partial(\delta\phi)}{r\,\partial\theta} = \frac{\Gamma}{4\pi} \frac{r\, d\zeta}{(r^2 + \zeta^2)^{3/2}}$$

from which the potential in the plane $z = 0$ is given by

$$\delta\phi(r, \theta) = \frac{-\Gamma\theta}{4\pi} \frac{r^2\, d\zeta}{(r^2 + \zeta^2)^{3/2}} \tag{6.4}$$

Exercise 26 Prove that

$$\int_{-\infty}^{\infty} \frac{r^2\, d\zeta}{(r^2 + \zeta^2)^{3/2}} = 2 \tag{6.5}$$

verification á postiori

and recover (6.3) by integrating (6.4) with respect to ζ for a straight vortex line of infinite length. HINT: Substitution $\zeta = r \tan \alpha$ reduces (6.5) to the integration of the cosine function.

Exercise 27 Show, from (6.2b), that the velocity at the center of a circular vortex ring of radius a is given by

circular vortex, flow on axis

$$q = \frac{\Gamma}{2a} \tag{6.6}$$

and find the velocity at a point on the ring axis distant h from the ring.

Exercise 28 Prove that the velocity induced by a circular vortex ring at an off-center point in its plane is given by

and at arbitrary point of its plane

$$q = \frac{\Gamma}{4\pi a} \int_0^{2\pi} \frac{(1 - k \cos \alpha)\, d\alpha}{(1 + k^2 - 2k \cos \alpha)^{3/2}} \tag{6.6a}$$

where k is the fractional distance of the point from the ring center [the integral in (6.6a) is not expressible in terms of elementary functions].

Exercise 29 Calculate the velocity at the center of a square vortex ring of dimension $2a$ on each side by summing the contributions of each of the sides separately and adding. Compare with (6.6).

7 GROUND EFFECT — MOMENTARY ENHANCEMENT OF AIRPLANE WING LIFT FORCE

When a uniform flow is superimposed on plane vortex flow as shown in Fig. V-2, some streamlines are nearly straight, others nearly circular. Just before takeoff occurs, the flow past an airplane wing is affected by the proximity of the flat ground surface beneath it, represented by streamlines taken as strictly straight. When allowance is made for the presence of the flat boundary by accordingly modifying the vortex potential in a suitable manner, an interesting feature of airplane dynamics is demonstrated that has at times proved hazardous.

idealized wing lift represented by line vortex

Airplane wing lift is, it is recalled, dependent on circulation, as shown, e.g., by (V.4.2). For present purposes the wing vortex of strength $\Gamma/2\pi$ can be taken as two-dimensional on the assumption that its distance h from the ground is small compared with the wing span. The wing vortex potential is now written as

$$\phi = \frac{\Gamma}{2\pi}\theta_1 \tag{7.1}$$

where the sense of circulation is clockwise as shown in Fig. VI-5 and θ_1 is the angle measured from the horizontal to the "field point" P as shown. According to (6.3) a vortex of equal strength and opposite sense of rotation situated at $x = 0, y = -h$, is represented by

image vortex, potential representation

$$\phi = \frac{-\Gamma}{2\pi}\theta_2 \tag{7.1a}$$

The line $y = 0$ is a straight streamline for the combined flow obtained by superposing (7.1) and (7.1a), by symmetry.

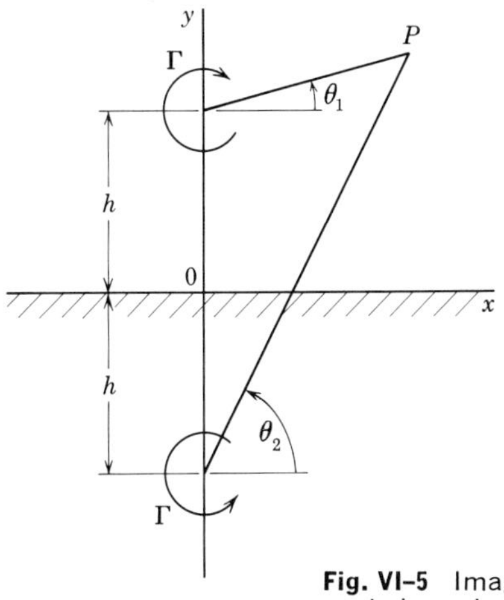

Fig. VI-5 Image vortex represents boundary condition.

complex potential

Exercise 30 Show that the real and imaginary parts, respectively, of the complex potential $W(z)$ given by

$$W(z) = \frac{-i\Gamma}{2\pi} \ln z \tag{7.2}$$

are the ordinary velocity potential and stream function of an infinite vortex at the origin [cf. (V.2.2) and (V.5.2)]. HINT: Recall that $z = re^{i\theta}$ and that the logarithm of a product is the sum of two logarithms.

Exercise 31 Using the final result of Exercise V.20 and resolving the velocity into its polar components, show that

and complex velocity

$$-\frac{dW}{dz} = (u_r - iu_\theta)e^{-i\theta} \tag{7.3}$$

in general.

Exercise 32 Show that the vortex pair (7.1) and (7.1a) superimposed and written in complex potential form as

$$W(z) = \frac{i\Gamma}{2\pi} \ln \frac{z + ih}{z - ih} \tag{7.4}$$

gives $y = 0$ as the streamline $\psi = 0$. Also demonstrate by direct calculation that $v(x, 0) = 0$.

The present method of accounting for the presence of a flat boundary by introducing a second vortex at the "image" point $z = y = -ih$ of the wing vortex at $z = y = +ih$ is an example of a commonly employed technique (called the method of images), extensively applied in hydrodynamics and electrostatics. The fictitious image vortex that provides the required correction of (7.1) at the boundary surface must also be taken into account when the wing lift force is estimated in the manner of Exercise V.19. The velocity associated with the image vortex, at the distance $2h$ that separates it from the wing, is

$$\frac{\Gamma}{2\pi} \frac{1}{2h}$$

horizontally directed. When this velocity increment is added to the flow speed U of section V.4 and the lift force is taken as the product of the velocity sum with the circulation as before, the modified lift force becomes

modified lift — temporary enhancement

$$Y' = \rho U \Gamma + \frac{\rho \Gamma^2}{4h} \tag{7.5}$$

The second term of (7.5) is recognized as an enhancement of wing lift; it is particularly significant when the wing is near the ground so that h is small, the forward speed U being also small. An airplane that is landing experiences a cushioning effect in this manner, delaying the end of airborne flight for some seconds. Owing to diminishing circulation, the lift increment is gradually lost as flight speed U decreases. The effect on takeoff is of course also a lift increment of special interest in very hot climates and at airports at considerable elevation. Under these circumstances the air density is reduced; thus takeoff weight may also require reduction. Pilots have been known to ignore this factor and succeed in leaving the ground under marginal conditions, profiting instantly from the lift augmentation represented by the ground effect term in (7.5). As this benefit diminishes with increasing height h, however, the airplane may continue in level flight at very low elevation to the end of the runway without achieving the anticipated height above the ground. The peculiar psychological sensation of flight under these conditions is rarely experienced more than once in a lifetime by daredevil airplane pilots.

8 STARTING RESISTANCE

As a final application of the dynamics of rectilinear vortex line elements, the modification of lift force (V.4.2) in the first instants of takeoff is modified according to the continuity requirements of Helmholtz' principles. If it is supposed that lift, and hence circulation, is suddenly generated in the vicinity of the leading edge of an airplane wing, an opposite circulation must be found nearby so that the circulation around the entire wing section remains zero, as it was before the appearance of the lift force. The vortex configuration can be approximated as a uniform lifting vortex just ahead of a "starting vortex" shed by the wing that is carried away by the stream at speed U, these two vortex elements being joined by two others to form a rectangular pattern similar to that of Exercise 29. When the starting vortex is still close to the wing (distance small compared with the total wing span), its influence on the flow may be regarded alone, neglecting the shorter trailing vortex elements that join it to the lifting vortex. If this distance is denoted Ut, where U is the relative speed of the stream as before, and t is the time elapsed since the lift force was (impulsively) generated, then the starting vortex "induces" a downward velocity

lifting vortex and starting vortex

$$v = -\frac{\Gamma}{2\pi}\frac{1}{Ut}$$

at the lifting vortex. When this velocity is small compared with U,

156 IRROTATIONAL FLOW IN THREE DIMENSIONS

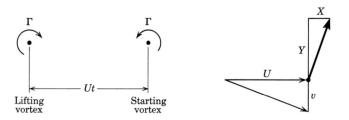

Fig. VI-6 Vortex interaction and starting resistance.

the approaching stream is deflected downward through an angle given by $v \div U$, and the resultant force corresponding to (V.4.2) is tilted so that it acquires a component X that can be estimated as the lift force Y multiplied by $v \div U$, i.e., as (Fig. VI-6)

drag component of tilted resultant force

$$X = \frac{\rho \Gamma^2}{2\pi U t} \qquad (8.1)$$

Although the *starting resistance* (8.1) is inappreciable except when the distance Ut is very small, the effect at takeoff is seen to be an additional but transient thrust requirement that must be provided by the propulsive units.

VII
Surface Gravity Waves

1 GRAVITY AND INERTIA FORCES COMPARED: FROUDE NUMBER

Pressures are balanced either by inertia or by gravity in most of the examples considered up to this point, but not by both together. In reality all three effects act simultaneously as a rule, the precise nature of any particular flow depending on the relative degrees of importance of inertia and gravity in opposing the pressure gradients Gravity *surface waves* introduce a distinctive aspect of fluid behavior when the static equilibrium of a liquid free surface is slightly disturbed, so that gravity predominates while effective inertia forces are small but nonvanishing. Apart from their intrinsic importance, the surface waves treated in this chapter are of interest as a preliminary to the study of fluid compressibility with its accompanying wave phenomena of a different sort.

pressure *and* gravity *and* inertia

Dimensionless forms provide a convenient manner of displaying the comparative importance of different effects represented by equations containing three or more types of terms. A classification of flows subject to the combined actions of gravity and inertia, including wave motions as one limit, is accordingly obtained by considering Euler's equations (II.5.7) for plane motion, written out as

$$\left.\begin{aligned}\underbrace{\frac{\partial u}{\partial t} + u\frac{\partial u}{\partial x} + v\frac{\partial u}{\partial y}}_{\text{inertia}} + \underbrace{\frac{1}{\rho}\frac{\partial p}{\partial x}}_{\text{pressure}} &= 0 \\ \underbrace{\frac{\partial v}{\partial t} + u\frac{\partial v}{\partial x} + v\frac{\partial v}{\partial y}}_{\text{inertia}} + \underbrace{\frac{1}{\rho}\frac{\partial p}{\partial y}}_{\text{pressure}} &= \underbrace{-g}_{\text{gravity}}\end{aligned}\right\} \quad (1.1)$$

when the y-axis is directed upward, opposite to the direction of the gravity force.

Only the gravity term g in (1.1) is known at the outset, and, even for steady flows, inertia terms like $u(\partial u/\partial x)$ vary not only from one term to another but also from point to point in a manner that remains to be determined by solving the entire set of flow equations. These terms are dimensionally equal to g, however; that is,

$$\left| u \frac{\partial u}{\partial x} \right| = |g| = LT^{-2}$$

and a preliminary basis for comparing gravity and inertia terms can thereby be found. Thus, if a length scale l could be found that characterizes the complete flow, as well as a time τ, these might be combined and compared with the standard value 32.2 ft sec^{-2} of g, for example. There is no time scale for a steady flow, and the combination LT^{-2} is therefore more advantageously formed in a slightly different manner as the product of a *characteristic length* l and a *characteristic velocity* U. A typical dimension of a solid boundary surface may be chosen as the length l, while the velocity of distant fluid particles can serve as U, as seen from flow examples already considered. The stream function (V.3.1) found to be proportional to speed U and a length a shows that these values are suitable in the case of uniform flow past a circular cylinder; in the case of flow past a wavy wall (V.6.16) shows that the waviness parameter h appears as an appropriate length scale.

<small>dimensional analysis: characteristic flow quantities</small>

Once any values of l and U have been selected, it becomes possible to rewrite the equations (1.1) in dimensionless form containing a ratio formed with these constants. For this purpose it is convenient to normalize the lengths and velocities appearing in (1.1) by introducing the dimensionless velocities and lengths denoted by primes and defined as follows:

<small>dimensionless variables</small>

$$u' = \frac{u}{U} \qquad v' = \frac{v}{U} \qquad x' = \frac{x}{l} \qquad y' = \frac{y}{l} \qquad (1.2)$$

When a dimensionless pressure p' is likewise defined for an incompressible fluid as

$$p' = \frac{p}{\rho U^2} \qquad (1.3)$$

and flow unsteadiness is momentarily ignored, the pair of equations (1.1) becomes

$$u' \frac{\partial u'}{\partial x'} + v' \frac{\partial u'}{\partial y'} + \frac{\partial p'}{\partial x'} = 0$$

$$u' \frac{\partial v'}{\partial x'} + v' \frac{\partial v'}{\partial y'} + \frac{\partial p'}{\partial y'} = -\frac{gl}{U^2} \qquad (1.4)$$

In the first equation of (1.4) only inertia and pressure terms are present, and they are therefore equal to each other in importance — the one effect exactly balances the other. A completely different state of affairs is found in the second of (1.4), since pressure and inertia effects are only of comparable magnitudes between themselves if the term on the right is small compared with either one separately. When gravity is important, however, corresponding to large values of the scale l or to small characteristic velocity U, three different cases present themselves, according to whether pressure predominates, or inertia, or both together more or less equally stand against the gravity term. The limit of predominant pressure effects is recognized as leading to fluid statics, but, in this chapter, interest centers on the effects of the relatively small inertia terms in the "perturbation" of fluid statics represented by the second of (1.4).

The preceding classification has been based on the magnitude of the parameter, or dimensionless group, on the right side of (1.4). It is customary to employ the square root of the reciprocal of this quantity, termed the *Froude number*,

"Froude number" — a measure of the insignificance of gravity forces

$$F \equiv \frac{U}{\sqrt{gl}} \qquad (1.5)$$

as the criterion for the retention or neglect of gravity effects in the dynamics of fluid motion.

In terms of the Froude number, hydrostatics corresponds to the limit of small values of F, i.e., $F \ll 1$ and the gravity force is of dominant importance with respect to pressure variations. $F \gg 1$, conversely, represents flows at sufficient speed and such scale that gravity forces are dwarfed by pressure and inertia; most aerodynamic and hydrodynamic flows of technical interest are in this category.

Exercise 1 Evaluate the Froude number and indicate whether gravity forces are important in each of the following cases: (a) surface wave with particle velocity of 1 ft sec^{-1} at the free surface of a pool of water of depth 8 ft; (b) flow past an airplane wing of chord (i.e., linear dimension) 8 ft at flight speed 400 mph; (c) winds in large-scale atmospheric motion, speed 200 mph, scale taken as one Earth diameter.

When fluid velocities are everywhere small as well as the Foude number ($F \ll 1$), the pressures in steady flow differ but slightly from the hydrostatic values; the nonlinear terms are regarded as small to the second order and therefore are neglected. In the unsteady case, however, only the *sum* of the first and last terms on the left side of the first equation of (1.1) can be set equal to zero, each term appearing as small to the first order only. Likewise in the second equation of (1.1), neglect of second-order terms alone leaves the velocity derivative $\partial v/\partial t$ as first-order smallness determined by the *difference* of the pressure gradient from the hydrostatic value determined by the remaining terms. Unsteady plane flow at low values of the Froude number is now represented by the Euler equations, obtained from (1.1) as

linearized "perturbation" flow equations

$$\left.\begin{aligned}\frac{\partial u}{\partial t}+\frac{1}{\rho}\frac{\partial p}{\partial x}&=0\\\frac{\partial v}{\partial t}+\frac{1}{\rho}\frac{\partial p}{\partial y}&=-g\end{aligned}\right\} \quad (1.6)$$

To return to the question of a characteristic time τ of the motion represented by the pair of equations (1.6): the validity of the first equation is unaffected by the magnitude of τ, when τ is large (as in motions of very long period), the horizontal pressure gradient is small accordingly, and conversely. But large values of τ have a different effect on the vertical motion and pressure gradient, since the term $\partial p/\partial y$ in this case does not become indefinitely smaller but approaches the hydrostatic value the more closely as τ increases. Quantitative meaning of this idea follows if it is noted that the parameters U and l define a time as the quotient of l divided by U. Two cases are now distinguished, depending on the magnitude of the characteristic time τ: (a) long-period motions for which $\tau \gg l/U$ and the term $\partial v/\partial t$ in (1.6) may be dropped [but not the corresponding term in the first of (1.6), as already explained]; (b) motions of either long or short period, when the first term in the second of (7.6) is *not* discarded — this is the general case that includes (a).

characteristic time

2 WAVES IN SHALLOW WATER

Because of the simplicity of the special case (a) of the preceding Section, a short and intuitive calculation will be carried through as a preliminary to (b), recognizing that the justification of the results is found as one limit of the general calculation given in the next section.

The flow considered is that of the disturbed equilibrium of liquid standing to a height H that is in some sense small (precise meaning to be specified at a later point of the discussion). When the origin of coordinates is taken at the undisturbed free surface, where the pressure is denoted zero, a horizontal boundary is

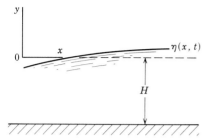

Fig. VII-1 Showing notation, shallow water waves.

located at $y = -H$. The condition of vanishing normal velocity there takes the form $v(x, -H) = 0$ (Fig. VII-1). It follows that the derivative $\partial v/\partial t$ is strictly zero there, and, if the free surface height $\eta(x, t)$ is not only small but also slowly varying with time, its variation $\partial v/\partial t$ does not affect the pressure gradient in (1.6). At points intermediate between the two fluid boundaries, $-H < y < \eta(x, t)$, it is assumed that the same quantity may be uniformly neglected, in other words, the quantity that is strictly zero at one end of a narrow range (over which it is espected to vary continuously) and nearly zero at the other end is supposed to be insignificant throughout the entire range of its values. In physical terms, since the horizontal boundary inhibits vertical motion in its proximity, while the total fluid thickness is small, the vertical motion is ignored completely. The absence of motion is also recognized as the limit of very long characteristic time, in accord with the earlier viewpoint.

justification of hydrostatic pressure approximation

The integration of the second of (1.6) now gives the pressure as the hydrostatic value, proportional to the depth $\eta(x, t) - y$ below the free surface:

$$p(x, y, t) = \rho g(\eta - y) \qquad (2.1)$$

Removing the pressure from the first of (1.6) by means of (2.1) furnishes the relationship between particle velocity component u and the disturbance height η as

$$\frac{\partial u}{\partial t} + g\frac{\partial \eta}{\partial x} = 0 \qquad (2.2)$$

A second equation between the same two quantities is obtained by noting that the vertical velocity $v(x, \eta, t)$ at the free surface is also the time rate of change of η itself, and the velocity component v at the same point is found by integrating the continuity equation:

$$v(x, \eta, t) = \frac{\partial \eta}{\partial t} = -\int_{-H}^{\eta} \frac{\partial u}{\partial x} dy$$
$$= -\frac{\partial u}{\partial x}\{\eta + H\} \doteq -\frac{\partial u}{\partial x} H \qquad (2.3)$$

approximation for vertical velocity components

Equation (2.2), which shows the y-derivative of $\partial u/\partial x$ to be zero, leads to the third expression of $v(x, \eta, t)$ in (2.3), while the last form follows from the fact that the disturbance height η is small compared with H. Hence

$$\frac{\partial \eta}{\partial t} = -H \frac{\partial u}{\partial x} \tag{2.4}$$

and the pair of equations (2.2) and (2.4) is solved for the disturbance height $\eta(x, t)$ by eliminating the velocity u:

free surface height disturbances satisfy wave equation

$$\frac{\partial^2 \eta}{\partial t^2} - gH \frac{\partial^2 \eta}{\partial x^2} = 0 \tag{2.5}$$

The disturbance height $\eta(x, t)$ is given by solutions of (2.5), which is termed the *one-dimensional wave equation*. This equation differs in form from Laplace's equation (IV.5.6), apart from a constant factor, only by having terms of different signs. The nature of its solutions, and hence also the free surface disturbance heights it represents, contrast sharply with the flow properties interior to a fluid that have been found to obey Laplace's equation.

For the interpretation of the solutions of (2.5), we write for conciseness gH as c^2, and verify that any differentiable function $f(\lambda)$, where $\lambda \equiv x - ct$, is a solution for

$$\frac{\partial f}{\partial t} = \frac{df}{d\lambda} \frac{\partial \lambda}{\partial t} = \frac{df}{d\lambda}(-c) \quad \text{and} \quad \frac{\partial f}{\partial x} = \frac{df}{d\lambda}$$

Hence the terms on the left side of (2.5) vanish identically:

$$\frac{\partial^2 f}{\partial t^2} - \frac{\partial^2 f}{\partial x^2} c^2 = (-c)^2 \frac{d^2 f}{d\lambda^2} - \frac{d^2 f}{d\lambda^2} c^2 \equiv 0$$

for arbitrary functions f. The same property is similarly verified for arbitrary functions $g(\sigma)$, where $\sigma \equiv x + ct$; thus the general solution of (2.5) can be written as

general solution given by two arbitrary functions

$$\eta(x, t) = f(x - ct) + g(x + ct) \tag{2.6}$$

The expression on the right side of (2.6) may be compared with the arbitrary complex function the real and imaginary parts of which give solutions of Laplace's equation in (V.5.2).

When $\eta(x, t)$ is represented by only the first term on the right side of (2.6), the disturbance height at any place x_0 and time t_0 is determined by the value of $x_0 - ct_0$. The value of η is the same at another point x_1 and time t_1 provided that

$$x_0 - ct_0 = x_1 - ct_1 \tag{2.7}$$

When the time interval $t_1 - t_0$ is taken arbitrarily small, the distance $x_1 - x_0$ is correspondingly small, and the x-displacement of points of given surface height is in the direction of $+x$, the speed given by the limit of the difference quotient $x_1 - x_0$ divided by $t_1 - t_0$, i.e., equal to c, according to (2.7). The velocity $c = \sqrt{gH}$ is referred to as the *phase velocity*, since in the *advancing wave* represented by $f(x - ct)$ points of constant "phase" (or amplitude) are found to move at this speed, independently of the form of the *wave* specified by the function f. In the same manner the disturbances represented by $g(x + ct)$ *propagate* in the opposite sense along the x-axis; such waves are termed *receding* waves. Both advancing and receding waves are called progressive waves to distinguish them from *standing waves*. It is emphasized that, for the assumed shallow water represented by small values of H, the phase velocity

$$c = \sqrt{gH} \tag{2.7a}$$

is also small.

advancing and receding waves

phase velocity

phase velocity depends on depth alone

Exercise 2 Show that the motion (2.6) is periodic in space and time if

$$f(x - ct) = \sin(x - ct) \quad \text{and} \quad g(x + ct) = \sin(x + ct)$$

and express its period in terms of the constants g and H. Note that the waveheight is permanently zero at certain places: $\eta(0,t) = \eta(2\pi,t) = 0$, so that (2.6) represents a standing wave in this case.

Exercise 3 Prove that every incompressible fluid motion described by the pair of equations (1.6) is irrotational. Determine whether the same is true of the motion represented by (2.5), which corresponds to the neglect of one term in (1.6).

3 WAVES IN WATER OF ARBITRARY DEPTH

Exercise 4 Derive the Euler equations for *unsteady* flow by modifying the development of Section II.4 to include the time rate of increase of x-momentum $(\partial/\partial t)(\rho u\, dx\, dy)$ so that (II.4.2) is extended as

$$\frac{\partial u}{\partial t} + u\frac{\partial u}{\partial x} + v\frac{\partial u}{\partial y} + \frac{1}{\rho}\frac{\partial p}{\partial x} = -\frac{\partial U}{\partial x}$$

Justify the corresponding vector form of the equations (VI.5.5) and show that for unsteady irrotational flow the Bernoulli integral (IV.5.3) is replaced by

$$-\frac{\partial \phi}{\partial t} + \frac{q^2}{2} + \frac{p}{\rho} + U = \text{const} \tag{3.1}$$

The motion of small disturbances at the free surface of a liquid of arbitrary depth is now investigated as a two-dimensional irrotational flow by solving Laplace's equation (IV.5.6)

flow at internal points given by solutions of Laplace's equation

$$\frac{\partial^2 \phi}{\partial x^2} + \frac{\partial^2 \phi}{\partial y^2} = 0 \qquad (3.2)$$

subject to conditions at the free surface $y = \eta(x, t)$ and, at the bottom, $y = -H$. The flow condition there is expressed in terms of the time-dependent velocity potential $\phi(x, y, t)$ by writing $v(x, -H, t) = 0$ as

$$\frac{\partial \phi}{\partial y}(x, -H, t) = 0 \qquad (3.3a)$$

and two boundary conditions

and the condition of constant pressure $p = 0$ at the free surface is obtained from (3.1) by writing the potential energy U as gy and neglecting the squared velocity term as before. Since, at the surface, $y = \eta(x, t)$ where the pressure is taken as zero, the first equation of (2.3) permits the time derivative of (3.1) to be expressed in terms of ϕ, this gives

$$\frac{\partial^2 \phi}{\partial t^2}(x, 0, t) + g \frac{\partial \phi}{\partial y}(x, 0, t) = 0 \qquad (3.3b)$$

for the *approximate* boundary condition applied, not at the free surface, but at the small distance η away from it, i.e., at $y = 0$ [cf. (V.6.12)]. That the constant on the right side of (3.1) becomes zero as in (3.3b) is seen by considering the case of zero disturbance. Both terms in (3.3b) are of first order in the velocity potential, the quantities neglected in imposing the condition at $y = 0$ being of higher order of smallness, as in the wavy wall flow of Section V.6.

The differential equation and its boundary conditions, (3.2) and (3.3a,b) are seen to be linear, i.e., of first degree, in the dependent variable $\phi(x, y, t)$ and also *homogeneous*, since the right side of each is zero. Product-type solutions are readily found, and the treatment of Section V.6 suggests taking a sinusoidal variation for the x-dependence, since an arbitrary wave form can be constructed as the sum of terms of this form. Guided also by the finding of Exercise 2, a periodic time dependence is also assumed, so that a solution is written in the form

$$\phi(x, y, t) = P(y) \cos \lambda x \cos \sigma t \qquad (3.4)$$

trial solution

where λ now represents the wave number $\lambda = 2\pi/L$ of a wave of length L and σ is the circular frequency of a motion of period T; hence $\sigma = 2\pi/T$ and the function $P(y)$ is now determined from (3.2).

Substitution in (3.2) gives

$$\left\{\frac{d^2P}{dy^2} - \lambda^2 P\right\} \cos \lambda x \cos \sigma t = 0$$

from which it is concluded that P satisfies the equation

$$\frac{d^2P}{dy^2} - \lambda^2 P = 0 \tag{3.5}$$

Equation (3.5) is familiar, having been found for the y-dependence in the wavy wall flow of Section 5.6; its solution is

$$P(y) = Ae^{\lambda y} - Be^{-\lambda y} \tag{3.6}$$

analogous to (V.6.13). It also follows from (3.3a) that

$$Ae^{-\lambda H} - Be^{\lambda H} = 0$$

(the integration constant B can be expressed in terms of A). By putting a new constant C equal to $2Ae^{-\lambda H}$ the potential (3.4) takes the form

$$\phi(x, y, t) = C \cosh [\lambda(y + H)] \cos \lambda x \cos \sigma t \tag{3.7}$$

where C is an integration constant and the parameters λ and σ remain unspecified.

flow potential established

Exercise 5 Show that the required boundary conditions are satisfied by the potential (3.7) only if

$$\sigma^2 = g\lambda \tanh \lambda H \tag{3.8}$$

frequency depends on wave length and on depth of liquid

That is, the frequency of *free oscillation* depends on the wavelength. [Recall the definitions (V.6.19) of the hyperbolic sine and cosine functions the quotient of which is represented in (3.8)].

The velocity potential (3.7) is now regarded as containing, in addition to the constant C which is expressible in terms of maximum disturbance amplitude, the parameter λ representing the length of a periodic disturbance, while the frequency σ is related to λ by (3.8). Particle velocities are obtained in the usual manner, and (3.1) permits the free surface height to be found directly from the potential, since this equation reduces to

$$\eta(x, t) = \frac{1}{g} \frac{\partial \phi}{\partial t} (x, 0, t) \tag{3.9}$$

free surface height...

when the potential energy at the free surface is again taken as $g\eta$. Hence the free surface height is given explicitly by

$$\eta(x, t) = \frac{-C\sigma}{g} \cosh \lambda H \cos \lambda x \sin \sigma t \tag{3.9a}$$

166 SURFACE GRAVITY WAVES

satisfies wave propagation equation

Exercise 6 Show that the disturbance height $\eta(x, t)$ given by (3.9a) is a solution of the one-dimensional wave equation

$$\frac{\partial^2 \eta}{\partial t^2} - c^2 \frac{\partial^2 \eta}{\partial x^2} = 0 \tag{3.10}$$

where the phase velocity c is given by

phase velocity is a function of wave length

$$c^2 = \frac{g}{\lambda} \tanh \lambda H \tag{3.11}$$

Equation (3.10) justifies referring to the small-disturbance motion of the free surface as a wave motion.

Exercise 7 Show that the velocity potential (3.7) of waves of (maximum) amplitude h is expressed in terms of this quantity by the equation

evaluation of constants

$$-\phi(x, y, t) = \frac{hg}{\sigma} \frac{\cosh\{\lambda(y + H)\}}{\cosh \lambda H} \cos \lambda x \cos \sigma t \tag{3.12}$$

identification of standing wave

The potential (3.12) shows that the free surface height is permanently zero at $x = L/4$ and at all half-wavelength intervals on both sides of this point, while the peak displacements h are attained periodically at $x = 0$, $x = L/2$, etc., halfway between the *nodes* of zero vertical displacement. The entire motion thus represented, in which the wave crests remain at fixed locations, therefore differs in appearance from the progressive waves in (2.6). The periodic standing wave represented by (3.12) is easily recognized as the superposition of progressive waves, as in Exercise 2. Since it is also seen from (3.12) that the x-velocities vanish at $x = 0$ and at $x = L$, vertical boundaries at these locations would serve to isolate a complete cycle of the wave motion, and it is not necessary to consider fluid of unlimited extent in the directions $-x$ and $+x$.

Exercise 8 Recall the definitions (V.6.19) and establish the relationships
(a) $\qquad 2 \sinh x \cdot \cosh x = \sinh 2x$
(b) $\qquad \cosh^2 x - \sinh^2 x = 1$
(c) $\qquad \sinh x \to x$ and $\cosh x \to 1$ as $x \to 0$
(d) $\qquad \sinh x \to e^x/2$ and $\cosh x \to e^x/2$ as $x \to \infty$

hyperbolic functions: identities and limiting values

Show that in the limit of shallow water waves ($H \ll L$) the wave speed c is independent of the wavelength and is correctly given by (2.7).

Exercise 9 Evaluate the potential energy of the fluid in one cycle of a standing wave by: (a) noting that $\rho g 2y$ is the work required to lift a unit mass from $-y$ to $+y$; (b) adding the energies of fluid particles in the column extending from $y = 0$ to $y = \eta$; (c) integrating the results of (b) over the half-cycle $0 < x < L/2$. Express the result in terms of wave amplitude h and wave length L. It may be noted that the kinetic energy of the wave motion varies periodically between the same limits as the potential energy; their sum is a constant when frictional effects are neglected, of course.

average values of potential and kinetic energies are equal

Ans. $\frac{1}{8}\rho g h^2 L$.

4 WAVES IN DEEP WATER: DISPERSION

When the depth H is large compared with the wavelength L of a surface wave, it follows from (3.11) and Exercise 5 that the phase velocity depends on the wavelength:

$$c = \sqrt{\frac{gL}{2\pi}} \qquad (4.1)$$

Since the long-wave components of an arbitrary surface disturbance represented as a sum of terms like (3.12) propagate more rapidly along the surface of deep water than do the shorter waves, an initial wave form is modified in the course of time [this contrasts with the shallow water waves for which (2.7a) shows all harmonics advancing at the same speed, so that the wave form is preserved in time]. The phenomenon of different harmonic components of a wave traveling with different speeds is familiar in optics, where the longer (red light) waves advance most rapidly. The relationship (4.1) exhibits a similar behavior for surface waves. This behavior is referred to in both cases as *normal dispersion*.

Exercise 10 Considering ocean waves on the Earth's surface, evaluate the phase velocity of a wave of length equal to one-half of the Earth's circumference by using (4.1). Calculate the time required for a disturbance of this scale to travel halfway around the Earth. Is it justified to apply (4.1) in this instance?

phase velocity is very great for large-scale disturbances

The periodic rise and fall of the free surface is represented in (3.9a) for a standing wave and is also verified for the progressive train of waves obtained by superposition, reversing the procedure of Exercise 2. From (3.12) it is seen that fluid particles below the free surface also share in the periodic movement, and this is exhibited in explicit form by eliminating the coordinates in either case. To illustrate for the points at the free surface, x is eliminated from

(3.9a) by noting that the second time derivative recovers η and hence, for a given x,

$$\frac{d^2\eta}{dt^2} + \sigma^2\eta = 0$$

or

$$\frac{d^2\eta}{dt^2} + g\lambda \tanh \lambda H \, \eta = 0 \tag{4.2}$$

according to (3.8). The coefficient of η in (4.2) is thus the square of the circular frequency σ of (3.9a), and the time-dependence of small displacements at a given location is described by the *harmonic oscillator* ordinary differential equation (4.2) the solutions of which are a sine function and a cosine. For deep water waves, $\lambda H \gg 1$ and the equation is approximated as

$$\frac{d^2\eta}{dt^2} + \frac{2\pi g}{L}\eta = 0 \tag{4.2a}$$

which shows that the frequency is great and the periodic time short when the wavelengths are short. For a wavelength equal to half the circumference of the Earth [when (4.2a) is not strictly applicable], a period of nearly 1 hour is obtained. The more complete calculation of the following Section will yield a value rather close to this one, indicating a surprising accuracy of (4.2a) even when applied beyond its proper domain.

5 OSCILLATION OF A SPHERICAL LIQUID MASS

The possibility of discussing surface disturbances as either progressive waves or standing waves (or "oscillations") interchangeably provides a basis for examining the dynamic response of the nearly spherical ellipsoidal form of a large liquid mass "disturbed" by the attractions of a tide-raising body such as the Moon. The Earth's radius is denoted by a as before, and a time-dependent velocity potential is written in the form

time-dependent potential of perturbed liquid sphere

$$\phi(r, \theta, t) = \left(\frac{r}{a}\right)^2 P_2(\theta) G(t) \tag{5.1}$$

as a product of three functions, as in (3.4) but with the difference that both coordinate-dependent factors are specified whereas the time-dependent function $G(t)$ remains to be determined. Comparison with (VI.1.12) assures that (5.1) represents an irrotational flow, since it is a solution of Laplace's equation (VI.1.5). Reference to (3.12) also shows that a constant factor, omitted for brevity from (5.1), relates to peak displacement. The free surface height at any point is now denoted $\eta(a, \phi, t)$ or, more simply, $\eta(\theta, t)$, and its time rate of change is again related to the potential by means of the radial velocity component at the free surface:

$$-\frac{\partial \phi}{\partial r}(a, \theta, t) = \frac{\partial \eta}{\partial t}(\theta, t) = -\frac{2}{a} P_2(\theta) G(t) \qquad (5.2)$$

boundary condition

The extended Bernoulli integral (3.1) again provides an additional relationship between the potential and the disturbance height when the gravity potential is expressed in terms of $\eta(\theta, t)$. For this purpose the potential is taken in the form

$$U(a + \eta, \theta, t) = \text{const} + \frac{2}{5} g\eta(\theta, t) \qquad (5.3)$$

where the second term on the right side of (5.3) can be shown to represent the modification due to the attraction of displaced water when the sphere of radius a is deformed as an ellipsoid of small eccentricity. Then (3.1) becomes

$$P_2(\theta) \frac{dG}{dt} = \frac{2}{5} g\eta(\theta, t) \qquad (5.4)$$

analogous to (3.9) for two-dimensional motion. The time-dependent factor is removed from (5.4) by differentiation of (5.2), so that

$$\frac{d^2\eta}{dt^2} + \frac{4}{5}\frac{g}{a}\eta = 0 \qquad (5.5)$$

time dependence

Comparison with (4.2a) shows how the more accurate expression of the gravity potential for the deformed sphere leads to a *natural frequency* $\sqrt{4g/5a}$ determined by surface gravity g and sphere radius a. When a is taken as the radius of the Earth, a half-period representing the time elapsed between a maximum positive displacement (peak) and the subsequent minimum surface height at a given point is found to be 47.2 minutes. Lord Kelvin, remarking on the relatively short period of gravity oscillations of a liquid sphere of the size of the Earth, noted that an elastic wave in even such a rigid material as steel would require a longer time to travel the straight-line distance equal to one Earth diameter.

Exercise 11 Take the phase speed of an elastic wave in a steel rod to be 10,140 ft sec^{-1} and calculate the time required to traverse the distance equal to one Earth diameter.

6 PARTICLE PATHS IN PROGRESSIVE WAVE MOTION

It is readily seen from (3.12) that the motion of fluid particles is rectilinear and vertical at all points under the crests and hollows in a standing wave pattern, rectilinear and horizontal at the nodes. A direct calculation also shows that the motion is rectilinear at the intermediate points, the path slopes varying continuously through all values. The trajectories, or orbits, are now examined in the motion that gives rise to an infinite train of progressive waves of sinusoidal form advancing in one direction.

standing waves: rectilinear particle paths

Waves of height h advancing in the $+x$-direction at speed σ/λ are represented by a combination of two potentials of the form of (3.12) as

surface disturbance representation as advancing wave

$$\phi(x, y, t) = \frac{hg}{\sigma} \frac{\cosh\{\lambda(y + H)\}}{\cosh \lambda H} \cos(\lambda x - \sigma t) \quad (6.1)$$

where the frequency σ is again given in terms of λ and H by (3.8). For waves in deep water $\lambda H \gg 1$, and part (d) of Exercise 5 shows that $\tanh \lambda H \to 1$ so that (3.8) is approximated by

$$\sigma^2 = g\lambda \quad (6.2)$$

while

$$\frac{\cosh \lambda(y + H)}{\cosh \lambda H} \to e^{\lambda y} \quad (6.2a)$$

Hence the potential for deep water waves becomes

$$\phi = \frac{hg}{\sigma} e^{\lambda y} \cos(\lambda x - \sigma t) \quad (6.3)$$

and (3.9) gives the wave height as

$$\eta(x, t) = h \sin(\lambda x - \sigma t) \quad (6.4)$$

The particle displacements in the vertical direction, relative to equilibrium levels, are represented by (6.4) and indicate that these are small (because h was so assumed) and completely unrelated to the finite phase speed of the surface wave. The particle velocities obtained as derivatives of the potential (6.3) are also small:

particle velocities

$$\frac{dx}{dt} = u(x, y, t) = \frac{\lambda hg}{\sigma} e^{\lambda y} \sin(\lambda x - \sigma t) \quad (6.5a)$$

$$\frac{dy}{dt} = v(x, y, t) = -\frac{\lambda hg}{\sigma} e^{\lambda y} \cos(\lambda x - \sigma t) \quad (6.5b)$$

Regarded as differential equations for the determination of particle positions $x(t)$ and $y(t)$, these equations are nonlinear because of the exponential and trigonometric functions of x and y which they contain. The small factor h on the right side of each one, however, implies the smallness of the x- and y-displacements from the equilibrium positions corresponding to $h = 0$; it follows that the right sides of these equations can each be written as the sum of a term independent of the displacements plus smaller terms that contain them. When it is seen that the latter terms are of a higher order of smallness than the first, the integration of (6.5) can be effected by treating x and y as constant on the right sides of the equations, while the displacements, denoted x' and y', are obtained as the integrals of the terms on the left. Thus

$$x'(t) = he^{\lambda y} \cos(\lambda x - \sigma t) \qquad (6.6a)$$

$$y'(t) = he^{\lambda y} \sin(\lambda x - \sigma t) \qquad (6.6b)$$

linearized approximation for particle paths

Exercise 12 Show that the particle orbits given by (6.6a) and (6.6b) are circles, with radii decreasing from h at the free surface to steadily and rapidly diminishing values at increasing depths below the surface. Can this motion be reconciled with the assumption of flow irrotationality?

*****Exercise 13** The trajectories of particles in wave motion of arbitary depth of water are obtained in the same manner as (6.6), but without the approximations (6.2) and (6.2a). Eliminate the time from the resulting expressions for x' and y' and show that the orbits are ellipses in general. Find the semimajor and semiminor axes a and b, and show that they reduce to the results of the preceding exercise when the depth h is great.

elliptic orbits

*****Exercise 14** Verify that the orbit eccentricity $e = \sqrt{a^2 - b^2}$ of the ellipses of the preceding exercise is independent of depth $-y$, so that the orbit ellipses are *confocal*.

VIII
Compressibility

1 ACOUSTIC PROPAGATION: SOUND WAVES

As long as fluid compressibility is ignored, both gravity and pressure gradient effects appear as *external* forces in Euler's equations. The reason is that gravity depends only on the mass and shape of the Earth and not on any property of the fluid on which it acts, whereas pressure gradient forces represent the influence of surrounding fluid on the motion of a particular volume element. Accounting for compressibility, i.e., for the volumetric reduction that accompanies even a momentary pressure rise, introduces particle displacements within the volume element. Although the displacements may be very small, the resistance offered by the fluid to volume reduction usually involves energy storage much the same as in a loaded coil spring. This energy is communicated to adjacent fluid by virtue of the *internal* forces of fluid compressibility, hence the description of the main features of acoustics and high-speed flows in terms of *internal waves*, which are analogous in many respects to the surface gravity waves of the preceding chapter.

<small>fluid elasticity: volumetric reduction is accompanied by restoring forces</small>

Before examining the modifications required to account for compressibility in the familiar Eulerian form of the equations of fluid motion, an indication of how internal compressive forces lead to wave-type behavior is presented. It is sufficient to consider one-dimensional motions, advantageously employing the Lagrangian description in which the instantaneous position $x(\xi, t)$ of a fluid particle is specified by the independent variable t and the initial (i.e., undisturbed) position ξ that identifies the particle in question, where ξ is written in place of the variable a appearing in (II.7.1). A one-dimensional volume element is taken as a slab of fluid of thickness $d\xi$ at the initial time denoted $t = 0$. Mass and momentum

<small>analysis of compression in Lagrangian variables...</small>

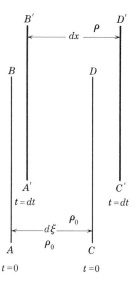

Fig. VIII-1 Illustrating Lagrangian form (1.1) of continuity equation.

conservation principles are applied directly to the fluid contained between the bounding planes AB and CD at $t = 0$ and between $A'B'$ and $C'D'$ at time dt, as indicated in Fig. VIII-1. If the initial density is denoted ρ_0 as before, and without subscript at the time dt, the continuity equation is written directly as

$$\rho_0 \, d\xi = \rho \, dx \qquad (1.1)$$ ... continuity

where dx is the separation distance of the planes $A'B'$ and $C'D'$. Each side of (1.1) gives the fluid mass contained in unit cross section of slab volume element, and the equation itself is a special case of (II.7.7) obtained by setting $y = b$.

If the pressures are p and $p + (\partial p/\partial x) \, dx$ at $A'B'$ and $C'D'$, respectively, the direct statement of Newton's second law of motion as the equality of force and the product of mass times acceleration gives the equation of momentum conservation as

$$\rho \frac{\partial^2 x}{\partial t^2} = -\frac{\partial p}{\partial x} \qquad (1.2)$$ ... Newton's second law

when the common factor dx is removed from both sides of the equation. The derivative on the left side of (1.2) is "Lagrangian" in the sense that ξ is regarded as constant, whereas the physical argument shows that the derivative on the right side considers t to be constant, so that it is "Eulerian." The latter term is transformed by supposing that the fluid is barotropic in the first instance, so that the pressure depends completely on the density $p = p(\rho)$. When

barotropic fluid

the pressure-density derivative is provisionally denoted c^2, the chain rule of differentiation gives

$$\frac{\partial p}{\partial x} = \frac{\partial p}{\partial \rho}\frac{\partial \rho}{\partial \xi}\frac{\partial \xi}{\partial x} = c^2 \frac{\partial \rho}{\partial \xi}\frac{\partial \xi}{\partial x} \qquad (1.3)$$

and the last factor on the right side of (1.3) is expressible in Lagrangian form as the reciprocal of the derivative $\partial x/\partial \xi$.* The density derivative is similarly found by writing (1.1) as

$$\rho = \rho_0 \frac{1}{\partial x/\partial \xi}$$

so that

$$\frac{1}{\rho_0}\frac{\partial \rho}{\partial \xi} = \frac{-1}{(\partial x/\partial \xi)^2}\frac{\partial^2 x}{\partial \xi^2} \qquad (1.4)$$

The derivative in the denominator of (1.4) is seen from (1.1) to be the ratio of densities ρ and ρ_0; hence, if the density increment is written $\delta\rho$ and $\delta\rho \ll \rho_0$, the compression being slight,

$$\left(\frac{\partial x}{\partial \xi}\right)^{-1} = \frac{\rho}{\rho_0} = \frac{\rho_0 + \delta\rho}{\rho_0} = 1 + \frac{\delta\rho}{\rho_0} \qquad (1.5)$$

then the last term on the right side of (1.5) is small compared with unity and $\partial x/\partial \xi$ can be replaced by unity in good approximation when the pressures (hence densities also) are *nearly* uniform. The density term in (1.2) is likewise replaced by ρ_0 when the acceleration is also small, so that substitution of (1.4) and (1.3) brings this equation to the form

compression generates internal waves

$$\frac{\partial^2 x}{\partial t^2} - c^2 \frac{\partial^2 x}{\partial \xi^2} = 0 \qquad (1.6)$$

which is recognized as identical with the one-dimensional wave equation (VII.2.5) used to describe the behavior of a surface disturbance in the gravity waves in a liquid. It follows that particle positions within a compressible fluid suffer displacements that are communicated to adjacent fluid at a definite phase speed, again denoted c, in a fashion analogous to the gravity surface waves. In the present instance, gravity plays no role and the wave propagation is entirely related to the resistance to compression that takes the form of increased pressure, entailing particle motions as indicated by (1.2).

*Note that $\partial \xi/\partial x = 1/(\partial x/\partial \xi)$ only when the second independent variable in one of the derivatives is the same as in the other, i.e., the two sets of variables are (x,t) and (ξ,t) in this case.

It is of interest to note some numerical values when (1.6) is interpreted in relation to plane acoustic (sound) waves in air at 0°C. Sir Isaac Newton evaluated the speed of sound, c, by supposing that the required barotropic condition in (1.3) represents isothermal compression. If sea level atmospheric pressure p_0 is taken as 1.013×10^6 dyne cm^{-2} and the density $\rho_0 = 0.00129$ gm cm^{-3}, the speed $c = 280$ m sec^{-1} is found by proceeding as in Section 1.4 for an ideal gas. This value is considerably below that found by direct observation, and it led Newton to conclude that the discrepancy is caused by fluid friction. The following sections consider the motion of a compressible fluid with fewer restrictions than above, and it is found that the more general treatment leads to a quite different estimate of the acoustic velocity — in much closer agreement with observation.

Newton's (erroneous) calculation of speed of sound

The preceding calculations furnish no indication of the magnitude of the particle displacements during the passage of a sound wave, nor of the particle velocities themselves. Measurements made nearly a century ago by Wien, Lord Rayleigh, and others indicate that the human ear is sensitive to sounds accompanying displacements as small as 10^{-7} cm for the standard laboratory frequency ("middle c"), 256 Hz.* Maximum particle velocities are then considerably less than 1 cm/sec and the several assumptions used in obtaining (1.6) appear to be amply justified (the reader can verify that the pressure and density increments are likewise sufficiently small to be approximated as above).

particle displacements in acoustic disturbances are very small

Exercise 1 If the smallness of particle displacements in a sound wave justifies replacing ξ by x after differentiation of (1.6) with respect to time, then the particle velocities also satisfy the one-dimensional wave equation

$$\frac{\partial^2 u}{\partial t^2} - c^2 \frac{\partial^2 u}{\partial x^2} = 0 \qquad (1.6a)$$

velocities also satisfy the wave equation

Establish (1.6a) by appealing to the Eulerian forms of the continuity and momentum conservation equations of small motion in the x-direction:

$$\frac{\partial \rho}{\partial t} + \rho \frac{\partial u}{\partial x} = 0$$
$$\frac{\partial u}{\partial t} + c^2 \frac{1}{\rho} \frac{\partial \rho}{\partial x} = 0 \qquad (1.7)$$

equivalent Eulerian formulation

where c^2 again represents the pressure-density derivative indicated in (1.3).

*In current notation, cycles per second are expressed as Hertz (Hz).

2 ENERGY CONSERVATION IN COMPRESSIBLE FLUID

Exercise 2 Extend the energy balance of Section II.5 to the case of rectilinear *unsteady* flow in the x-direction, so that (II.5.3) is replaced by

$$\rho \frac{DE}{Dt} + \frac{\partial}{\partial x}(pu) = 0 \tag{2.1}$$

Write pu as $(p/\rho)\rho u$ and notice that

$$\frac{\partial}{\partial x}(pu) = \rho u \frac{\partial}{\partial x}\left(\frac{p}{\rho}\right) - \frac{p}{\rho}\frac{\partial \rho}{\partial t}$$

in order to prove that (II.5.3a) is replaced by

general form of energy principle when friction and heat addition are unimportant

$$\frac{D}{Dt}\left(\frac{p}{\rho} + E\right) = \frac{1}{\rho}\frac{\partial p}{\partial t} \tag{2.2}$$

as the energy equation of unsteady frictionless fluid motion, *not* restricted to one-dimensional flow.

Exercise 3 When the energy E of an ideal gas is taken as the sum of kinetic energy $q^2/2$ and internal energy $c_v T$ per unit mass, show that for steady flow (2.2) reduces to

specialization: steady flow of ideal gas

$$h + \frac{q^2}{2} = \text{const} \tag{2.3}$$

where the enthalpy h is given by $h = c_p T$ in terms of the absolute temperature and the specific heat at constant pressure.

When flow is steady and one-dimensional, differentiations with respect to the flow coordinate direction x are expressed in compact form by suppressing the dx; hence the differential form of (2.3) is

$$c_p\, dT + q\, dq = 0 \tag{2.3a}$$

The energy equation (2.3a), in conjunction with the corresponding form of the momentum relationship,

$$dp + \rho q\, dq = 0 \tag{2.4}$$

can now be applied to evaluate the pressure-density derivative introduced in (1.3) and to study one-dimensional motions at unrestricted flow speeds.

The appearance of the differentials dT, dp, etc., in (2.3a) and (2.4) limits the application of these equations to *continuous* flows, i.e., to motion without abrupt changes in these quantities. In this case the elimination of the flow speed q and replacement of the temperature by means of (I.3.8) leads to the equation

adiabatic relationship; differential form ...

$$\gamma \frac{d\rho}{\rho} - \frac{dp}{p} = 0 \tag{2.3b}$$

where γ denotes the ratio of specific heats, $\gamma \equiv c_p/c_v$.

Exercise 4 Show that the first law of thermodynamics, when applied to an ideal gas, namely,

$$dQ = d(c_v T) + p\, d\!\left(\frac{1}{\rho}\right) \qquad (2.5)$$

is equivalent to (2.3b) in the absence of heat addition, $dQ = 0$, all state changes occurring reversibly. Integration of (2.5) then furnishes the *isentropic* pressure-density relationship expressible as

$$\frac{p}{\rho^\gamma} = \frac{p_0}{\rho_0^\gamma} \qquad (2.5a)$$

... and integral form

Exercise 5 Calculate the speed of sound, c, by using the isentropic pressure-density derivative obtained from (2.3b) or (2.5a), and evaluate for the conditions that led to Newton's value of 280 m sec^{-1} in the preceding section. (Take $\gamma = 1.4$ and compare with the directly measured speed of 332 m sec^{-1}). The reconciliation of theory and observation in this manner is due to Laplace.   Ans. $c^2 = \gamma RT$.

When the subscript zero refers to stagnation conditions and the enthalpy of an ideal gas is expressed in terms of the pressure and density, the steady-flow energy equation (2.3) can be written as

$$\frac{\gamma}{\gamma - 1}\frac{p}{\rho} + \frac{q^2}{2} = \frac{\gamma}{\gamma - 1}\frac{p_0}{\rho_0} \qquad (2.6)$$

Bernoulli's Equation, ideal gas ...

and recognized as the compressible-flow couterpart of the Bernoulli integral (II.3.11) when gravity is neglected:

$$\frac{p}{\rho} + \frac{q^2}{2} = \frac{p_0}{\rho_0} \qquad (2.7)$$

... and incompressible fluid limit

Exercise 6 Prove that (2.6) reduces to the incompressible form of Bernoulli's equation (2.7) when the velocities are so small that the pressure is very slightly different from the stagnation value p_0, i.e., so that

$$\frac{p}{p_0} = 1 - \epsilon \qquad \text{where} \quad \epsilon \ll 1$$

HINT: Use (2.5a) to express the density ρ in terms of ϵ and substitute for p and ρ in (2.6), with the approximation

$$(1 - \epsilon)^{(\gamma - 1/\gamma)} \doteq 1 - \frac{\gamma - 1}{\gamma}\epsilon$$

Exercise 7 Use (2.3b) to estimate the temperature drop in air accompanying a pressure reduction of 5 percent ($dp/p = 0.05$) and thus explain the phenomenon of automobile carburetor icing in near-freezing weather conditions.

3 NORMAL SHOCK WAVES IN IDEAL GAS

The conservation principles of fluid mass, momentum, and energy in steady flow parallel to the x-axis can be written with the aid of (2.3) and (2.4) in differential form as

conservation principles; steady, continuous, one-dimensional flow (differential form)

$$\left. \begin{array}{l} d(\rho u) = 0 \\ d(p + \rho u^2) = 0 \\ d\left(h + \dfrac{u^2}{2}\right) = 0 \end{array} \right\} \quad (3.1)$$

When the enthalpy is expressed in terms of pressure and density as

$$h = \frac{\gamma}{\gamma - 1} \frac{p}{\rho}$$

the set of equations (3.1), three in number, is seen to permit any one of the flow quantities p, u, ρ to be found by elimination of the other two. If conditions p_1 u_1, ρ_1 are known at one place, the mass flux $\rho_1 u_1$ there must also be equal to the mass flux $\rho_2 u_2$ at another where the flow quantities are identified by the subscript 2. The same types of statements are indicated about the momentum and energy flux totals in the second and third of equations (3.1), and the *integral* form of the three statements is accordingly

...without restriction to continuous flow (integrated form)

$$\left. \begin{array}{l} \rho_2 u_2 = \rho_1 u_1 \\ p_2 + \rho_2 u_2{}^2 = p_1 + \rho_1 u_1{}^2 \\ \dfrac{\gamma}{\gamma - 1} \dfrac{p_2}{\rho_2} + \dfrac{u_2{}^2}{2} = \dfrac{\gamma}{\gamma - 1} \dfrac{p_1}{\rho_1} + \dfrac{u_1{}^2}{2} \end{array} \right\} \quad (3.2)$$

The set of equations (3.2) differs essentially from the differential forms (3.1) or their more explicit representations like (2.3a) and (2.4). The absence of derivatives, or of differentials, removes the restriction to continuous flows (i.e., to flow with continuously varying velocity, pressure, etc.), leaving only the more gross statements of the equality of the respective flux represented by each equation at the two locations identified by subscripts 1 and 2. Uniform flow, $u_2 = u_1$, $p_2 = p_1$, etc., is evidently one possible condition under which the equations are all satisfied, but the presence of quadratic terms in the velocities opens the possibility of nonuniform and even discontinuous flow as well. Direct algebraic solution of the equations shows in fact that a velocity $u_2 \neq u_1$ is possible; this represents an abrupt, i.e., discontinuous, change in each of the flow quantities p, u, and ρ, while their respective combinations forming the mass, momentum, and energy fluxes all remain constant consistent with the conservation principles.

Exercise 8 Use the first two equations of (3.2) to eliminate the density ρ_2 and pressure p_2 from the last, and show that the velocity discontinuity across a *normal shock wave* is given by

$$\frac{u_2}{u_1} = 1 - \frac{2}{\gamma + 1}\left(1 - \frac{\gamma p_1}{\rho_1 u_1^2}\right) \quad (3.3)$$

normal shock wave: velocity discontinuity ...

Also verify that the pressure rise is given by

$$\frac{p_2}{p_1} = 1 + \frac{2\gamma}{\gamma + 1}\left(\frac{\rho_1 u_1^2}{\gamma p_1} - 1\right) \quad (3.4)$$

... and pressure

The preceding equations furnish no indication of the distance $x_2 - x_1$, for example, over which the transition of flow speed from the value u_1 to the value u_2, given by (3.3) may be realized, and there is in fact nothing here to indicate that any finite distance is required, when the flow principles are written as in (3.2). The actual occurrence of very abrupt or "shock" transitions in high-speed flows is proof that the mathematical solution represented by (3.3) or (3.4) is at least closely approximated in reality. Frictional stresses not included in (3.2) permit estimates to be made of the small but finite thickness of the region of variation of flow properties between the limits identified by the different subscripts in each of the equations (3.2).

The pressure jump (3.4) is seen to increase from very small values when the flow speed u_1 exceeds but slightly the sound speed $\sqrt{\gamma p_1/\rho_1}$ obtained in Exercise 5 to large values when the dimensionless speed ratio is also large. Since u_1 is interpreted as a given initial flow condition that defines (along with p_1 and ρ_1) both the speed and the thermodynamic state of the fluid compressed in passing through the shock wave, both (3.3) and (3.4) show that the final flow state is determined by the same ratio. The velocity decrease $u_1 - u_2$ and the pressure rise (sometimes termed the strength of the shock wave) $p_2 - p_1$ also approach zero as the same ratio tends to the value unity.

Exercise 9 By calculating the density ratio with the aid of (3.3) and the first of (3.2), evaluate the four quantities

$$\frac{p_2}{p_1} - 1 \quad \frac{p_2}{p_1} + 1 \quad \frac{\rho_2}{\rho_1} - 1 \quad \frac{\rho_2}{\rho_1} + 1$$

and establish Hugoniot's relationship between the state variables ahead of a shock wave and those behind it, namely,

$$\frac{p_2 - p_1}{\rho_2 - \rho_1} = \gamma \frac{p_2 + p_1}{\rho_2 + \rho_1} \quad (3.5)$$

Hugoniot's equation

In what manner are (2.3b) and (3.5) interrelated?

Exercise 10 Note that

$$2(p_2\rho_1 - p_1\rho_2) = (p_2 - p_1)(\rho_2 + \rho_1) - (p_2 + p_1)(\rho_2 - \rho_1) \quad (3.6)$$

and show that the energy equation in (3.2) may be written as

$$u_1^2 - u_2^2 = \frac{\gamma}{\gamma - 1} \frac{(\rho_2 + \rho_1)(\rho_2 - \rho_1)}{\rho_1 \rho_2} \left(\frac{p_2 - p_1}{\rho_2 - \rho_1} - \frac{p_2 + p_1}{\rho_2 + \rho_1} \right) \quad (3.7)$$

Combine the continuity equation in (3.2) with (3.5) and (3.7) to obtain

$$u_1 \frac{u_1 + u_2}{2} = \frac{\gamma}{\rho_1} \frac{p_1 + p_2}{2} \quad (3.8)$$

From (3.8) it is seen that in the limit of very weak shock waves, i.e., when

$$\frac{u_1 + u_2}{2} \doteq u_1 \qquad \frac{p_1 + p_2}{2} \doteq p_1$$

the speeds u_1 and u_2 ahead and behind a shock wave are both given by

$$\sqrt{\gamma \frac{p_1}{\rho_1}} \quad (3.9)$$

acoustic speed more accurately given by isentropic pressure-density variation

Expression (3.9) therefore represents the "sonic" or *acoustic* velocity, since sound waves are recognized as very weak pressure disturbances. But (3.9) has also been found as the isentropic pressure-density derivative; hence this is the appropriate value of c to use in (1.6), in accord with Laplace's determination. The physical interpretation is that the passage of a sound wave at a given point in a fluid occurs so quickly that heat conduction from the compressed region to adjacent fluid is insignificant, the process being much more nearly isentropic than isothermal.

4 MACH NUMBER: THE DYNAMIC MEASURE OF FLUID COMPRESSIBILITY

The fluid compressibility (I.4.1) corresponding to the isentropic condition that holds for very weak shock waves can be written as

isentropic compressibility

$$K_S = \frac{1}{\rho} \left(\frac{\partial \rho}{\partial p} \right)_S = \frac{1}{\rho_1 c^2} \quad (4.1)$$

where the subscript S indicates that the entropy is constant in the compression described by the partial derivative. When the density is identified with the value in the initial flow, the denominator of (4.1) is seen to be proportional to the pressure p_1, so that a dimensionless parameter is obtained from (4.1) by introducing another quantity having the same dimension. The product of the density and the square of the flow speed u_1 is suitable for this purpose and is in fact the conventional parameter of compressiblity in discussions of high-speed flows:

$$M_1^2 = \frac{u_1^2}{c_1^2} \qquad (4.2)$$

Mach number: the dimensionless flow speed

Comparison with (3.9) shows that the *Mach number* M_1 is simply the velocity normalized with respect to the characteristic speed defined by the pressure and density in an ideal gas in adiabatic flow. At each point in a flow the Mach number M is similarly determined by the flow speed and the thermodynamic state variables. The normal shock relationships (3.3) and (3.4) are seen to be functions of M_1^2 alone.

Exercise 11 Use (3.2), (3.3), and (3.4) to evaluate the Mach number M_2 behind a normal shock wave. Show that, in the limit $M_1 \to 1$, $M_2 \to 1$ also, and that $M_1 > 1$ implies $M_2 < 1$. Prove that, as $M_1 \to \infty$, $M_2^2 \to \frac{1}{7}$ in a diatomic gas for which $\gamma = 1.4$.

$$\text{Ans.} \quad M_2^2 = \frac{1 + \dfrac{\gamma - 1}{\gamma + 1}(M_1^2 - 1)}{1 + \dfrac{2\gamma}{\gamma + 1}(M_1^2 - 1)} \qquad (4.3)$$

Exercise 12 (a) Take the entropy variation between adjacent states in an ideal gas as

$$dS = c_v \frac{dp}{p} - c_p \frac{d\rho}{\rho} \qquad (4.4)$$

to establish for the entropy rise across a normal shock wave

$$S_2 - S_1 = c_v \ln\left[\frac{p_2}{p_1}\left(\frac{u_2}{u_1}\right)^\gamma\right] \qquad (4.4a)$$

(b) Letting $\theta \equiv M_1^2 - 1$ for conciseness, show that (3.4) is expressible as

$$\frac{p_2}{p_1} = 1 + \frac{2\gamma\theta}{\gamma + 1} \qquad (4.5)$$

and that the velocity ratio (3.3) can be written as a series of ascending powers of θ when $2\theta < 1$. Obtain the first four terms in the expansion of $(u_2/u_1)^\gamma$. (c) Recall that $\ln(1 + x) \doteq x$ when $x \ll 1$ and show that the leading term in the expansion of (4.4a) gives the entropy variation across a normal shock wave as

$$c_v \frac{2\gamma(\gamma - 1)}{3(\gamma + 1)^2}(M_1^2 - 1)^3 \qquad (4.6)$$

shock waves occur in supersonic flows only

to the first order. Explain why shock waves are observed only in *supersonic* ow, $M_1 > 1$. HINT: Recall the second law of thermodynamics.

5 HEAT SHOCK; HEAT PULSE

The discontinuous normal shock wave flow solution of the equations (3.2) has appeared as an alternate to simple uniform flow, its possibility having been signaled by the quadratic character of the algebraic equations for the flow speed. The spontaneous appearance of shock waves in this manner raises the question whether different means can be found to "trip" flow discontinuities. Heat addition in a localized region, as by chemical combustion or detonation for example, is now considered a direct extension of adiabatic flow, accounted for by a modification of the third equation only of the set (3.2).

When it is assumed that the amount of heat, Q, in energy units, is acquired by unit mass of fluid in passing from flow state 1 to state 2, without the introduction of forces, the third equation of (3.2) becomes

modified energy condition

$$\frac{\gamma}{\gamma-1}\frac{p_2}{\rho_2} + \frac{u_2^2}{2} = \frac{\gamma}{\gamma-1}\frac{p_1}{\rho_1} + \frac{u_1^2}{2} + Q \tag{5.1}$$

in an ideal gas.

Exercise 13 Eliminate p_2 and ρ_2 from (5.1) and the first two of (3.2) in order to find the velocity ratio as

thermal and dynamic shock effects combined

$$\frac{u_2}{u_1} = 1 + \frac{1}{\gamma+1}\left\{\frac{1-M_1^2}{M_1^2} \pm \sqrt{\left(\frac{1-M_1^2}{M_1^2}\right)^2 - \frac{2Q}{u_1^2}(\gamma^2-1)}\right\} \tag{5.2}$$

Select the appropriate sign to determine the effect of heat addition on discontinuous flow by examining the limiting case $Q = 0$ when (5.2) reduces to (3.3). Show that in this case the addition of heat in a subsonic flow ($M_1 < 1$) causes an increase in speed, with the opposite effect on supersonic flow.

Exercise 14 Show that the velocity ratio u_2/u_1 obtained from the continuity and momentum relations alone in (3.2) gives

$$\frac{u_2}{u_1} = \frac{\gamma}{\gamma+1} + \frac{1}{M_1^2(\gamma+1)} \tag{5.3}$$

when $M_2 = 1$. HINT: Express c_2 in terms of p_2 and ρ_2 and equate to u_2 to show that

$$u_2 = \frac{\gamma p_1}{\rho_1}\frac{1}{u_1} + \gamma(u_1 - u_2) \tag{5.3a}$$

Exercise 15 Demonstrate that when heat is added in the amount

$$Q = \frac{c_p T_1}{2(\gamma+1)}\frac{(1-M_1^2)^2}{M_1^2} \tag{5.4}$$

an initially subsonic flow is brought to sonic speed $M_2 = 1$ by an *increase* in velocity, and an initially supersonic flow is brought to the same speed by a *decrease* in velocity accompanying the same quantity of heat addition. HINT: Substitute (5.4) in (5.2) and compare the result with (5.3).

a remarkable property

The influence of heat addition on *continuous* one-dimensional flow is examined from the extended form of the equations expressing conservation principles in differential terms. Thus equations (3.1) are modified to allow for the incremental heat "pulse" dQ by writing, in place of (3.1),

$$d(\rho u) = 0$$
$$d(p + \rho u^2) = 0$$
$$d\left(h + \frac{u^2}{2}\right) = dQ$$

(5.5)

differential heating

When the enthalpy of an ideal gas is expressed in terms of the pressure and the density with the aid of the thermodynamic equation of state, the three equations (5.5) can be regarded as a system of linear algebraic equations for the differentials dp, $d\rho$, and du. In matrix notation they are

$$\begin{vmatrix} 0 & \dfrac{1}{\rho} & \dfrac{1}{u} \\ 1 & 0 & \rho u \\ \dfrac{\gamma}{\gamma - 1}\dfrac{1}{\rho} & -\dfrac{\gamma p}{(\gamma - 1)\rho^2} & u \end{vmatrix} \begin{vmatrix} dp \\ d\rho \\ du \end{vmatrix} = \begin{vmatrix} 0 \\ 0 \\ dQ \end{vmatrix} \quad (5.5a)$$

matrix form of the three conservation equations

Exercise 16 Use Cramer's rule or any other method to solve the equations (5.5a) for the velocity differential du. Show that the flow speed is increased in continuous subsonic flow when heat is added, but decreased in supersonic flow. From the form of the solution for du, show that the signs of both dp and $d\rho$ also change in passing from subsonic to supersonic flow. Explain the density variation caused by heat addition to a subsonic flow and compare with the case of no motion.

Exercise 17 Show that the last equation of (5.5a), in the absence of heat addition, $dQ = 0$, is equivalent to (2.3b). Verify that (2.3b) is recovered by multiplying each element of the second row in the 3 × 3 matrix of (5.5a) by $1/\rho$ and subtracting the new elements from the respective terms in the third row.

6 LINEARIZED SMALL-DISTURBANCE FLOW IN TWO DIMENSIONS

The principle of mass conservation in a compressible fluid, expressed in stream tube form as (II.2.1), is written in terms of Cartesian coordinates and velocity components by summing the mass flux terms for an infinitesimal element in the same manner that led to (II.2.4) and (II.2.4a) for incompressible fluid. In this case each of the velocity components is multiplied by the density, regarded as a variable quantity, and examination of Fig. II-4 then leads to the continuity equation for steady flow in the x, y plane as

continuity equation for compressible flow

$$\frac{\partial(\rho u)}{\partial x} + \frac{\partial(\rho v)}{\partial y} = 0 \tag{6.1}$$

The total mass of fluid within a small element is determined by its instantaneous density, and it is easily shown that (6.1) is extended to unsteady flow by the addition of the time derivative term shown in (1.7). Because the present discussion is confined to steady flow, the continuity equation (6.1) is supplemented by the Euler equations (II.4.3) which become

Euler equations unaffected by compressibility

$$\left. \begin{array}{l} u\dfrac{\partial u}{\partial x} + v\dfrac{\partial u}{\partial y} + \dfrac{1}{\rho}\dfrac{\partial p}{\partial x} = 0 \\[6pt] u\dfrac{\partial v}{\partial x} + v\dfrac{\partial v}{\partial y} + \dfrac{1}{\rho}\dfrac{\partial p}{\partial y} = 0 \end{array} \right\} \tag{6.2}$$

when gravity is neglected (cf. Exercise II.8).

The two velocity components, the pressure and the density now form four dependent variables in equations (6.1) and (6.2), three in number, while either (2.3a) or (2.5a) provides the fourth relationship

adiabatic energy equation

$$\frac{p}{\rho^\gamma} = \text{const} \tag{6.3}$$

making the mathematical problem determinate.

The adiabatic pressure-density relationship (6.3) may now be used to express the pressure derivatives in (6.2) in terms of the sonic speed

$$c = \sqrt{\frac{\gamma p}{\rho}} \tag{6.4}$$

and the density derivatives that are also present in (6.1). When subscripts are adopted as in (3.9) to identify a particular value of c, (2.6) shows how this quantity is determined by the velocity. If the subscript zero now refers to conditions in a region of uniform flow U, this equation becomes

$$\frac{c^2}{\gamma - 1} + \frac{q^2}{2} = \frac{c_0^2}{\gamma - 1} + \frac{U^2}{2} \qquad (6.5)$$

equivalent to (2.6)

showing that the sound speed at any point is determined entirely by the flow speed, and not by the pressure or density separately.

Exercise 18 Write the pressure derivative terms in (6.2) in the form

$$\frac{1}{\rho} c^2 \frac{\partial \rho}{\partial x}$$

and eliminate the density derivatives by combining the momentum and continuity equations in the form

$$\frac{\partial u}{\partial x}\left\{1 - \frac{u^2}{c^2}\right\} + \frac{\partial v}{\partial y}\left\{1 - \frac{v^2}{c^2}\right\} - \frac{uv}{c^2}\left(\frac{\partial u}{\partial y} + \frac{\partial v}{\partial x}\right) = 0 \qquad (6.6)$$

elimination of density: an intractable equation

containing only velocity components and their derivatives. In what sense can (II.2.4a) be regarded as a limiting case of (6.6)?

Equation (6.6) is not in a form suitable for solution purposes, both because of the appearance of the two quantities u and v in one equation and because of the nonlinear character of the terms containing products of velocities and their derivatives. An approximation removes the second difficulty when the flow is everywhere only very little different from uniform and constant, in the same manner as in Section V.6. In this case of considerable practical importance, (6.6) is greatly simplified if the velocity is expressed as the sum of a constant term plus a small disturbance the second and higher powers of which are negligible compared with the disturbance velocities (and their derivatives) themselves.

basis for approximation

When the principal flow is denoted U as before and

$$u = U + u' \qquad v = v' \qquad (6.7)$$

where the *perturbation velocities* u' and v' are very small compared with U, each of the factors outside the braces in (6.6) is small to the first order (derivatives of the perturbation velocities also being assumed small). Thus the retention of terms of "normal" order in the expressions in braces gives the approximate equation

$$\frac{\partial u'}{\partial x}(1 - M^2) + \frac{\partial v'}{\partial y} = 0 \qquad (6.8)$$

linearized equation for the velocities

where M^2 denotes the square of the velocity ratio U divided by c_0 [(6.5) shows $c_0 = c$ plus small terms only, when q^2 is the sum of the squares of velocity components (6.7)].

The barotropic relationship (6.3) meets the requirements of Kelvin's theorem in Section IV.2, the proof of which is now completed by substituting from (II.5.7) in (IV.2.2) and proceeding as before to establish the constancy of circulation as (IV.2.3). Because the argument for the absence of fluid rotation that was used for incompressible flow to introduce the velocity potential is equally valid in this case, the velocity perturbations are again given as in (V.6.6) by

<small>introduction of a "perturbation potential" for velocities</small>

$$u' = -\frac{\partial \phi}{\partial x} \qquad v' = -\frac{\partial \phi}{\partial y} \tag{6.9}$$

and (6.8) becomes finally

<small>linearized potential equation</small>

$$\frac{\partial^2 \phi}{\partial x^2}(1 - M^2) + \frac{\partial^2 \phi}{\partial y^2} = 0 \tag{6.10}$$

which is known as the linearized potential equation for plane compressible flow.

The constant M is the free-stream Mach number, and (6.10) shows that, when this quantity is less than unity, *subsonic* flow ($M < 1$) is described by an equation having two terms of the same sign and therefore being of the same form as Laplace's equation (IV.5.6). *Supersonic* flow ($M > 1$) contrasts with this by having two terms of opposite signs, corresponding to the equation of wave propagation in one dimension (1.6a).

Exercise 19 A scale transformation of the independent variable x given by

$$x' = \alpha x \tag{6.11}$$

with unspecified transformation parameter α is introduced in (6.10) to recover the classical equations in two distinct cases. Determine the value of the "stretching" factor α for subsonic flows such that $\phi(x', y)$ is a solution of Laplace's equation. Also determine α for supersonic flow, interpreting y in (6.10) as t in (1.6a), and express the phase velocity c in terms of $M^2 - 1$.

7 SUPERSONIC FLOW PAST WAVY WALL

When the uniform flow speed U is supersonic and $M^2 - 1$ is denoted m^2 for conciseness, the perturbation potential is determined as a solution of (6.10) in the form

<small>equivalent to (6.10)</small>

$$\frac{\partial^2 \phi}{\partial x^2} - \frac{1}{m^2}\frac{\partial^2 \phi}{\partial y^2} = 0 \tag{7.1}$$

which is equivalent to the one-dimensional wave equation. It follows as in the incompressible case considered in Section (V.6) that the pressure is found from ϕ as

$$p(x, y) - p_0 = \rho_0 U \frac{\partial \phi}{\partial x} \tag{7.2}$$

in linearized approximation. By considering again the wavy wall defined by

$$y = h \sin \lambda x \qquad h \ll \frac{2\pi}{\lambda} \tag{7.3}$$

the distinctive features of supersonic flow are contrasted with the typical properties of incompressible fluid motion determined earlier.

The condition of tangential flow based on (6.7) and (6.9) and the boundary (7.3) is

$$-\frac{\partial \phi}{\partial y}(x, 0) = Uh\lambda \cos \lambda x \tag{7.4}$$

boundary condition, consistent approximation

exactly as in (V.6.12), so that (7.4) serves as a boundary condition that supplements (7.1). Recalling the form of the general solution of the wave equation as given by (VII.2.6) ϕ is taken as a function of $x - my$

$$\phi(x, y) = F(x - my) \tag{7.5}$$

assumed solution in general form

and the form of the function is sought on the basis of (7.4). It follows from (7.4) that

$$mF'(x) = Uh\lambda \cos \lambda x \tag{7.6}$$

where the prime denotes differentiation and the argument $x - my$ reduces to x when the condition (7.4) is applied at the axis of the wavy wall as before. Equation (7.6) is readily integrated as

$$F(x) = \frac{Uh}{m} \sin \lambda x$$

while (7.5) indicates that the potential is obtained by replacing x by $x - my$:

$$\phi(x, y) = \frac{Uh}{(M^2 - 1)^{1/2}} \sin \{\lambda(x - my)\} \tag{7.6a}$$

satisfies all conditions

The potential is constant along each line, $x - my = $ const, and hence the velocities and pressure are constant also. The equation of this family of lines is

$$y = \frac{x}{(M^2 - 1)^{1/2}} + \text{const} \tag{7.7}$$

and all evidently have the same positive slope

$$\frac{1}{(M^2 - 1)^{1/2}}$$

measured from the x-axis. The interpretation of the solution (7.6a) is therefore that the flow disturbance at any point x on the wall (7.3) near the y-axis "propagates" in the downstream direction so that the same disturbance is found for points distant from the wall ($y > 0$) at x-values greater than the coordinate of the point of the wall from which the disturbance emanates as (7.7). The sharp wave fronts of an unstationary plane compressive wave (1.6a) are thus also found in steady two-dimensional motion. Comparison with (VII.2.6) shows that in the present case the y-axis plays the same role as did the time variable t in the one-dimensional unsteady wave propagation process considered in Chapter VII. In the same way that the waves formed at the bow of a ship spread downstream only, the choice of solution (7.5) with the minus sign as shown implies that flow (compressibility) disturbances are carried downstream only. This means that solutions of the form $G(x + my)$ which also satisfy the equation (7.1) are excluded [cf. (VII.2.6)] on physical grounds. The supplementary boundary condition

$$\phi(0, y) = 0 \qquad (7.8)$$

provides the appropriate mathematical restriction on solutions of (7.1) if, for example, the wavy wall corresponds only to $x > 0$ so that $x < 0$ is the upstream region in which the flow must be uniform, $u = U, v = 0$.

margin note: flow disturbance affects only part of complete flow field

margin note: justification of assumed solution form

Exercise 20 Calculate the force components F_y and F_x defined by (V.6.20) for the supersonic flow (7.6a) past one complete cycle of wall waviness. Show that the supersonic *wave drag*, reduced to coefficient form by dividing by $(\rho_0 U^2/2)L$ can be written as

$$c_x = \frac{(\lambda h)^2}{(M^2 - 1)^{1/2}} \qquad (7.9)$$

margin note: wave drag occurs in supersonic flow

margin note: drag is small for slender-boundary configurations

The drag force (7.9), which has no counterpart in incompressible or subsonic flow, represents an "in phase" relationship between the pressure and the boundary wall slope. Both are proportional to the $\cos \lambda x$, in supersonic flow, whereas in subsonic flow the pressure dependence $\sin \lambda x$ gives rise to forces that cancel each other when a complete wave cycle is considered. Two features of (7.9) are typical of linearized supersonic flow calculations. The Mach number dependence, which indicates large forces at speeds only slightly in excess of the sonic value (the "sonic barrier") also shows that at high supersonic speeds the force varies nearly inversely as the Mach number. The appearance of the wave height h to the second power is also characteristic of supersonic airplane wing

margin note: Mach number dependence—the sound "barrier"

drag and indicates the importance of thin wing profiles to "cut" through the air with the least disturbance to the surrounding fluid. It is of interest to note that, in incompressible fluid also, the drag force on each separate portion of one wave cycle is proportional to the square of the lateral dimension h, although these components cancel for each complete wave cycle.

contrast with incompressible flow

Exercise 21 Define $F_{y/4}$ and $F_{x/4}$ as the forces on the quarter of the wave cycle extending from $x = 0$ to $x = \pi/2\lambda$ and calculate each one for the incompressible flow (V.6.16). Indicate on a sketch of the quarter-wave cycle the magnitude and direction of each component, recalling the assumption in (7.3).

Exercise 22 Repeat the preceding calculation for the case of supersonic flow, showing that the force is now reversed in direction. Demonstrate that the larger (transverse) force component cancels when the integration is extended over a complete cycle, but the drag force contribution of each separate quarter-wave is equal to the others, recovering (7.9).

		$M = 0$	$M > 1$
Ans.	$F_{y/4}$	$\rho U^2 h$	$\dfrac{\rho_0 U^2 h}{(M^2 - 1)^{1/2}}$
	$F_{x/4}$	$-\dfrac{\rho U^2 h^2 \lambda}{2}$	$+\dfrac{\rho_0 U^2 h^2 \lambda}{(M^2 - 1)^{1/2}} \dfrac{\pi}{4}$

8 STARS HEATED BY RADIATIVE ENERGY LOSS

Conditions within a gaseous star are studied by considering the balance (I.6.8) of pressure and gravitational forces, the very wide range of density variations preventing the use of the simplifying assumption that was useful in the case of planets idealized as liquids. Extreme values of the temperature inside a star are accompanied by strong convective actions, and they justify the adoption of a barotropic relationship that permits the pressure to be expressed in terms of density alone. It will be shown in this section that, even when dynamic effects are completely neglected and the equation of stellar statics cannot be solved analytically, it is still possible to infer certain features of star structure and evolution directly from the equation (i.e. without attempting to integrate it).

The fundamental equation (I.6.8)

$$\frac{1}{r^2}\frac{d}{dr}\left(\frac{r^2}{\rho}\frac{dp}{dr}\right) + 4\pi k^2 \rho = 0 \tag{8.1}$$

fundamental equation of stellar statics

containing both pressure and density regarded as functions of radial distance r is reduced by assuming a barotropic relationship containing two constants, in the form

barotropic assumption

$$p = C\rho^n \qquad (8.2)$$

Equation (8.1) then becomes

a difficult equation

$$\frac{Cn}{r^2}\frac{d}{dr}\left(r^2\rho^{n-2}\frac{d\rho}{dr}\right) + 4\pi k^2\rho = 0 \qquad (8.3)$$

which is an ordinary differential equation for the density variation with radial distance $\rho(r)$. Its solution is hindered by the nonlinearity in the first term, which raises the question whether there might be particular values of the constants C and n for which the equation can be solved in closed form (i.e., without recourse to numerical methods).

Exercise 23 When $n = 2$, (8.3) is a linear equation; verify that a solution corresponding to finite central density $\rho(0)$ is

an analytic solution

$$\rho(r) = \frac{A \sin \lambda r}{r} \qquad (8.4)$$

and evaluate the constant A in terms of the density $\rho(0)$. One requirement of a physically meaningful solution is that the density derivative with respect to r should vanish at $r = 0$; is this condition met by (8.4)? The fact that barotropic relationships corresponding to $n = 2$ are not found in Nature reduces the importance of the solution (8.4) studied by Laplace.

Exercise 24 A particular solution of the nonlinear equation (8.3) corresponding to the value $n = 1.2$ has been given by Sir A. Schuster as

another, also not meaningful

$$\rho(r) = \rho_0\left\{1 + \frac{r^2}{a^2}\right\}^{-5/2} \qquad (8.5)$$

Substitute (8.5) in (8.3) in order to evaluate the constant C in terms of the central density ρ_0 and the constant a. An important inconvenience of (8.5) is that the polytropic index $n = 1.2$ is not physically meaningful, as the conditions within a star are very sensitive to the value chosen, and only values of $\frac{4}{3}$ or greater are considered realistic.

another basis of approximation

In the absence of solutions of (8.1) and even of its more idealized form (8.3), it is still possible to introduce the mass and energy conservation principles and draw certain conclusions from (8.1) which are grossly consistent with Newtonian mechanics. Instead of applying these laws strictly at each point of the fluid mass, however, as is done when solutions of the equations give each fluid property at every point, a more modest goal is adopted by requiring only the overall, or average, satisfaction of each of the principles applied to the entire gas sphere as a whole. Precision is sacrificed, in other words, for the sake of expediency when proper mathematical solutions are unavailable.

The entire gaseous sphere is therefore considered as a unit, and its gradual contraction (or expansion) is related to the energy processes that occur within it. A particle is located at distance r from the center at one instant of time and at distance αr at a later time, and the parameter α also relates the density and total mass at both instants in a manner now to be determined. The equation

$$r' = \alpha r \tag{8.6}$$

Homologous configurations — similarity transformation

is written and r' is regarded as the position at a standard reference time, so that it is a constant while position r at an arbitrary time is determined by the value of α regarded as a function of time. Logarithmic differentiation of (8.6) gives

$$0 = \frac{d\alpha}{\alpha} + \frac{dr}{r} \tag{8.7}$$

while the fact that the total mass remains constant indicates that the average density ρ is related to r so that $\rho r^3 = $ const. Differentiation and comparison with (8.7) then show that

$$\rho' = \alpha^{-3}\rho \quad \text{and} \quad 0 = -3\frac{d\alpha}{\alpha} + \frac{d\rho}{\rho} \tag{8.8}$$

implications for density ...

Exercise 25 Set

$$p' = \beta p \tag{8.9}$$

... and for pressure ...

and determine β in terms of α by substituting from (8.7), (8.8), and (8.9) for r, ρ, and p in (8.1). What is the physical justification of the requirement that (8.1) be satisfied when all its variables are replaced by the prime quantities?

Exercise 26 Assume an ideal gas and use (8.8) and (8.9) to show that temperature variations are given by

$$\frac{dT}{T} = \frac{d\alpha}{\alpha} \tag{8.10}$$

... and for temperature

so that contraction ($dr < 0$) corresponds to increased temperature $dT > 0$.

In order to determine whether a star becomes cooler or hotter when it loses energy by radiation to surrounding space, (8.10) shows that it is necessary to know only whether the process corresponds to expansion or contraction of the entire mass. Since the balance

(8.1) of pressure and gravity forces has interrelated the temperature and density increments appearing in (2.5), the sum of these terms being negative when energy is removed $dQ < 0$, the first law of thermodynamics permits the determination in a given case of whether or not a contraction occurs. Substitutions in (2.5) now give

$$dQ = \frac{d\alpha}{\alpha} \{c_v T - 3RT\}$$
$$= \frac{d\alpha}{\alpha} c_v T (4 - 3\gamma) \qquad (8.11)$$

energy consequence of assumed similarity

and show that the sign of $d\alpha$ depends entirely on the value of the ratio γ of specific heats. It will be recalled that γ in (2.5a) is a particular value of the polytropic index n of (8.2); therefore when

$$\gamma > \tfrac{4}{3} \qquad (8.12)$$

loss of heat, $dQ < 0$, implies $d\alpha > 0$ and, according to (8.7), a contraction $dr < 0$ accompanied by a temperature rise $dT > 0$. This gives the curious result that the more a star radiates heat from its surface, the hotter it becomes. The explanation is, of course, simply that contraction of the star liberates enough gravitational energy to account for both the radiated energy *and* the heating of the star. Enormous amounts of energy are released by stellar radiation [represented by the integral of (8.11) over the entire mass of each star], and the continual rise of entropy which this radiation entails is taken to indicate the inevitability of the "heat death" of the universe after a sufficient length of time.

It is interesting to compare the value shown on the right side of (8.12) with the value $\tfrac{5}{3}$ of the ratio of specific heats for an ideal monatomic gas. The high temperatures within a star assure that polyatomic molecules are completely dissociated, and the question whether the *homologous configurations* represented by the complete range of values of α from zero to infinity correspond to expansion or contraction must be decided by more delicate arguments than have been considered here.

delicate criterion

9 ACOUSTIC DISPERSION: GRAVITATIONAL COLLAPSE AND THE BIRTH OF GALAXIES

If the ultimate heat death of the universe is announced by the second law of thermodynamics, there is at least greater uncertainty about explaining the formation of these vast concentrations of heat and energy in the starry entropy machines as they are presently known. According to one view the "zero" state of our universe

VIII. 9 / GRAVITATIONAL COLLAPSE AND THE BIRTH OF GALAXIES

was an extremely tenuous cosmic dust cloud of hydrogen atoms in which the weight of matter contained in a volume the size of the Sun was no greater than that of a teaspoon of water. One possible explanation of how the stars could have formed from an initially homogeneous primeval gas employs an extension of the mechanics of wave propagation in the acoustic limit examined in Section 1. A key element in establishing the instability of very large fluid masses is shown to be the *dispersive* behavior of acoustic waves when gravitational effects are included and disturbances of large scale are considered.

evolution of tenuous primeval dust cloud

The gravity force in a large spherical fluid mass is taken again as in static equilibrium with pressure forces as analyzed in Chapter I. Thus, from (I.7.1) and (I.6.7),

$$\frac{\nabla p}{\rho} = -\nabla U \qquad (9.1)$$

pressure-gravity balance

and

$$\frac{\nabla p}{\rho} = -\frac{4\pi k^2}{r^2} \int_0^r \rho r^2 \, dr \, \mathbf{e}_r \qquad (9.2)$$

Eliminating the pressure between (9.1) and (9.2) in the manner that led to (I.6a.8) then shows that the gravitational potential U at an internal point of a large fluid mass satisfies Poisson's equation:

elimination of pressure

$$\frac{1}{r^2}\frac{\partial}{\partial r}\left(r^2 \frac{\partial U}{\partial r}\right) = 4\pi k^2 \rho \qquad (9.3)$$

The left side of (9.3) is the already familiar "Laplacian" operator in a spherically symmetric configuration [cf. (VI.1.5)]. By taking the case of plane disturbances as in Section 1 for greater simplicity, the corresponding form of (9.3) is seen by comparing with (VI.1.5a) to be

gravitational potential determined by local conditions

$$\frac{\partial^2 U}{\partial x^2} = 4\pi k^2 \rho \qquad (9.3a)$$

The equations of continuity and momentum, including the gravity force $-\partial U/\partial x$ are now written by extending (1.7) as

$$\left.\begin{array}{l} \dfrac{\partial \rho}{\partial t} + \rho \dfrac{\partial u}{\partial x} = 0 \\[6pt] \dfrac{\partial u}{\partial t} + c^2 \dfrac{1}{\rho}\dfrac{\partial \rho}{\partial x} = -\dfrac{\partial U}{\partial x} \end{array}\right\} \qquad (9.4)$$

Exercise 27 Defining the "condensation" s as the infinitesimal fractional density increment, i.e., such that

$$\rho = \rho_0(1 + s) \qquad (9.5)$$

where ρ_0 is a constant density corresponding to the undisturbed state $u = 0$, show that, when gravity is neglected, the condensation satisfies the same equation as the velocity, viz., (1.6a). When gravity is included, however, show that

disturbances exhibit wave-like behavior

$$\frac{\partial^2 s}{\partial t^2} - c^2 \frac{\partial^2 s}{\partial x^2} - 4\pi k^2 \rho_0 s = 4\pi k^2 \rho_0 \quad (9.6)$$

When an acoustic disturbance is represented as a sinusoidal density variation of wavelength $L = 2\pi/\lambda$ at time $t = 0$, such as

$$s = A \sin \lambda x \qquad t = 0 \quad (9.7)$$

and gravity is neglected, (9.6) reduces to the ordinary one-dimensional wave equation, and the advancing wave corresponding to (9.7) is found as

$$s(x, t) = A \sin (\lambda x - \sigma t) \quad (9.8)$$

where σ is the frequency as previously employed. Substitution in the simplied form of (9.6) then shows that

$$-\sigma^2 + c^2 \lambda^2 = 0 \quad (9.9)$$

and the fact that the phase velocity c is determined by the compressibility of the fluid alone, i.e., as

nondispersive when gravity is neglected: disturbances are propagated away

$$\frac{\sigma}{\lambda} = c = \sqrt{\frac{\partial p}{\partial \rho}}$$

indicates the ordinary nondispersive character of sound waves (contrast with deep water waves; Section VII.4).

In the more general case represented by (9.6), when the gravity terms are *not* ignored, the wave propagation characteristics are determined by the terms on the left side of the equation, i.e., by the homogeneous equation obtained by discarding the constant on the right side of the equation. A wave of the form (9.8) now gives the relationship between frequency and wavelength as

$$-\sigma^2 + c^2 \lambda^2 = +4\pi k^2 \rho_0 \quad (9.9a)$$

in place of (9.9), while the phase velocity is given by

$$\left(\frac{\sigma}{\lambda}\right)^2 = c^2 - \frac{4\pi k^2 \rho_0}{\lambda^2} \quad (9.10)$$

gravity is essential for large-scale disturbances

in a form that depends on the wave length of the disturbance (9.7). The importance of this feature is seen directly from (9.10): the phase velocity σ/λ is no longer independent of the wave length but is diminished for long waves (λ small). Because, for sufficiently large disturbances, the phase velocity becomes arbitrarily small, disturbances are not transmitted as propagated waves throughout an unlimited region but remain in the region where they are generated. Then the energy associated with the disturbance also accumulates and increases steadily, in sharp contrast with conditions in adjacent regions. It is now a commonly accepted cosmogonic principle that disperse matter in the universe collected into the discrete concentrations recognized as the stars by a mechanism of the type just indicated.

dispersion related to gravitation: energy accumulates

S1 PRANDTL-GLAUERT COMPRESSIBILITY CORRECTION IN SUBSONIC FLOW

Effects of compressibility in subsonic flows are determined solely by M^2, the square of the Mach number, according to the linearized equation for the disturbance velocity potential (6.10). It follows that, even at flow speeds as great as a few hundred miles per hour in the lower atmosphere, the main features of a flow differ very little from the corresponding flow of an incompressible fluid. Because of the great practical importance of the fluid pressure, however, quantitatively accurate estimates of the compressibility pressure correction have been sought by a variety of means. The simplest and most useful approximation is obtained by an appropriate scale transformation of (6.10), to show that the pressure is increased, as compared with the value in incompressible flow, by the factor

$$\frac{1}{\sqrt{1-M^2}} \qquad (S1.1)$$

Prandtl-Glauert factor to estimate influence of compressibility

The result will be demonstrated for flow past a wavy wall, and the same argument will be seen to apply to an arbitrary boundary having small inclinations of its tangents relative to the distant flow. Wing profiles employed on modern military and commercial aircraft fit into this category, and were evolved with the aid of intensive theoretical studies as well as painstaking experimental test programs. The procedure here consists in modifying (6.10) in a manner that permits the compressible flow potential to be inferred directly from the corresponding incompressible case (V.6.16).

It is convenient to retain for the symbols U, x, y, λ, h, ϕ, etc., their meanings in the earlier discussion of incompressible flow, and to denote flow quantities in subsonic flow at Mach number M by $\tilde{x}$, $\tilde{y}'$, $\tilde{\phi}$, etc., so that (6.10) is replaced by

$$\frac{\partial^2 \tilde{\phi}}{\partial \tilde{x}^2}(1-M^2) + \frac{\partial^2 \tilde{\phi}}{\partial \tilde{y}'^2} = 0 \qquad (S1.2)$$

notation for comparison flow

and the equation of the boundary surface is
$$\tilde{y}' = \tilde{h} \sin \tilde{\lambda}\tilde{x} \tag{S1.3}$$

If, for conciseness, we put
$$m^2 \equiv 1 - M^2$$

where m is positive, a stretching of the $\tilde{y}'$ coordinate given by
$$m\tilde{y}' = \tilde{y}$$

reduces (S1.2) to the form of Laplace's equation

scale (similarity) transformation

$$\frac{\partial^2 \tilde{\phi}}{\partial \tilde{x}^2} + \frac{\partial^2 \tilde{\phi}}{\partial \tilde{y}^2} = 0 \tag{S1.2a}$$

and the boundary surface to

reduction to familiar case

$$\tilde{y} = m\tilde{h} \sin \tilde{\lambda}\,\tilde{x} \tag{S1.3a}$$

Exercise 28 Note that repetition of the argument of Section V.6 leads to a solution of (S1.2a) in the form
$$\tilde{\phi}(\tilde{x}, \tilde{y}) = \tilde{A} \cos \tilde{\lambda}\tilde{x}\, e^{-\lambda \tilde{y}}$$

and evaluation of integration constant A by means of the tangential flow boundary condition obtained from (S1.3). Thus determine A and establish that

potential of subsonic flow

$$\tilde{\phi}(\tilde{x}, \tilde{y}') = \frac{\tilde{h}U}{m} \cos \tilde{\lambda}\tilde{x} \exp(-\lambda \tilde{y}'/m) \tag{S1.4}$$

When the same scale is employed for measuring distances in the mean-flow direction for compressible and for incompressible flows, $\tilde{x} = x$, and then equivalent wall wavelengths in the two cases is assured by setting $\tilde{\lambda} = \lambda$. If the free stream pressure is taken as $\tilde{p}_0 = 0$ for convenience in the compressible flow and also in the incompressible flow (V.6.17a), the pressure $\tilde{p}(x, 0)$ at a boundary point in the compressible flow is found as

pressure in subsonic flow

$$\tilde{p}(x, 0) = -\frac{\tilde{h}\rho\lambda U^2}{m} \sin \lambda x \tag{S1.5}$$

Two types of comparison with incompressible flow can now be made:

(a) If it is desired to determine the form of the wall waviness $\tilde{h}$ such that the compressible flow pressure (S1.5) shall be equal to that of the incompressible flow (V.6.17a) at corresponding points x, the wall waviness must be given by

two types of similarity ...

... flow pressures are matched, boundary wall modified ...

$$\frac{\tilde{h}}{m} = h \quad \text{or} \quad \tilde{h} = hm < h$$

and the boundary that disturbs the uniform flow must be *more slender* than in the incompressible comparison flow.

(b) For a single boundary waviness $\bar{h} = h$ the magnitude of the pressure disturbance $\bar{p}(x, 0)$ in compressible flow is *increased* as compared with its incompressible flow counterpart $p(x, 0)$:

$$\frac{\bar{p}(x, 0)}{p(x, 0)} = \frac{1}{\sqrt{1 - M^2}} \qquad (S1.6)$$

...given the same boundary, *higher* pressures occur in compressible flow

The result expressed by (S1.6) is called the Prandtl-Glauert correction for compressibility in subsonic flow. The intensification of pressures in this case may be compared with the effect of a contiguous plane boundary, as in Exercise V.23. In compressible subsonic flows the more general result (S1.1) is inferred from the *local* features of the flow in the vicinity of the point where the pressure is calculated.

*S2 APPLICATION OF THE HODOGRAPH EQUATIONS — CHAPLYGIN'S APPROXIMATION

When it is specifically required to predict compressibility effects starting with a knowledge of the incompressible flow pattern in a given case, such as might be determined with the aid of primitive experimental facilities, the hodograph form of the equations provides a more fundamental approach than the preceding calculations. Extension and refinement of the Prandtl-Glauert correction formula (S1.6) are now obtained *without* the restriction of small disturbances from uniform basic flow, the chief approximation being thermodynamic in character. The essential feature of the method in question is that Laplace's equation is recovered again, notwithstanding the greater generality of the streamline patterns.

linearization *without* restriction to slightly disturbed uniform flow

Streamline coordinates are again employed as a starting point, and the continuity equation for a compressible fluid in steady motion, given by (II.8.1) as

$$\frac{\partial}{\partial s} (\rho q \, \Delta n) = 0 \qquad (S2.1)$$

introduction of streamline coordinates and hodograph variables

is now considered without supposing the density to be constant.

Exercise 29 Neglect gravity and let s again denote distance measured along a flow streamline, so that (II.3.10) can be written

$$\rho q \frac{\partial q}{\partial s} + c^2 \frac{\partial \rho}{\partial s} = 0$$

and show that (II.8.2) is extended to the case of compressible flow as

$$\left(1 - \frac{q^2}{c^2}\right)\frac{\partial q}{\partial s} + q\frac{\partial \theta}{\partial n} = 0 \tag{S2.2}$$

where the term in parentheses contains the square of the Mach number, henceforth written in the more concise form as M^2, as before.

The condition of irrotationality in streamline form, obtained as (IV.8.2) requires no modification in the case of compressible flow; therefore (S2.2) and the equation

$$\frac{\partial q}{\partial n} - q\frac{\partial \theta}{\partial s} = 0 \tag{S2.3}$$

form a pair of simultaneous equations for the determination of the flow speed q and inclination θ.

Exercise 30 Extend the finding of Exercise IV.17 to evaluate the derivatives appearing in (S2.2), in order to show that

$$\frac{\partial q}{\partial s}:\frac{\partial n}{\partial \theta} = \frac{\partial \theta}{\partial n}:\frac{\partial s}{\partial q} = \frac{\partial q}{\partial n}:-\frac{\partial s}{\partial \theta} = \frac{\partial \theta}{\partial s}:-\frac{\partial n}{\partial q} \tag{S2.4}$$

where the independent variables on the left side of each proportionality symbol ":" are s and n, and those on the right are the hodograph variables q and θ.

By using (S2.4) the hodograph form of the continuity and irrotationality conditions, (S2.2) and (S2.3), is obtained as

$$\frac{1-M^2}{\rho q}\frac{\partial \psi}{\partial \theta} + \frac{\partial \phi}{\partial q} = 0 \tag{S2.5a}$$

linear equations in hodograph-coordinate independent variables

$$\frac{\rho}{q}\frac{\partial \phi}{\partial \theta} - \frac{\partial \psi}{\partial q} = 0 \tag{S2.5b}$$

when n and s are replaced by the stream function and velocity potential, respectively, according to the relationships

$$-\frac{\partial \psi}{\partial n} = \rho q \qquad -\frac{\partial \phi}{\partial s} = q \tag{S2.6}$$

Exercise 31 Justify the first of (S2.6) by considering the forms of the continuity equation (6.1) and (II.8.2), and also by evaluation of fluid *mass* flux in the manner of (V.1.2).

A single equation for either ϕ or ψ alone is obtained by selective differentiations of the pair of equations, (S2.5a,b). When it is observed that the coefficients are functions of flow speed q but not dependent on flow inclination θ, the elimination of ϕ appears to afford an advantage of simplicity because the resulting equation can be written as the sum of two terms equated to zero:

$$\frac{1-M^2}{\rho^2}\frac{\partial^2\psi}{\partial\theta^2} + \frac{q}{\rho}\frac{\partial}{\partial q}\left(\frac{q}{\rho}\frac{\partial\psi}{\partial q}\right) = 0 \qquad (S2.7)$$

equation satisfied by the stream function

A transformation of the speed variable in the manner of Exercise V.15 permits the second term on the left side of (S2.7) to be written as a second derivative with coefficient unity. If, in addition,

$$\frac{1-M^2}{\rho^2} = \text{const} \qquad (S2.8)$$

then a scale transformation of θ, analogous to that used to obtain (S1.2a), would reduce (S2.7) to the simplest form (IV.5.6) of Laplace's equation. The fact that compressible flow solutions might then be obtained by well-known classical means prompted the Russian mathematician Chaplygin to study a fictitious gas for which the condition (S2.8) is true exactly. The manner in which (S2.8) permits compressibility effects to be estimated without restriction to small perturbations of uniform flow to extend the validity of the Prandtl-Glauert correction term of (S1.6) will now be indicated. Physical justification for the adoption of (S2.8) will be left as exercises for the reader.

fictitious gas with polytropic index $\gamma = -1$

When (S2.8) is used in (S2.7), the latter equation can be written as

$$\frac{\partial^2\psi}{\partial\theta^2} + \left(\frac{q}{\sqrt{1-M^2}}\frac{\partial}{\partial q}\right)^2\psi = 0 \qquad (S2.9)$$

reduction of (S2.7)

where the notation implies that the differentiation operation

$$\frac{q}{\sqrt{1-M^2}}\frac{\partial}{\partial q}$$

involving multiplication after differentiation with respect to q, is repeated so that the left side of (S2.9) is of the Laplace form with two second-derivative terms. The limit of incompressible flow, $M = 0$, may now be used to provide an estimate of compressibility effects, $M \neq 0$, by establishing a correspondence between the two flows in the manner that has led to (S1.6) for a more restricted class of flows. If the speed of an incompressible flow is now denoted by w and q is retained as the speed of the comparison flow of the same

200 COMPRESSIBILITY

form with compressibility taken into account, the relationship between w and q is, from (S2.9),

$$\frac{q}{\sqrt{1-M^2}}\frac{\partial}{\partial q} = \frac{w}{1}\frac{\partial}{\partial w}$$

or

a useful transformation

$$\frac{dw}{w} = \sqrt{1-M^2}\,\frac{dq}{q} \tag{S2.10}$$

When (S2.10) is regarded as a first-order differential equation, and it is seen that M^2 depends only on q, the variables w and q are separated. It follows that for each (incompressible flow) speed w, a value of q is determined that satisfies the conditions of a steady, irrotational, *compressible* flow having at each point the same direction as the incompressible comparison flow. For the sake of simplicity, only the case of small compressibility effects in subsonic flow, are considered; that is, i.e.,

$$M^2 \ll 1$$

so that fourth and higher powers of the Mach number can be neglected compared with M^2. Within this order of approximation, (2.6) written in the form

$$\frac{c^2}{\gamma-1} + \frac{q^2}{2} = \frac{c_0^2}{\gamma-1}$$

shows that (S2.10) can be expressed as

establishment of comparison

$$\frac{dw}{w} = \sqrt{1-M_0^2}\,\frac{dq}{q} \tag{S2.10a}$$

where M_0 is formed by using the stagnation value of the speed of sound:

$$M_0 \equiv \frac{q}{c_0}$$

Under the present assumption, (S2.10a) can be integrated to give

the velocities related...

$$\left(\frac{q}{w}\right)^2 = \frac{1}{\sqrt{1-M_0^2}} \tag{S2.11}$$

after some tedious but elementary reductions. The variation of density with Mach number in adiabatic flow is similarly found from (2.5a) and (2.6) as

...and the densities...

$$\frac{\rho}{\rho_0} = \left(1 - \frac{\gamma-1}{2}M^2\right)^{1/(\gamma-1)} \doteq \sqrt{1-M_0^2} \tag{S2.12}$$

where the final expression is based on the present approximation.

The pressure variation in compressible flow can now be written with the aid of the preceding relationships as

$$d\bar{p} = -\bar{\rho}q\,dq \doteq -\rho_0\sqrt{1-M_0^2}\left(\frac{q}{w}\right)^2 \frac{w\,dw}{\sqrt{1-M_0^2}} = dp\,\frac{1}{\sqrt{1-M_0^2}}$$

where dp is the pressure variation in incompressible flow, given by $= -\rho_0 w\,dw$. The ratio of pressure variations in the two flows is therefore

$$\frac{d\bar{p}}{dp} = \frac{1}{\sqrt{1-M^2}} \tag{S2.13}$$

... and the pressures

to the present order of approximation, in agreement with (S1.6). As mentioned earlier, the present development is not restricted to slightly disturbed parallel flows, an assumption used in deriving the Prandtl-Glauert pressure correction (S1.6). It should also be noted that, when the Mach number is not limited to small values, extension of the present calculation, due to Kármán and Tsien, shows that a slightly greater correction than (S2.13) occurs in regions of under-pressure $dp < 0$.

*Exercise 32 (a) Obtain the equation satisfied by the velocity potential by elimination of ψ from (S2.5). (b) Show that the assumption (S2.8) permits simplification of the preceding equation, and that in this case ϕ satisfies the same equation (S2.7) that determines ψ.

*Exercise 33 Prove that

$$\frac{1-M^2}{\rho^2} = \frac{1}{\rho_0^2}\left(1 - \frac{\gamma+1}{2}\frac{q^2}{c_0^2}\right)\left(1 - \frac{\gamma-1}{2}\frac{q^2}{c_0^2}\right)^{-(\gamma+1)/(\gamma-1)} \tag{S2.14}$$

and hence that the condition (S2.8) is satisfied provided $\gamma = -1$, the value of the constant then being ρ_0^{-2}.

The numerical value of the polytropic index γ in an actual gas is found in the range $1 < \gamma < 1.67$ that does not contain the value -1, so that the physical significance of (S2.8) requires explanation. It should first be pointed out that the more accurate isentropic relationship (2.5a) is considered valid over a wide range of values of the variables p and ρ, whereas a much narrower range of values is ordinarily encountered in a given flow problem. Hence a different function may serve as a useful physical representation provided that it closely approximates (2.5a) in the neighborhood of the pressure and density values relevant to a given situation. That this is in fact the case here provides the justification of Chaplygin's imaginative argument.

the fiction $\gamma = -1$ has a physical justification: close approximation of actual conditions within a limited range of variables

The question is to show how a given function $F(\rho)$, such as (2.5a),

$$p = F(\rho) = p_0\left(\frac{\rho}{\rho_0}\right)^\gamma \tag{S2.14a}$$

can be approximated by another function $f(\rho)$ of the form

Can suitable constants be chosen?

$$f(\rho) = A + B\rho^m \tag{S2.14b}$$

for example, where m is prescribed but the constants A and B are to be so adjusted as to assure a "good" approximation in a sense dictated by the physical problem. For present purposes two criteria are adopted: (a) that f shall assume the same value as F at a given value of the independent variable denoted ρ_0; and (b) that the first derivative of the approximating function shall equal that of the given function at the same point. In analytical form, then,

$$f(\rho_0) = F(\rho_0) \tag{S2.15a}$$

conditions to be satisfied

$$\left.\frac{df}{d\rho}\right|_{\rho_0} = \left.\frac{dF}{d\rho}\right|_{\rho_0} \tag{S2.15b}$$

Exercise 34 Determine the values of A and B in (S2.14b) according to the conditions (S2.15a,b).

$$\text{Ans.} \quad A = F(\rho_0) - \frac{\rho_0}{m}\left.\frac{dF}{d\rho}\right|_{\rho_0}$$

Exercise 35 Take F in the final form shown on the right side of (S2.14a), denote the derivative at ρ_0 by c_0^2 as before, and assign the value -1 to the constant m to obtain for the approximating function (S2.14b) the explicit form

adiabatic energy equation for Chaplygin's fictitious gas

$$p = p_0 + \rho_0 c_0^2 - \frac{\rho_0^2 c_0^2}{\rho} \tag{S2.14c}$$

representing the "tangent approximation" to the adiabatic curve in the plane of cartesian coordinates $1/\rho$, p, as shown in Fig. VIII-2. Also show that the right side of (S2.14c) represents the first two terms of the Taylor series expansion of (2.5a) in positive powers of $(1/\rho) - (1/\rho_0)$. See that Chaplygin's procedure is accurate for sufficiently small variations of density.

S3 ENTROPY JUMP ACROSS NORMAL SHOCK WAVE, NONIDEAL GAS

The fact that compression shock waves occur only in supersonic flow ($M_1 > 1$) is established on the basis of the second law of thermodynamics in the case of an ideal gas by the entropy expression (4.6). The same is true for nonideal gases, and it is of

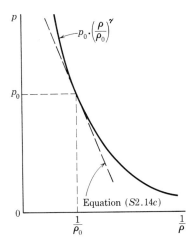

Fig. VIII-2 Illustrating Chaplygin's approximation.

interest to exhibit by means of an alternative entropy calculation the essential thermodynamic characteristic on which the argument turns.

When the restriction to ideal gas is removed, the enthalpy is expressed in terms of internal energy, pressure, and density by the defining relationship

$$h \equiv e + \frac{p}{\rho} \tag{S3.1}$$

while the first law of thermodynamics (2.5) assumes the form

$$T\,dS = dh - \frac{dp}{\rho} \tag{S3.2}$$

entropy related to enthalpy and pressure variations, arbitrary gas

and the integrated form of the energy equation in (3.2) is written as

$$h_2 + \frac{u_2^2}{2} = h_1 + \frac{u_1^2}{2} \tag{S3.3}$$

Exercise 36 Eliminate the velocities from (S3.3) and the first two of equations (3.2) in order to express the enthalpy change in the steady one-dimensional flow of an arbitrary gas in terms of the pressures and densities:

mass and momentum equations permit elimination of velocities

$$h_2 - h_1 = \left\{ \frac{1}{\rho_2} + \frac{1}{\rho_1} \right\} \frac{(p_2 - p_1)}{2} \tag{S3.4}$$

When the enthalpy is expressed in terms of pressure and density, (S3.4) is termed the dynamic pressure-density relationship for an arbitrary gas.

The entropy change from flow condition 1 to condition 2 is obtained by the integration of (S3.2) between corresponding limits.

Since the absolute temperature T is always positive, the positive or negative quality of the entropy variation is determined by the sign of the integral of the expressions on either side of this equation:

integration of (S3.2)

$$\int_{S_1}^{S_2} T\, dS = h_2 - h_1 - \int_{p_1}^{p_2} \frac{dp}{\rho} \tag{S3.5}$$

An approximation to the value of the integral on the right side of (S3.5) is evidently given by (S3.4). This expression is recognized as the area of a trapezoid of width $p_2 - p_1$; hence the linear approximation of variation of $1/\rho$ as a function of pressure. The sign of the entropy variation $S_2 - S_1$ is therefore to be found from a more detailed examination of the pressure-density relationship, obtained with the aid of the *trapezoidal rule of integration*:

a close estimate

$$\int_{x_1}^{x_2} f(x)\, dx = \frac{f(x_1) + f(x_2)}{2}(x_2 - x_1)$$
$$- \frac{f''(x_1)}{12}(x_2 - x_1)^3 + 0[(x_2 - x_1)^4] \tag{S3.6}$$

The meaning of (S3.6) is expressed as follows: The value of a definite integral over a small range $x_2 - x_1$ of the integration variable is given by the trapezoidal area term plus a term depending only on the *second* derivative of the integrand function, with an error that vanishes with the fourth power of the integration interval. Taking for f the reciprocal of the density $1/\rho$ and interpreting the independent variable x as the pressure p gives the entropy integral (S3.5) as

$$\int_{S_1}^{S_2} T\, dS = \frac{(p_2 - p_1)^3}{12} \frac{d^2}{dp^2}\left(\frac{1}{\rho}\right) \tag{S3.7}$$

argument turns on a fine-detail feature of Hugoniot's equation

the fourth and higher powers of the pressure difference $p_2 - p_1$ being neglected and the derivative in (S3.7) being evaluated at p_1. It follows that compression shocks $p_2 > p_1$ correspond to increased entropy $S_2 > S_1$ when the pressure-density derivative in (S3.7) is positive. For an arbitrary gas the relationship that determines this derivative is obtained by eliminating the velocities from the conservation equations, as was done by (3.5) in the case of an ideal gas.

In the particular case of an ideal gas and an initial state defined by a given pair of values p_1 and ρ_1, the final states p_2 and ρ_2 are given by (3.5), which may now be written as

$$\frac{p - p_1}{\rho - \rho_1} = \gamma \frac{p + p_1}{\rho + \rho_1} \tag{S3.8}$$

where the subscript 2 is suppressed to emphasize the consideration of a range of final states so that p and ρ are regarded as variables. The resemblance of (S3.8) to the adiabatic relationship (2.5a), now written as

$$\frac{p}{p_1} = \left(\frac{\rho}{\rho_1}\right)^\gamma \qquad (S3.9)$$

has already been noted. In a graphical representation of the two functions (S3.8) and (S3.9), both curves pass through the point p_1, ρ_1 with the same slope.

superficial resemblance

Exercise 37 (a) Solve (S3.9) to express $1/\rho$ as a function of the pressure p, and calculate the first, second, and third derivatives of $1/\rho$ with respect to the pressure. (b) Repeat for the pressure-density relationship (S3.8), which can be written as

$$\frac{\rho_1}{\rho} = \frac{(\gamma - 1)p + (\gamma + 1)p_1}{(\gamma + 1)p + (\gamma - 1)p_1}$$

in order to show that the two curves have not only the same slope but also the same curvature, the distinction between their forms appearing only in the third derivatives. (c) Substitute from part (b) in (S3.7) to evaluate the right side of that equation in the case of an ideal gas, and show that the result is identical with (4.6) by using (3.4).

osculating curves

***Exercise 38** Prove the trapezoidal rule of integration (S3.6) by writing $f(x)$ as one-half of the sum of the two Taylor series expansions about the points x_1 and x_2 and integrating between these two limits. HINT: Note that the difference of first derivative terms $f'(x_2) - f'(x_1)$ can be written as the second derivative $f''(x_1)$ plus terms of order $x_2 - x_1$.

Exercise 39 Show that the occurrence of compression shocks in supersonic flow is a general property of all fluids for which the dynamic pressure-density relationship [corresponding to (S3.4)] p vs. $1/\rho$ has a positive curvature in the plane defined by these variables as rectangular coordinates.

IX
Viscosity—Fluid Friction

1 NEWTON'S FRICTION LAW: PIPE FLOW RESISTANCE

Simple flow experiments performed by Sir Isaac Newton led him to two conclusions concerning fluid friction which are fundamental to all that is known of the mechanics of real fluids. The observation that fluid does not slide along a solid boundary surface, but rather adheres to it in all cases, implies an extension of the boundary condition of zero normal flow written as (II.6.1) or (II.6.3a). Because of the "no-slip" condition, the tangential flow component at a solid boundary vanishes, as well as the normal component. Practical calculations of frictional flow resistance therefore consider the relative flow velocity to be zero at every point of a solid boundary surface.

The second fact deduced by Newton relates to the force exerted by fluid and solid boundary surface on each other. For a pair of plane surfaces separated by a thin layer of fluid and in uniform relative motion parallel to each, Newton found that the force is directly proportional to the relative velocity U, inversely proportional to the separation distance D. The force on unit area of either plane surface then becomes

$$\tau = \mu \frac{U}{D} \qquad (1.1)$$

Newton's empirical findings:
(i) "no slip" at solid boundary
(ii) shear stress proportional to velocity shear

and is recognized as representing a *shear stress*. The factor μ appearing as proportionality factor is a property of the fluid, and it is independent of the flow geometry. According to Newton's third law of motion relating to action and reaction, the stress (1.1) is regarded either as acting on the fluid in the direction opposing motion, or on the boundary surface in the reverse direction.

If the distance of a fluid particle from one plane surface is

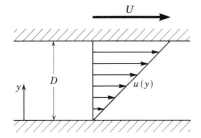

Fig. IX-1 Viscous flow between parallel walls in relative motion.

denoted y, the tangential velocity component $u(y)$ at that surface is zero, $u(0) = 0$. The no-slip condition at the second boundary is similarly expressed by writing $u(D) = U$, when it is considered that this surface is moving while the surface, or wall, at $y = 0$ is at rest (Fig. IX-1).

From the fact that only the *relative* motion of the two walls is of importance in determining the fluid motion in the space between them, it follows by symmetry that the velocity variation in the fluid is uniform:

$$u(y) = U \frac{y}{D} \qquad (1.2)$$

from many particular statements

Equation (1.2) is a statement of the velocity at all points located by y, and, since y takes all values in the interval $0 < y < D$, the equation represents a large number of separate statements, each referring to a distinct value of y. A single statement concerning the flow at *all* points of the interval is obtained by differentiating:

$$\frac{\partial u}{\partial y} = \frac{U}{D} \qquad (1.2a)$$

to a general conclusion

The right side of (1.2a) no longer contains the variable y, and the meaning of (1.2a) is therefore that the velocity *derivative* has the same value throughout the flow. When this is combined with (1.1), the fluid stress at the boundary is given by the flow shear $\partial u/\partial y$ as

$$\tau = \mu \frac{\partial u}{\partial y} \qquad (1.3)$$

and it is apparent that this is also the shear stress at any internal point of the flow, i.e., at any value of y where (1.2) is valid (consider the equilibrium of the fluid element extending from y to one of the boundary surfaces, for example). Equation (1.3) relates the stresses in a fluid to the fluid motions which are regarded as giving rise to them; the resemblance to the corresponding stress-strain relationships in solids is apparent. In the present instance it is not the strain (spatial variation of *displacement*), but the time rate of strain (derivative of *velocity*) that determines the stresses, although the formal similarity with the elasticity relationships of solid bodies remains an important and often useful property in the study of real fluids. The linear stress-strain relationship of the form (1.3) characterizes fluids termed Newtonian.

The factor μ in (1.3) is called the *coefficient of shear viscosity* and is evaluated experimentally for a large number of fluids under a wide range of conditions. Kinetic theory provides its interpretation in terms of momentum variations of individual molecules in shear flow and indicates the dependence of the viscosity coefficient on the molecular dimensions and speeds. Extension of (1.3) to two- or three-dimensional flows gives the complete representation of the effects of fluid friction in modifying the balance of inertia, pressure, and gravity forces as described by the Euler equations. The partial derivative notation of (1.3) is appropriate for showing how the more general relationships obtained below reduce in the case of the one-dimensional parallel flow of Fig. IX-1.

An example of a technically important fluid motion determined entirely by applying (1.3) in conjunction with the no-slip boundary condition is provided by viscous flow in a circular pipe, illustrated by Fig. IX-2. In the simplest case of steady incompressible flow the frictional resistance at the pipe wall is balanced by the pressure drop in the direction of flow. When the pressures p_1 and $p_2 < p_1$ are uniform across two sections of the pipe normal to its axis and separated by distance l, the cylindrical fluid volume of radius r experiences a net pressure force proportional to πr^2. The shear stress τ on the lateral surface of area $2\pi r l$ gives for the condition of dynamic equilibrium

$$(p_1 - p_2)\pi r^2 + \tau 2\pi r = 0 \tag{1.4}$$

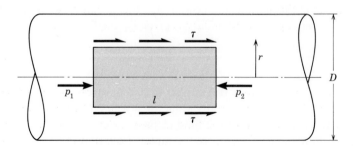

Fig. IX-2 Viscous flow in circular pipe.

where the shear stress in the flow, $u(r)$, assumed axially symmetric, is given by

$$\tau = \mu \frac{\partial u}{\partial r} \tag{1.3a}$$

the distance r being measured from the center of the pipe.

Introduction of (1.3a) into (1.4) leads to the first-order ordinary differential equation

$$\frac{p_1 - p_2}{2l\mu} r \, dr + du = 0 \tag{1.5}$$ differential equation

each of the terms of which is an exact differential. When the pipe diameter is denoted D, the no-slip condition is

$$u\left(\frac{D}{2}\right) = 0 \tag{1.6}$$ boundary condition

and it shows that the integration constant of (1.5) is most easily evaluated by integrating the equation between limits r and $D/2$. Thus the velocity "profile" $u(r)$ is parabolic and is given by the equation

$$u(r) = \frac{p_1 - p_2}{4l\mu}\left(\frac{D^2}{4} - r^2\right) \tag{1.7a}$$

or by writing

$$u_M \equiv \frac{p_1 - p_2}{4l\mu} \frac{D^2}{4} \tag{1.8}$$

for the maximum velocity which occurs on the pipe axis $r = 0$, $u_M \equiv u(0)$,

$$u(r) = u_M\left[1 - \left(\frac{r}{D/2}\right)^2\right] \tag{1.7b}$$ velocity "profile"

Exercise 1 (a) Denoting the average velocity over the circular pipe cross section by V, prove that $V = u_M/2$. (b) Defining the discharge Q as the total quantity of flow across any section in unit time so that

$$Q = V\pi\left(\frac{D}{2}\right)^2 \tag{1.9}$$ discharge Q

show that the discharge through a circular pipe is proportional to the fourth power of the pipe diameter (Poiseuille's law).

Exercise 2 (a) Use (1.4) to show that the frictional resistance to flow in unit length of pipe is given either as the fluid stress at the pipe wall,

$\tau(D/2)$, multiplied by the perimeter πD, or by the pressure gradient $(p_1 - p_2)/l$ multiplied by the pipe cross section area. (b) Define the dimensionless resistance coefficient (or friction factor) f by dividing the friction force per unit length by

$$\rho \frac{V^2}{2} \pi \frac{D^2}{4} \frac{1}{D} \qquad (1.10)$$

and show that the frictional resistance to steady flow in a circular pipe is given by

friction factor

$$f = \frac{64}{R} \qquad (1.11)$$

where the *Reynolds number* R is the dimensionless group defined as

Reynolds number: measure of the unimportance *of friction compared with inertia effects*

$$R = \frac{\rho V D}{\mu} \qquad (1.12)$$

Exercise 3 Poiseuille's flow (1.7) in a circular pipe is generalized by considering the motion in the annular region $D_1/2 < r < D_2/2$. The balance of forces (1.4) is no longer valid but can be correctly obtained by considering an elementary cylindrical shell volume of radius r and thickness dr. Show that (1.4) is replaced by

generalization of the preceding

$$(p_1 - p_2) 2\pi r\, dr + \frac{\partial}{\partial r} \{\tau 2\pi r l\}\, dr = 0 \qquad (1.13)$$

and deduce the form of the velocity profile as

$$u(r) = Ar^2 + B \ln r + C \qquad (1.14)$$

Evaluate the three constants in (1.14) by using (1.13) and the two conditions that now correspond to (1.6). Show that the flow reduces to (1.7a) when $D_1 \to 0$.

2 VISCOUS STRESSES: THE NAVIER-STOKES EQUATIONS

The stress (1.3) is tangential to flow in the x-direction, and, if axes are so chosen that there are both x- and y-components of velocity at any point, the stress must also be dependent on the x-variation of the y-component of velocity. The stress itself then has components in the respective coordinate directions, and these components can be distinguished by introducing subscripts. Thus (1.3), representing a stress in the x-direction on a surface parallel to the y-axis, can be written τ_{xy}, and the same notation convention means that τ_{yx} is the stress in the y-direction acting on a surface parallel to the x-axis. A convention with regard to signs is also adopted such that τ_{xy} is directed in the positive sense of the $+x$-axis on the surface element at the greater distance from the x-axis, as indicated in Fig. IX-3, when the shear $\partial u/\partial y$ is greater than zero. A relationship of

stress in two-dimensional flow

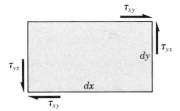

Fig. IX-3 Convention for denoting viscous stresses.

fundamental importance is obtained by considering the effect of the stresses τ_{xy} and τ_{yx} on the rotation of the fluid element on which they act.

Exercise 4 Stresses τ_{xy} and τ_{yx}, respectively, act on the sides BD and CD of a rectangular fluid volume of unit depth, where the lengths of the sides are given by $|BD| = l$ and $|CD| = h$. Evaluate the angular acceleration of the fluid volume by calculating both the resultant force moment about the fourth vertex A and the moment of inertia of the rectangle about the axis through A and normal to the rectangle.

the order of the subscripts is unimportant

$$\text{Ans: } 3\frac{\tau_{xy} - \tau_{xy}}{\rho(h^2 + l^2)}.$$

The result of Exercise 4 shows that reducing the lengths of both sides of the rectangle by a factor of *two* leads to an increase in the angular acceleration by a factor of *four*, if the stresses are unequal and thus there is a net force moment. This conclusion becomes physically absurd, however, in implying unlimited (i.e., infinite) acceleration of infinitesimal fluid elements as continual reduction of scale permits l to be replaced by dx, h by dy, which tend to zero. It therefore necessarily follows that $\tau_{yx} = \tau_{xy}$ and that (1.3) is generalized by writing

$$\tau_{yx} = \tau_{xy} = \mu\left\{\frac{\partial u}{\partial y} + \frac{\partial v}{\partial x}\right\} \quad (2.1)$$

shear stress, general expression

for *both* of the tangential stresses. [The formulae (2.1) are derived in the supplement S1 at the end of this chapter.] It can also be shown that normal viscous stresses, additive to the fluid pressure stress, are given by*

$$\tau_{xx} = 2\mu\left\{\frac{\partial u}{\partial x} - \frac{\theta}{3}\right\} \quad \tau_{yy} = 2\mu\left\{\frac{\partial v}{\partial y} - \frac{\theta}{3}\right\} \quad (2.2)$$

normal viscous stresses

where

$$\theta \equiv \frac{\partial u}{\partial x} + \frac{\partial v}{\partial y} + \frac{\partial w}{\partial z}$$

*The customary assumption is followed here, i.e., viscous actions do not affect fluid compression in general, a conclusion strictly supported only for monatomic gases. The implied neglect of *bulk viscosity* in other cases is justified solely by reason of fundamental uncertainties of fluid physics.

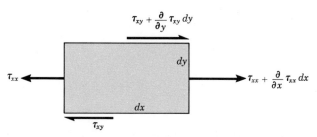

Fig. IX-4 Viscous stresses in x-direction.

is the divergence of the velocity, equal to zero in an incompressible fluid [cf., equation (II.2.4)]. From the forms of (2.1) and (2.2) it is clear that the order of subscripts is unimportant, and in three-dimensional flows the tangential stresses τ_{xz} and τ_{zx} are obtained from (2.1) by simple cyclic interchange of variables u, v, w and x, y, z.

Fluid friction effects on the motion of an *incompressible* fluid in general two-dimensional motion are now accounted for by including the stresses (2.1) and (2.2) in the force balance of Section II.4. The resultant of the stresses in the x-direction on all four edges of an element of dimensions dx and dy is seen from Fig. IX-4 to be expressible as

summation of viscous stresses...

$$\left\{ \frac{\partial}{\partial x} \tau_{xx} + \frac{\partial}{\partial y} \tau_{xy} \right\} dx\, dy \qquad (2.3)$$

Exercise 5 When the viscosity coefficient μ is assumed constant and $\nu \equiv \mu/\rho$ denotes the kinematic viscosity, show that the viscous stress components in the x-direction, per unit mass of fluid, are obtained from (2.3) as

...in the x-direction

$$\nu \left\{ \frac{\partial^2 u}{\partial x^2} + \frac{\partial^2 u}{\partial y^2} \right\} \qquad (2.3a)$$

Demonstrate that the principle of fluid momentum conservation (II.5.7) applied to a viscous incompressible fluid motion in two dimensions is expressed by

Navier-Stokes equations, incompressible fluid

$$\left. \begin{array}{l} \dfrac{Du}{Dt} + \dfrac{1}{\rho} \dfrac{\partial p}{\partial x} = -\dfrac{\partial U}{\partial x} + \nu \left\{ \dfrac{\partial^2 u}{\partial x^2} + \dfrac{\partial^2 u}{\partial y^2} \right\} \\[2mm] \dfrac{Dv}{Dt} + \dfrac{1}{\rho} \dfrac{\partial p}{\partial y} = -\dfrac{\partial U}{\partial y} + \nu \left\{ \dfrac{\partial^2 v}{\partial x^2} + \dfrac{\partial^2 v}{\partial y^2} \right\} \end{array} \right\} \quad (2.4)$$

which are called the Navier-Stokes equations of viscous fluid motion.

Exercise 6 Prove the vector identity

$$\nabla^2 \mathbf{A} = \nabla(\nabla \cdot \mathbf{A}) - \nabla \times (\nabla \times \mathbf{A}) \qquad (2.5)$$

for an arbitrary vector **A** having Cartesian components α, β, γ by evaluating the gradient of the divergence of **A** [first term on the right side of (2.5)] as well as the curl of the curl of **A**. Then identify α with the velocity component u in (2.4) to show that the vector form of the Navier-Stokes equations corresponding to (II.5.7a) is

$$\frac{D\mathbf{q}}{Dt} + \frac{1}{\rho}\nabla p = -\nabla U + \nu \nabla^2 \mathbf{q} \qquad (2.4a)$$

vector form of the preceding

when the fluid is incompressible. Demonstrate that viscous stresses in incompressible fluid motion are entirely related to the flow vorticity $\boldsymbol{\omega} \equiv \nabla \times \mathbf{q}$. What can be said concerning viscous stresses in regions of flow that are irrotational?

Exercise 7 By introducing a characteristic velocity V and a characteristic length l in the manner of Section 7.1, show that the dynamical importance of fluid friction terms in (2.4) is determined by the magnitude of the Reynolds number R of (1.12), where l now replaces D. Note that large values of the Reynolds number imply the comparative *un*importance of fluid friction in the balance of inertia, pressure, and gravity terms. Justify the neglect of viscosity in the flows of slightly viscous fluids like air and water, in regions of flow that are not close to solid boundary surfaces.

Air at 68°F = 20°C
μ_A = 0.00018 gm cm^{-1} sec^{-1}
     = 3.76 × 10^{-6} slug ft^{-1} sec^{-1}

Water at 20°C
μ_W = 0.010 gm cm^{-1} sec^{-1}
     = 2.086 × 10^{-5} slug ft^{-1} sec^{-1}

3 SWIRLING FLOW: THE IDEALLY STIRRED COFFEE CUP

Viscous flow in a circular cylinderical region is now considered when radial and axial velocities are both zero and the motion is everywhere transverse, $u_\theta(r) = q(r)$ at each instant. The effect of frictional stresses at the boundary $r = a$, in the absence of pressure gradient forces, is to damp out an initial flow, and equation (2.4a) can be applied to determine how the decay of the motion in time depends on its initial form. Gravity is neglected, and the unsteady axially symmetric flow is represented by the velocity u_θ as a function of radial distance and time as

idealization of flow

$$\mathbf{q} = q(r, t)\mathbf{e}_\theta \qquad (3.1)$$

and this flow is seen to consist of the transverse component u_θ alone.

Exercise 8 (a) Apply the definition of the total derivative (II.1.11a), adapted for cylindrical coordinates r, θ, z and corresponding unit vectors $\mathbf{e}_r, \mathbf{e}_\theta, \mathbf{e}_z$, to (3.1) to show that

$$\frac{D\mathbf{q}}{Dt} = \frac{\partial q}{\partial t}\mathbf{e}_\theta - \frac{q^2}{r}\mathbf{e}_r \qquad (3.2)$$

the acceleration

(b) Take the expression for the curl of a vector, referred to its cylindrical coordinate components, as

$$\nabla \times \mathbf{A} = \frac{1}{r}\begin{vmatrix} \mathbf{e}_r & r\mathbf{e}_\theta & \mathbf{e}_z \\ \frac{\partial}{\partial r} & \frac{\partial}{\partial \theta} & \frac{\partial}{\partial z} \\ A_r & rA_\theta & A_z \end{vmatrix} \tag{3.3}$$

and calculate the vorticity by putting $A_r = A_z = 0$ and taking A_θ as q in (3.1) to check the result of Exercise IV.7. (c) Calculate the curl of the vorticity and use (2.5) and (3.2) to obtain the scalar form (2.4a) in this case as

the equations for unsteady axi-symmetric flow

$$\frac{-q^2}{r} + \frac{1}{\rho}\frac{\partial p}{\partial r} = 0$$
$$\frac{\partial q}{\partial t} = \nu\left\{\frac{\partial^2 q}{\partial r^2} + \frac{1}{r}\frac{\partial q}{\partial r} - \frac{q}{r^2}\right\} \tag{3.4}$$

Consequently the velocity is determined by the second equation of (3.4), and the radial pressure variation is given in terms of this quantity by the first equation of (3.4).

Because the time variable t does not appear explicitly in (3.4), an exponential time dependence is inferred as in previous examples. Then the velocity $q(r, t)$ is determined as a solution of an ordinary differential equation obtained by setting q equal to the product of an exponential and an unknown function $R(r)$:

assumed product-type solution

$$q(r, t) = e^{-c^2\nu t}R(r) \tag{3.5}$$

where c^2 is an as yet unspecified *separation constant* of the type employed in Exercise V.22.

When the product cr is written as x for brevity, substitution of (3.5) in the second of (3.4) yields the equation for $R(x/c)$, denoted $R(x)$:

$$x^2\frac{d^2R}{dx^2} + x\frac{dR}{dx} + (x^2 - 1)R = 0 \tag{3.6}$$

radial variation determined by Bessel's equation

the solutions of which are known as Bessel functions of order unity. These functions are named for Friedrich Wilhelm Bessel, who in 1817 introduced them in an investigation on Kepler's problem in orbital dynamics. Daniel Bernoulli in 1732 obtained the functions of order zero, corresponding to a different value of a constant in (3.6), in analyzing the motion of a chain suspended at one end. Euler's and Fourier's use of the same equations in problems of vibrating membranes and heat conduction in solid cylinders are

inspection of the solution

but two instances, among many more, of the importance of Bessel functions in physics and engineering.

For large values of the independent variable x, equation (3.6) differs but slightly from the harmonic oscillator equation given as the first equation of (V.5.4) and known to have sine and cosine functions as solutions. The oscillatory character of solutions of (3.6) for smaller values of x is also easily demonstrated, and the vanishing of the function at each "zero" of the Bessel function provides a convenient means for satisfying the no-slip condition at the cylinder wall $r = a$:

boundary conditions

$$q(a, t) = 0 \quad \text{or} \quad R(ca) = 0 \qquad (3.7)$$

where ca is the value of x corresponding to $r = a$. The Bessel function of interest vanishes at $x = 0$ and also at other values of x, here denoted j_1, j_2, j_3, etc., that appear in tables of functions tabulated to high degrees of accuracy (the intervals $j_n - j_{n-1}$ approaching π as n becomes large, as for the ordinary sine function). The constant c is determined by requiring $ca = j_1$; greater values of c corresponding to j_2, j_3, etc., also give solutions of (3.5) any linear combination of which is still a solution of this equation.

Each value of c determines a distinct damping or *attenuation* rate, since the velocity is decreased by a factor $1/e$ for the corresponding "mode" when the time $1/c^2\nu$ has elapsed. The slowest decay therefore corresponds to the smallest value of c, i.e., to the value

$$c = \frac{j_1}{a}$$

for which the complete motion is, according to (3.5),

$$q(r, t) = C \exp\left\{\frac{-j_1^2 \nu t}{a^2}\right\} J_1\left(\frac{r}{a} j_1\right) \qquad (3.8)$$

solution of equation satisfying required conditions

where the standard notation J_1 is written for the first-order Bessel function, and an arbitrary amplitude constant C is included. The last factor in (3.8) shows that the most slowly decaying motion corresponds to the radial variation of J_1 given by the product of r and j_1/a which vanishes both at $r = 0$ and at $r = a$, having a positive maximum at $r = 0.628a$. Hence, if a cup is idealized as cylindrical and it is desired to impart a persistent motion to fluid in its interior, the most effective (i.e., the most slowly damped) mode is that which has a peak at a distance slightly greater than halfway from the axis to the wall of the cup, in the sense that the most prolonged motion is assured by stirring at this location. To the extent that the best stirring provides maximum time for mixture or for melting of a sugar cube, (3.8) indicates where a given amount of energy might ideally by applied to initiate the flow.

stir at one definite radial distance

4 THE FRICTIONAL WIND

Friction forces near the Earth's surface alter the balance of the Coriolis and pressure gradient forces leading to geostrophic winds in the direction of the isobars, as described in Section III.5. The fact that the horizontal velocities at the Earth's surface vanish on account of a no-slip condition like (1.6) shows that a more complete form of (III.5.2) is required, and it is obtained by including the viscous stress terms. These terms are simplified by noting that the thickness of the atmosphere is very small compared with the continental dimensions of the pressure systems that give rise to geostrophic winds; the layer within which friction effects are important is still thinner by at least two orders of magnitude. It follows that the vertical variations of wind speed are much greater than the horizontal variations, so that, when z denotes the distance above the Earth's surface, terms like $\partial^2 u/\partial z^2$ in (2.4) greatly exceed the derivative terms with respect to x and y. In purely horizontal motion the gravity terms may be omitted, and the balance of Coriolis, pressure, and frictional stresses gives, in place of (III.5.2)

$$-2\omega \sin\lambda\, v + \frac{1}{\rho}\frac{\partial p}{\partial x} = \nu \frac{\partial^2 u}{\partial z^2}$$
$$+2\omega \sin\lambda\, u + \frac{1}{\rho}\frac{\partial p}{\partial y} = \nu \frac{\partial^2 v}{\partial z^2} \qquad (4.1)$$

subject to the no-slip conditions on $u(x, y, z)$ and $v(x, y, z)$ that appear as

$$u(x, y, 0) = 0 \qquad v(x, y, 0) = 0 \qquad (4.2)$$

If the level where friction forces are reduced to zero is $z = h$ and the x-axis is taken parallel to the geostrophic wind at $z \geq h$, two additional conditions on the velocities are expressed as

$$u(x, y, h) = U \qquad v(x, y, h) = 0 \qquad (4.3)$$

where U now denotes the geostrophic wind speed at $(0, 0, z)$ for $z \geq h$. When the conditions (4.3) are inserted in (.41), the vanishing of the friction stresses at the level $z = h$ shows that the pressure is given by

$$\frac{\partial p}{\partial x} = 0 \quad \text{and} \quad 2U\omega \sin\lambda + \frac{1}{\rho}\frac{\partial p}{\partial y} = 0 \qquad (4.4)$$

and thus (4.1) become

$$\frac{\partial^2 u}{\partial z^2} + \frac{2\omega \sin\lambda}{\nu} v = 0$$
$$\frac{\partial^2 v}{\partial z^2} + \frac{2\omega \sin\lambda}{\nu} (U - u) = 0 \qquad (4.5)$$

for the range $0 < z < h$ of the only remaining independent variable.

By noticing again that the independent variable does not appear explicitly in the equations, and in the light of preceding examples, solutions are expected to take the form of exponential functions and/or trigonometric functions (which are, of course, also recognized as exponential functions with imaginary arguments). The solutions of (4.5) are in fact verified to be of this form and are written in full as

$$u(z) = U(1 - e^{-\beta z} \cos \beta z)$$
$$v(z) = U e^{-\beta z} \sin \beta z \qquad (4.6)$$

tentative solution

where the dependence on x and y is now omitted and the constant β is seen, by substitution in (4.5), to be given by

$$\beta^2 = \frac{\omega \sin \lambda}{\nu} \qquad (4.7)$$

Exercise 9 Show that each of the equations (4.5) separately establishes the value of β according to (4.7) when u and v are of the forms (4.6). What conclusion would be drawn if two distinct values β_1 and β_2 were implied by the two equations (4.5)?

verification

Exercise 10 Recall that, in the limit $x \to 0$,

$$\sin x \to x \qquad \tan^{-1} x \to x$$
$$\cos x \to 1 - \frac{x^2}{2} \qquad e^{-x} \to 1 - x$$

and prove that the wind direction at the Earth's surface forms an angle of 45° with the local isobar (i.e., x-)direction by evaluating

$$\lim_{z \to 0} \left\{ \tan^{-1} \frac{v(z)}{u(z)} \right\}$$

general conclusion

Note that the result is independent of the thickness h of the friction layer, independent of the viscosity ν and also of the wind velocity U. Wind markings on sea level weather maps generally confirm that surface winds are inclined toward low pressure zones in approximately this manner.

Exercise 11 The velocities (4.6) do not conform strictly to the conditions (4.3), for any finite height h but only asymptotically as $z \to \infty$. Calculate the height z_1, where the error becomes less than 0.01, by solving the equation

$$e^{-\beta z_1} = 0.01$$

for z_1 when the Earth's rotation is $\omega = 0.728 \times 10^{-4}$ sec^{-1}, latitude = 30°, and the ordinary kinematic viscosity of air at 0°F is 1.27×10^{-4} ft^2 sec^{-1}.

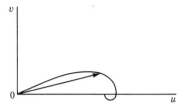

Fig. IX-5 Hodograph representation of frictional wind.

estimation of friction layer thickness

(The actual *turbulent* state of winds in the atmosphere requires the use of a larger effective value of ν, and somewhat larger values of friction height h result. The components (4.6) of frictional wind are shown by a single curve for all values of z in the range $(0, h)$ by employing the hodograph variables u and v of Section IV.8 as Cartesian coordinate axes. The vanishing velocity at the surface $z = 0$ corresponds to the origin, while the velocities at different heights are given in both magnitude and direction by successive points of the spiral-like curve approaching $u = U$, $v = 0$ as $z \to h$. The wind direction angle diminishes continuously from 45° to zero for the wind Fig. IX-5.

5 ANOMALOUS POSITIVE FRICTIONAL ACCELERATION IN ONE-DIMENSIONAL FLOW: FRICTIONAL COOLING

frictional resistance represented one-dimensionally

The one-dimensional treatment of various fluid motions considered earlier has capitalized on the resulting computational simplicity in demonstrating the principal features of flows that are in reality multi-dimensional. It is of interest now to show that the same type of idealization, in flows affected by frictional stresses like (2.2) indicates flow characteristics that are just the opposite of what is found by more complete analysis and by observation. Specifically the variations of flow speed and temperature produced by a one-dimensional flow resistance force in the uniform motion of an ideal gas are considered. Force of magnitude X per unit volume is taken to act in the $+x$-direction, so that $X < 0$ corresponds to flow resistance when $u > 0$. Only the second equation of (VIII.3.1) is modified by the inclusion of a term $X\,dx$. According to the result of Exercise VIII.17, the last equation of the set can be written as (VIII.2.3b), and the matrix form of the three equations analogous to (VIII.5.5a) is now

$$\begin{vmatrix} 0 & u & \rho \\ 1 & 0 & \rho u \\ \rho & -\gamma p & 0 \end{vmatrix} \begin{vmatrix} dp \\ d\rho \\ du \end{vmatrix} = \begin{vmatrix} 0 \\ X\,dx \\ 0 \end{vmatrix} \tag{5.1}$$

Solving for the velocity increment du yields

$$\frac{du}{u} = \frac{(-X\,dx)/\rho c^2}{1 - M^2} \tag{5.2}$$

which shows that a resisting force in subsonic flow (i.e., $-X > 0$, $M < 1$) corresponds to an *increase* in flow speed, $du > 0$. The opposite effect in supersonic flow is evident; hence a one-dimensional friction force produces flow changes of the same qualitative form as the addition of heat, as shown in Section VIII.5. Although positive flow acceleration caused by friction forces in subsonic flow are not ordinarily observed (*tangential* stresses related to two-dimensional flow being usually dominant), the finding (5.2) has been confirmed in laboratory tests in which the resistance force is provided by placing a fine screen across the flow inside a vertical pipe open at both ends. Convective motion is established by a heating effect of the screen not considered here.

subsonic: acceleration
supersonic: deceleration

Exercise 12 Use the first and third of equations (5.1) to express the density and pressure variations in terms of du, and the differential form of the equation of state of an ideal gas, viz.,

$$\frac{dT}{T} = \frac{dp}{p} - \frac{d\rho}{\rho} \quad (5.3)$$

to prove that the friction force represented as in (5.1) causes the temperature to *drop*, $dT < 0$, in subsonic flow, the constant γ being positive. Does this finding tend to vitiate the result of Exercise VIII.7?

6 VISCOUS BOUNDARY LAYER; BLASIUS' CALCULATION

Exercise 13 Show that the dimensionless form of the Navier-Stokes equations of steady plane flow, obtained by introducing a characteristic velocity V and a single characteristic length l in the manner of Section VII.1 is

$$\left. \begin{array}{l} u'\dfrac{\partial u'}{\partial x'} + v'\dfrac{\partial u'}{\partial y'} + \dfrac{\partial p'}{\partial x'} = \dfrac{1}{R}\left\{\dfrac{\partial^2 u'}{\partial x'^2} + \dfrac{\partial^2 u'}{\partial y'^2}\right\} \\[2ex] u'\dfrac{\partial v'}{\partial x'} + v'\dfrac{\partial v'}{\partial y'} + \dfrac{\partial p'}{\partial y'} = \dfrac{1}{R}\left\{\dfrac{\partial^2 v'}{\partial x'^2} + \dfrac{\partial^2 v'}{\partial y'^2}\right\} \end{array} \right\} \quad (6.1)$$

dimensionless form of equations

when gravity is neglected and the Reynolds number based on V and l is defined by the dimensionless ratio

$$R \equiv \frac{Vl}{\nu} \quad (6.2)$$

so that viscous stresses play a minor role in the balance of inertia pressure and friction forces when $R \gg 1$.

Exercise 14 Justify the reference to air as a fluid of small viscosity by calculating the numerical value of the Reynolds number for air in the flow

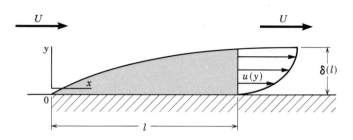

Fig. IX-6 Flat-plate boundary layer growth.

in part (b) of Exercise VII.1 (use the value of kinematic viscosity given in Exercise 11 above).

When a fluid of small viscosity like air or water flows past a solid boundary only a comparatively narrow zone adjacent to the surface is significantly affected by frictional stresses. The calculation of flow resistance depends on the detailed flow in this friction layer, commonly termed the viscous *boundary layer*. An estimate of the extent of this flow region, or *boundary layer thickness*, is obtained by a refinement of the normalization technique that led to equations (6.1) and that now forms the starting point for the calculation of frictional resistance forces.

viscous boundary layer; estimation of thickness

For the sake of definiteness a uniform flow U is considered parallel to the axis of x, modified in the vicinity of this axis by supposing that a plane boundary extends along the positive half of the axis, for which the no-slip velcity condition is written as (Fig. IX-6)

no-slip at boundary surface

$$u(x, 0) = 0 \qquad (6.3)$$

Then U and l are characteristic velocity and length, respectively, in the flow direction at distance $x = l$ from the "flat plate" leading edge that marks the beginning of the boundary layer region, and the boundary layer thickness $\delta(l)$ itself is taken as a characteristic length in the direction transverse to the uniform flow. When derivative terms like $\partial u'/\partial x'$ are considered to be of ordinary (normal) order of magnitude, the actual velocity derivative terms in (2.4) are indicated with respect to order by writing, for example,

order-of-magnitude estimates starting with one characteristic length and one velocity

deduction of order-of-magnitude of flow quantities...

$$\frac{\partial u}{\partial x} \sim \frac{U}{l}$$

IX. 6 / VISCOUS BOUNDARY LAYER; BLASIUS' CALCULATION

It is necessary to determine also the order of magnitude of the velocity component v. The continuity equation

$$\frac{\partial u}{\partial x} + \frac{\partial v}{\partial y} = 0 \tag{6.4}$$

can be used for this purpose, since integration across the boundary layer gives

$$v(x, \delta) = -\int_0^\delta \frac{\partial u}{\partial x} dy \sim \frac{U}{l}\delta \tag{6.4a}$$

This equation shows that the transverse velocity component is small by comparison with U, since the friction layer itself is small compared with l:

$$\frac{\delta}{l} \ll 1 \tag{6.5}$$

Because both u and v increase from zero values on the plate to their characteristic values U and $v(x, \delta)$ in the small distance δ, y-derivatives generally much larger than the corresponding derivatives in the x-direction. Thus

$$\frac{\partial v}{\partial y} \sim \frac{U}{l} \quad \text{and} \quad \frac{\partial v}{\partial x} \sim \frac{U}{l}\frac{\delta}{l}$$

$$\frac{\partial^2 u}{\partial x^2} \sim \frac{U}{l^2} \quad \text{and} \quad \frac{\partial^2 u}{\partial y^2} \sim \frac{U}{\delta^2} \tag{6.6}$$

... and their derivatives ...

and all the other derivatives appearing in (2.4) may be similarly "ordered." When pressure variations in the flow direction are omitted, the external flow adjacent to the boundary layer being assumed uniform, the equations corresponding to (6.1) are

$$\left.\begin{aligned} u\frac{\partial u}{\partial x} + v\frac{\partial u}{\partial y} &= \nu\left\{\frac{\partial^2 u}{\partial x^2} + \frac{\partial^2 u}{\partial y^2}\right\} \\ u\frac{\partial v}{\partial x} + v\frac{\partial v}{\partial y} + \frac{1}{\rho}\frac{\partial p}{\partial y} &= \nu\left\{\frac{\partial^2 v}{\partial x^2} + \frac{\partial^2 v}{\partial y^2}\right\} \end{aligned}\right\} \tag{6.7}$$

and the order estimates (6.6) applied to the first equation of (6.7) now furnish the desired relationship to determine δ. Both terms on the left are of order U^2/l, and the terms in braces on the right side are seen by (6.6) to be of order U/δ^2, use being made of (6.5). Thus the extent of the region where x-stresses and x-inertia terms are in balance is obtained by writing this equation as

$$\frac{U^2}{l} \sim \nu\frac{U}{\delta^2}$$

which is solved for δ to give

$$\frac{\delta}{l} \sim \frac{1}{R^{1/2}} \tag{6.8}$$

...and of the boundary layer thickness itself

where $R \equiv Ul/\nu$, the external flow speed U replacing the characteristic velocity V of (6.2). The boundary layer thickness is therefore inversely proportional to the square root of the Reynolds number based on flow velocity U and distance l from the leading edge of the flat plate. For a flow in which the Reynolds number is of the order of one million, for example, the boundary layer thickness becomes of the order of one-tenth of one percent of the distance l.

Exercise 15 Because of the smallness of δ it follows that flow within the boundary layer is nearly parallel to the plate, and the nearly straight streamlines indicate that the lateral pressure gradient must be very small. Confirm this conclusion by applying the ordering arguments (6.6) and the evaluation (6.8) to the second equation of (6.7), to show that the lateral pressure gradient term is small compared with each of the terms in the first equation of (6.7). Justify the approximate form of the boundary layer equations (6.7) as

simplified equations

$$\left. \begin{aligned} u\frac{\partial u}{\partial x} + v\frac{\partial u}{\partial y} &= \nu\frac{\partial^2 u}{\partial y^2} \\ \frac{\partial p}{\partial y} &= 0 \end{aligned} \right\} \tag{6.9}$$

When the pressure is constant in the boundary layer as indicated by the second equation of (6.9), the velocity components are determined by the continuity equation (6.4) together with the first of (6.9). They are reduced to a single equation when the stream function is employed as in (V.1.4), although the resulting equation is nonlinear because it contains products of the derivatives of the stream function. The subscript notation is employed now for partial derivatives, and this equation is

reduction to a single equation, nonlinear in partial derivatives

$$\psi_y \psi_{yx} - \psi_x \psi_{yy} = -\nu \psi_{yyy} \tag{6.10}$$

and the ordering relationships (6.4a) and (6.8) suggest the manner by which (6.10) can be reduced to an ordinary differential equation.

Substituting (6.8) in (6.4a) shows that the velocity component v varies in the flow direction like the inverse square root of the distance l, here denoted x. Then the stream function that gives v by an x-differentiation must be proportional to the square root of x. Also, by writing (6.8) as

$$\frac{\delta^2}{l} \sim \frac{\nu}{U} \tag{6.8a}$$

and associating δ and l, respectively, with the y- and x-coordinates, a *similarity transformation* that replaces y and x by a particular combination of the two variables is chosen as y to the second power divided by x to the first power. The square root of this ratio is equally useful, and the transformations are written as

$$\psi(x, y) = -\sqrt{\nu U x} f(\eta) \quad \text{where} \quad \eta \equiv \frac{y}{2}\sqrt{\frac{U}{\nu x}} \qquad (6.11)$$

similarity transformation...

and the constant factors are chosen with a view to simplicity of resulting equations. The new *similarity variable* η is well chosen by the definition (6.11) if it does in fact lead to the desired simplification of (6.10). By direct differentiation of (6.11) it is found that

$$u = \frac{U}{2} f'(\eta) \quad \text{and} \quad v = -\frac{1}{2}\sqrt{\frac{\nu U}{x}} \{f - f' \cdot \eta\} \qquad (6.12)$$

where f' stands for the derivative of the function f with respect to the similarity variable η on which alone it depends. The further differentiations indicated in (6.9), or equivalently in (6.10), furnish an equation for determining the stream function in terms of $f(\eta)$.

Exercise 16 Take the stream function ψ and the similarity variable η as given in (6.11) and show that (6.10) is satisfied when the function f is a solution of the ordinary differential equation

$$ff'' - f''' = 0 \qquad (6.13)$$

...leads to an ordinary differential equation

derived first and solved numerically by H. Blasius in 1908.

Exercise 17 The three boundary conditions used to evaluate integration constants arising in the solution of the third-order equation (6.13) are taken as the conditions of zero normal and zero tangential velocity at the boundary

$$v(x, 0) = 0 \quad \text{and} \quad u(x, 0) = 0 \qquad (6.14a)$$

the transformation is appropriate for the particular flow in question

and the condition that the tangential velocity at the edge of the boundary layer is equal to U:

$$u(x, \delta) = U \qquad (6.14b)$$

Express each of these conditions in terms of the function f and its first derivative evaluated at the appropriate η-value obtained from (6.11).

Exercise 18 Use (1.3) and (6.12) to evaluate the viscous stress τ_0 at a boundary point $y = 0$ at distance x from the leading edge of the plate. Determine the total friction force F on the length x as

$$F = \int_0^x \tau_0(x)\, dx \qquad (6.15)$$

flat-plate friction coefficient

Defining the dimensionless skin friction coefficient c_f as the force on unit breadth of plate divided by $\rho(U^2/2)x$, show that

$$c_f = \frac{f''(0)}{\sqrt{Ux/\nu}} = \frac{1.328}{R^{1/2}} \qquad (6.15a)$$

where the numerical value found by Blasius is inserted in the last expression of (6.15a) and R again denotes Reynolds number, based in this case on the length x as shown in the preceding expression.

The friction coefficient (6.15a) has been used widely in practical calculations of viscous drag (resistance) on surfaces adjacent to uniform flow when the speed U (better, when the Reynolds number R) is not too large. The chief qualitative difference from the case of laminar pipe flow (1.11) is the decrease of friction coefficient as the inverse square root of R for boundary layer flow, in contrast with the simpler and more rapid decrease indicated in (1.11).

7 NONZERO PRESSURE GRADIENT — INTEGRAL OF MOMENTUM

When the flow outside the boundary layer is not uniform along the plate, $dU/dx \neq 0$ and the pressure increases or decreases depending on whether the flow is decelerating or accelerating:

external flow specifies the pressure gradient

$$U\frac{dU}{dx} + \frac{1}{\rho}\frac{dp}{dx} = 0 \qquad (7.1)$$

The same gradient of pressure affects the boundary layer flow, the equations corresponding to (6.9) now taking the form

equations no longer reducible in the manner of (6.10)

$$\left. \begin{array}{c} u\dfrac{\partial u}{\partial x} + v\dfrac{\partial u}{\partial y} + \dfrac{1}{\rho}\dfrac{\partial p}{\partial x} = \nu\dfrac{\partial^2 u}{\partial y^2} \\[6pt] \dfrac{\partial p}{\partial y} = 0 \end{array} \right\} \qquad (7.2)$$

where the pressure gradient term can be regarded as a given function of x determined by the form of the external flow as (7.1). On account of this term, introduction of the stream function in the manner of preceding section would lead to an equation like (6.10) but containing one additional term independent of ψ. In this circumstance the former reduction to an ordinary equation by a transformation like (6.11) is not possible and an entirely different procedure is adopted.

Confronted with an equation like the first of (7.2), or its equivalent in terms of the stream function, a further approximation is based on abandoning the search for the individual particle velocities in favor of a more modest goal that presents fewer mathematical complications. Hence, instead of requiring the balance of x-momentum with the pressure and viscous forces at every point in accordance with the first equation of (7.2), this requirement can be met *on the average* by insisting only that the x-momentum of the entire boundary layer conform to the requirements set by an arbitrary external pressure gradient and the frictional force that acts on the boundary layer at the surface where the stress is τ_0. The viewpoint in this case is similar to that in the analysis of stellar statics in Section (VIII.8) where, because of a differential equation for which useful solutions are not known, a broader but less detailed analysis was adopted which also conforms with the physical laws only on an overall average basis.

approximation comparable to analysis in section VIII.8

The forces and momentum flux through a section of the boundary layer of length dx, height $\delta(x)$, and unit breadth are analyzed in the manner of Section (II.5). Thus the x-forces on the boundary elements AC and CD are

$$\left(-\tau_0 + p\frac{d\delta}{dx}\right)dx \qquad (7.3)$$

as indicated in Fig. IX-7, where the first term of (7.3) is shown by the negative sign to be directed opposite to the flow and the last term represents the x-projection of the pressure force directed at right angles to the edge of the boundary layer. The pressure force on AB is $p\delta$ and, when this is added to the corresponding term for the edge BD and to (7.3), the total force in the x-direction is

forces on a section of the boundary layer . . .

$$-\left(\tau_0 + \delta\frac{dp}{dx}\right)dx \qquad (7.4)$$

The net outward flux of x-momentum, equated to (7.4) when the flow is steady, is obtained by integrating the momentum flux $-\rho u^2$ across the boundary layer thickness δ at AB and adding to the

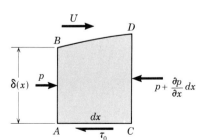

Fig. IX-7 Forces on boundary layer element of thickness $\delta(x)$.

corresponding term for the side CD, as well as by accounting for the flux across BD. The former pair of terms is expressed as

$$\frac{d}{dx}\int_0^\delta \rho u^2 \, dy \, dx \tag{7.5a}$$

and the momentum flux

while the velocity components U and $v(\delta)$ carry fluid momentum across BD. The first of these contributions gives

$$\rho U^2 \frac{d\delta}{dx} \, dx \tag{7.5b}$$

because the flow is inward across the control surface AC and the second may be written as

$$\rho U v(\delta) \, dx \tag{7.5c}$$

Exercise 19 When the boundary layer *displacement thickness* δ_* is defined by the integral of the boundary layer velocity deficit, i.e., as

displacement thickness δ_* defined

$$U\delta_* \equiv \int_0^\delta \{U - u(y)\} \, dy \tag{7.6}$$

show that the normal component of velocity at the edge of the boundary layer is given by

$$v(\delta) = -\int_0^\delta \frac{\partial u}{\partial x} \, dy = U \frac{d\delta_*}{dx} - (\delta - \delta_*) \frac{dU}{dx} \tag{7.7}$$

Give the physical interpretation of the velocity $v(\delta)$ for flow with zero pressure gradient.

momentum thickness θ defined

Exercise 20 When the boundary layer *momentum thickness* θ is defined by the integral of the momentum deficit across the boundary layer, i.e., as

$$U^2\theta = \int_0^\delta u(U - u) \, dy \tag{7.8}$$

show that

$$\int_0^\delta u^2 \, dy = U^2 (\delta - \delta_* - \theta) \tag{7.9a}$$

and that

$$2\int_0^\delta u \frac{\partial u}{\partial x} \, dy = 2U \frac{dU}{dx} (\delta - \delta_* - \theta) - U^2 \frac{d}{dx}(\delta_* + \theta) \tag{7.9b}$$

Exercise 21 Equate (7.4) and (7.5), using definitions (7.6) and to obtain the boundary layer *momentum integral equation* as

wall stress in terms of integral-flow quantities

$$\frac{\tau_0}{\rho U^2} = \frac{1}{U^2} \frac{d}{dx}(U^2\theta) + \frac{\delta_*}{U} \frac{dU}{dx} \tag{7.10}$$

IX. 7 / NONZERO PRESSURE GRADIENT — INTEGRAL OF MOMENTUM

When the external flow $U(x)$ is known, and hence the pressure gradient, as well as the displacement and momentum thicknesses, (7.10) provides an expression for the friction stress τ_0 exerted by the fluid on the boundary surface.

Exercise 22 Derive equation (7.10) by integrating the boundary layer equation (7.2) across the boundary layer. HINT: Show that

$$\int_0^\delta v \frac{\partial u}{\partial y} dy = v(\delta) \cdot U + \int_0^\delta u \frac{\partial u}{\partial x} dy$$

by integrating by parts and using the continuity equation (6.4).

Exercise 23 (a) Show that, when the external flow is uniform, $U = $ const, the friction coefficient c_f defined as before from (6.15) is related to the momentum thickness θ as

$$c_f = \frac{2\theta}{x} \qquad (7.11)$$

a special case, interpretation of θ

(b) Calculate the wall shear stress τ_0, the displacement thickness δ_*, and the momentum thickness for the velocity profile

$$u = U \sin\left(\frac{y}{\delta}\frac{\pi}{2}\right) \qquad (7.12)$$

trial profile

that satisfies the conditions (6.14) and also blends smoothly with the external flow, since

$$\left.\frac{\partial u}{\partial y}\right|_\delta = 0$$

Ans. $\tau_0 = \dfrac{\mu U \pi}{2\delta}$, $\delta_* = \delta\left(1 - \dfrac{2}{\pi}\right)$.

(c) Substitute the values of τ_0 and θ just obtained into (7.10) and integrate the resulting equation for δ as a function of x, to show that

comparison with "exact" solution

$$c_f = \frac{\sqrt{2\pi}\sqrt{(4-\pi)/\pi}}{R^{1/2}} \qquad (7.13)$$

(d) Obtain the numerical value of the coefficient on the right side of (7.13) and compare with Blasius' result (6.15a). How do you explain the close agreement?

Exercise 24 Repeat the preceding calculations, adopting the velocity profile

$$\frac{u}{U} = \frac{3}{2}\frac{y}{\delta} - \frac{1}{2}\left(\frac{y}{\delta}\right)^3 \qquad (7.14)$$

instead of (7.12).

The calculation of frictional resistance when U is a given function of x proceeds, in the same manner as in Exercise 18, by integration of the shear stress τ_0 along the plate surface.

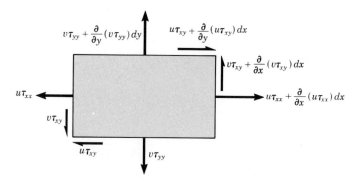

Fig. IX-8 Viscous stress energy flux.

8 VISCOUS DISSIPATION OF FLOW ENERGY

When viscous stresses are taken into account in the energy balance of an incompressible fluid in the manner of Section II.5, the normal stress terms τ_{xx} and τ_{yy} add to the normal pressures of (II.5.2) and the tangential stresses also participate in the exchange of energy between an element and its surroundings. The relevant effects are shown in Fig. IX-8 for the rectangular fluid element in the x, y plane having dimensions dx and dy. When the flow is steady, the preceding argument leads to the equation

energy equation, including effects of viscous stresses

$$\rho\left\{\frac{\partial(Eu)}{\partial x} + \frac{\partial(Ev)}{\partial y}\right\} + \frac{\partial(pu)}{\partial x} + \frac{\partial(pv)}{\partial y} = \frac{\partial(u\tau_{xx})}{\partial x} + \frac{\partial(v\tau_{xy})}{\partial y} + \frac{\partial(u\tau_{yx})}{\partial x} + \frac{(v\tau_{yy})}{\partial y} \quad (8.1)$$

where the viscous stress terms on the right side of (8.1) also enter into the momentum balance, as seen by comparing with (2.3) and (2.4). It is important to note that, contrary to the case of inviscid and incompressible fluid (II.5.5), (8.1) does not reduce to the Bernoulli integral.

Exercise 25 Multiply the first of (2.4) by u, the second by v and add, and thus evaluate the streamline derivative of the energy sum $q^2/2 + p/\rho + U$ in terms of the viscous stresses appearing in (8.1).

If the energy E is now regarded as the sum of the internal energy of the fluid per unit mass, here denoted e, plus the kinetic, pressure, and potential energies just enumerated, i.e., as

$$E = e + \frac{q^2}{2} + \frac{p}{\rho} + U \quad (8.2)$$

where $q^2 = u^2 + v^2$, the result of subtracting the equation of Exercise 25 from (8.1) is

$$\frac{De}{Dt} = \frac{\mu}{\rho}\left\{2\left[\left(\frac{\partial u}{\partial x}\right)^2 + \left(\frac{\partial v}{\partial y}\right)^2\right] + \left[\frac{\partial u}{\partial y} + \frac{\partial v}{\partial x}\right]^2\right\} \qquad (8.3)$$

mechanical energy "loss" is thermodynamic energy gain

Equation (8.3) shows that the internal energy is increased owing to the action of viscosity and motion involving either of the two types represented by the bracketed terms. These terms have particular physical interpretations. Although internal energy of a fluid is independent of flow speed, (8.3) indicates that flow energy is converted to internal (thermal) form through the mechanism of viscosity. While total energy is of course preserved, the reduction of kinetic (i.e., flow) energy, termed viscous dissipation, is demonstrated. The quantity on the right side of (8.3), called the specific *dissipation function*, provides a measure of the rate at which flow energy is converted to heat. In an ideal gas the internal energy is a function of temperature only (cf. Ex. VIII.3).

dissipation function

Exercise 26 Show that viscous flow between plane parallel boundaries separated by distance $2h$ from each other is represented by

$$u(y) = \frac{p_1 - p_2}{2\mu l}[h^2 - y^2]$$

$$= u_M\left[1 - \left(\frac{y}{h}\right)^2\right] \qquad (8.4)$$

dissipation localized

where the notation corresponds to that for the case of pipe flow, Section 1. (b) Evaluate the dissipation function for the flow (8.4) and determine in which region of the flow the fluid is the most intensely heated.

9 DECAY OF A PLANE SOUND WAVE

The dissipation of flow energy shown by loss of pressure in a pipe and that shown by the reduction of flow speed as seen by the growth of the boundary layer in flow at uniform pressure are two instances in which viscous stresses give rise to increases of entropy in heating the fluid. Even in the very small amplitude motion represented by a plane acoustic wave in the total absence of shear the action of fluid friction is also shown as an attenuation of flow.

acoustic dissipation

The strictly one-dimensional motion in the x-direction is accompanied by the longitudinal viscous stress (2.2) in the form

$$\tau_{xx} = \frac{4}{3}\mu\frac{\partial u}{\partial x} \qquad (9.1)$$

instead of the form used in (2.4) (why?). When the fluid is again regarded as barotropic and particle displacements are small, as before, instead of (VIII.1.7) the momentum condition is obtained in the form

$$\frac{\partial u}{\partial t} + c^2 \frac{1}{\rho} \frac{\partial \rho}{\partial x} = \frac{4}{3} \nu \frac{\partial^2 u}{\partial x^2} \qquad (9.2)$$

The continuity equation remains as shown in (VIII.1.7), and, when it is used to eliminate the density from (9.2) as before, the equation for the fluid velocity is

modified wave equation

$$\frac{\partial^2 u}{\partial t^2} - c^2 \frac{\partial^2 u}{\partial x^2} = \frac{4}{3} \nu \frac{\partial^3 u}{\partial x^2 \partial t} \qquad (9.3)$$

which differs from (VIII.1.6a) only in the viscous term on the right side of (9.3). It is apparent that, when the kinematic viscosity ν is very small, the motion closely approximates the free plane wave in inviscid fluid as found in Section VIII.1 [cf. equation (VIII.1.6a)]

The action of viscosity is now studied for a *forced* wave corresponding to a periodic motion that is maintained at $x = 0$:

forced wave

$$u(0, t) = A \cos \sigma t \qquad (9.4)$$

for example, where A is an unspecified amplitude constant and the frequency σ measures the pitch of the sound produced. Since an arbitrary function of $x - ct$ satisfies (9.3) in the limit $\nu \to 0$, while the absence of independent variable t from the equation implies exponential solutions, the function

assumed solution

$$u(x, t) = A e^{-bx} \cos \left\{ \sigma \left(t - \frac{x}{c} \right) \right\} \qquad (9.5)$$

may be tentatively considered. If a value of the constant b can be found such that the function (9.5) satisfies (9.3), then that function may justifiably be denoted $u(x, t)$ and regarded as representing a *damped acoustic wave*. The procedure is clearly analogous to that of Section 4 above.

Exercise 27 a) Substitute (9.5) in (9.3) and determine b by the condition that the coefficient of the sine terms must vanish. (b) Notice that the vanishing of the coefficient of the cosine terms does *not* give the same value of b, and justify the adoption of the value found in part (a) (cf., Exercise 9).

It is clear from the form of (9.5) that b has the dimensions of (length)$^{-1}$, and, when x is equal to this length, i.e., when

$$x = \frac{1}{b}$$

the amplitude of the motion is reduced by the factor e^{-1}, equal to $(2.718...)^{-1}$ times the amplitude A at $x = 0$. Since

$$\frac{1}{b} = \frac{3c^2}{2\nu\sigma^2} \qquad (9.6)$$

decay depends on pitch

it is seen that the decay of a sound wave depends on the pitch. The effect of viscosity on the amplitude of the motion is very slight, according to (9.6), except for sounds of very short wavelength.

Exercise 28 Evaluate the length given by (9.6) for air in which the acoustic speed is given by $c = 332$ m sec^{-1}, the kinematic viscosity $\nu = 0.133$ cm^2 sec^{-1}, and the frequency σ corresponds to middle C (256 Hz). Does the result seem reasonable?

10 VISCOUS DISSIPATION RELATED TO DEFORMATION OF FLUID ELEMENT

A physical interpretation of the individual terms appearing in the dissipation function (8.3) is obtained by considering the rate at which viscous stresses perform work on a rectangular element of incompressible fluid. It will be seen that the particular combination of velocity derivatives in (8.3) corresponds to the forces exerted by surrounding fluid and to motion parallel to the stresses, the appropriate force-velocity products representing the rate of energy increase attributable to fluid friction.

It is convenient to consider first a solid material element of dimensions dx and dy, of unit depth, subject to surface tractions τ_{xy}, τ_{yx}, τ_{xx}, and τ_{yy} which distort the initially rectangular form of the element. When the stresses and the deformations (strains) are sufficiently small, Hooke's law indicates the proportionality of stress and strain, and the material is regarded as elastic in its behavior. Shear force τ_{xy} that produces deformation shown by the angle θ_1 in Fig. IX-9 is therefore zero when $\theta_1 = 0$, and the work done by this stress in developing strain θ_1 is proportional to one-

dissipation and mechanical working

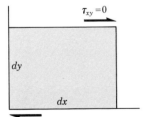
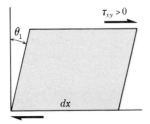

Fig. IX-9 Shear distortion of solid material element.

half of their product. Since the force is given by $\tau_{xy}\,dx$ and the displacement of the elementary surface on which it acts is $\theta_1\,dy$, the work performed is

$$\frac{\tau_{xy}}{2}\theta_1\,dx\,dy \qquad (10.1a)$$

In the same manner it is shown that the stress τ_{yx} that produces strain displacements parallel to the y-axis represents an amount of energy

shear work

$$\frac{\tau_{yx}}{2}\theta_2\,dx\,dy \qquad (10.1b)$$

where θ_2 is the angular displacement of the elementary edges initially parallel to the x-axis, as shown.

If the x- and y-components of strain displacements are denoted, respectively, by u and v, the angles θ_1 and θ_2 are expressed in terms of them as

$$\theta_1 = \frac{\partial u}{\partial y} \qquad \theta_2 = \frac{\partial v}{\partial x} \qquad (10.2)$$

Adding (10.1a) and (10.1b) after substituting the values (10.2) then gives the strain energy associated with the stresses τ_{xy} and τ_{yx}, per unit volume of material, as

$$\frac{1}{2}\left\{\tau_{xy}\frac{\partial u}{\partial y} + \tau_{yx}\frac{\partial v}{\partial x}\right\} \qquad (10.3)$$

Two modifications of (10.3) are required to obtain the corresponding expression for the effects of frictional stresses in a viscous fluid: First, the terms u and v are recognized as time derivatives of displacements (viz., velocity components); thus the expression (10.3) represents the *rate* at which neighboring fluid performs work on the element $dx\,dy$. Secondly the stresses τ_{xy} and τ_{yx} transmit energy in an amount given directly by their respective products with the velocities, and thus the factor $\frac{1}{2}$ in (10.3) should now be suppressed. Finally, taking the stresses in the form (2.1) gives the expression of the rate at which viscous shear stresses perform work on unit volume of fluid as

shear dissipation rate

$$\mu\left[\frac{\partial u}{\partial y} + \frac{\partial v}{\partial x}\right]^2 \qquad (10.4a)$$

Exercise 29 (a) Consider the work done by the normal shear stress τ_{xx} in stretching the solid material element from dimensions dx and dy to $(1 + \partial u/\partial x)\,dx$ and dy as shown in Fig. IX-10. Show that this is

stretching work

$$\frac{\tau_{xx}}{2}\frac{\partial u}{\partial x}\,dx\,dy$$

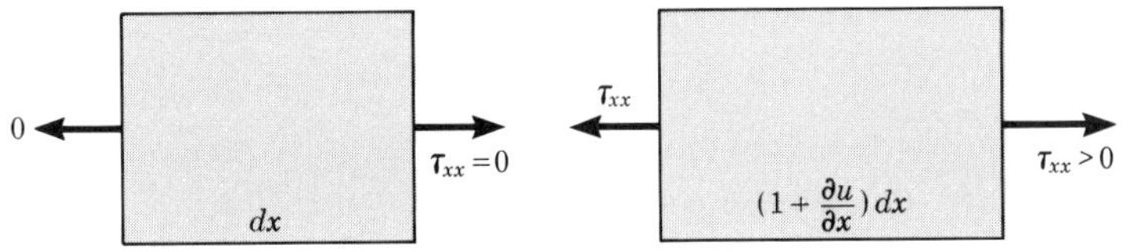

IX. 10 / VISCOUS DISSIPATION RELATED TO DEFORMATION OF FLUID ELEMENT

Fig. IX-10 Stretching of solid material element.

(b) Adapt the result of part (a) to represent the rate of working of normal viscous stress τ_{xx} on the corresponding fluid element. By considering the normal stress τ_{yy} also, show that the normal stresses furnish energy to unit volume of fluid at the rate

$$2\mu\left[\left(\frac{\partial u}{\partial x}\right)^2 + \left(\frac{\partial v}{\partial y}\right)^2\right] \qquad (10.4b)$$

(c) Combine (10.4a) and (10.4b) to establish that the normal and tangential viscous fluid stresses together account for energy production per unit mass given by

$$\frac{\mu}{\rho}\left\{2\left[\left(\frac{\partial u}{\partial x}\right)^2 + \left(\frac{\partial v}{\partial y}\right)^2\right] + \left[\frac{\partial u}{\partial y} + \frac{\partial v}{\partial x}\right]^2\right\} \qquad (10.5)$$

total dissipation: shear and stretching

which is identical with the dissipation function of (8.3).

Finally it should be noted that each of the terms in (10.5) has a separate kinematical significance. For this purpose it is convenient to return to the analysis of solid displacements, and to consider a decomposition of the general relative displacement of a point from initial position x, y to nearby location $x + dx, y + dy$. Its x-component is given by

$$u(x + dx, y + dy) = u(x, y) + \frac{\partial u}{\partial x}dx + \frac{\partial u}{\partial y}dy \qquad (10.6)$$

when dx and dy are small, and the last term on the right side is recognized as containing part of the expression (IV.1.2) for the rotation of the fluid element. Now the conclusion (2.1) of the equality of tangential stresses indicates that no flow energy is spent in contributing to the rotation of a fluid element, and it is therefore appropriate to rewrite (10.6) in a manner that isolates the rotation. This is evidently accomplished by expressing the differential x-displacement as

$$du = \frac{\partial u}{\partial x}dx + \frac{1}{2}\left(\frac{\partial u}{\partial y} - \frac{\partial v}{\partial x}\right)dy + \frac{1}{2}\left(\frac{\partial u}{\partial y} + \frac{\partial v}{\partial x}\right)dy \qquad (10.7)$$

decomposition of displacement: stretching, rotation, and shear

where the general displacement is exhibited as the sum of three distinct parts. The first term represents linear extension or stretching, and the corresponding first term on the right side of (10.5) is the associated energy rate. The second group on the right side of (10.7) represents rotation, which does not enter into our energy balance, and the third is identified as the total shearing distortion, the energy of which appears also in (10.5). Both types of terms in (10.5) represent genuine distortion (rates) resisted by viscous friction and therefore maintained only by the expenditure of mechanical energy. The argument leading to (8.3) shows that it is the flow energy of the surrounding fluid, which is steadily "dissipated" (more accurately, transformed) to internal energy appearing in the form of heat, that gives the dissipation function.

S1 PRECISE FORMULAS FOR VISCOUS STRESSES

The viscous stress formulas (2.1) and (2.2) are obtained by elementary methods based on the assumption that the fluid medium is *isotropic*, i.e., has the same properties in all directions. For then the stresses acting on a fluid element are of the same mathematical form regardless of the orientation of coordinate axes in space, and the stresses in general three-dimensional flow are found when a small number of seemingly plausible hypotheses are also introduced.

Just as Hooke's Law in elasticity takes each stress proportional to strain in a solid body, viscous fluid stresses are likewise supposed to depend *linearly* on the rate of change of strain (rate-of-strain, for short), so that we write in the first instance

Hooke's Law: linear stress and strain-rate dependence

$$\left. \begin{array}{l} \tau_{xy} = \tau_{yx} = a_{11}\dfrac{\partial u}{\partial x} + a_{12}\dfrac{\partial u}{\partial y} + a_{13}\dfrac{\partial v}{\partial x} + a_{14}\dfrac{\partial v}{\partial y} \\[6pt] \tau_{xx} = a_{21}\dfrac{\partial u}{\partial x} + a_{22}\dfrac{\partial u}{\partial y} + a_{23}\dfrac{\partial v}{\partial x} + a_{24}\dfrac{\partial v}{\partial y} \\[6pt] \tau_{yy} = a_{31}\dfrac{\partial u}{\partial x} + a_{32}\dfrac{\partial u}{\partial y} + a_{33}\dfrac{\partial v}{\partial x} + a_{34}\dfrac{\partial v}{\partial y} \end{array} \right\} \quad (S1.1)$$

In somewhat more compact matrix form, the same equations are

$$\begin{vmatrix} \tau_{xy} \\ \tau_{xx} \\ \tau_{yy} \end{vmatrix} = \begin{vmatrix} a_{11} & a_{12} & a_{13} & a_{14} \\ a_{21} & a_{22} & a_{23} & a_{24} \\ a_{31} & a_{32} & a_{33} & a_{34} \end{vmatrix} \begin{vmatrix} \dfrac{\partial u}{\partial x} \\ \dfrac{\partial u}{\partial y} \\ \dfrac{\partial v}{\partial x} \\ \dfrac{\partial v}{\partial y} \end{vmatrix} \quad (S1.1a)$$

and the matrix of coefficients a_{ij} is denoted for convenience by $|A|$. Among the twelve coefficients appearing in $|A|$ it is to be expected that various relationships are available that permit the stresses to be expressed in terms of the three distinct rates-of-strain appearing, for example, in (10.5). Appropriate physical and mathematical arguments are now sought.

For this purpose we impose the condition (hypothesis) that shear stress is purely a function of the shear rate-of-strain, independent of stretching (normal) rate-of-strain, so that

introduction of physical hypotheses

$$a_{11} = a_{14} = 0 \qquad (S1.2)$$

Likewise supposing that normal stresses are independent of shear rate-of-strain gives

$$a_{22} = a_{23} = a_{32} = a_{33} = 0 \qquad (S1.3)$$

The equality of τ_{xy} and of τ_{yx}, moreover, implies the equality of the two remaining coefficients, and when these are again denoted μ in conformance with Newton's finding, we have

$$a_{12} = a_{13} = \mu \qquad (S1.4)$$

The normal stress in the x-direction, viz., τ_{xx} must also depend on the y-component of normal rate-of-strain $\partial v/\partial y$ in the same manner that τ_{yy} depends on $\partial u/\partial x$. Accordingly we have, upon introducing constants a and b,

$$\begin{aligned} a_{21} = a_{34} \equiv a \\ a_{24} = a_{31} \equiv b \end{aligned} \qquad (S1.5)$$

and only the three constants μ, a and b remain, of which the first may be considered known. Two additional restrictions are next examined which permit a and b to be expressed in terms of μ.

According to (S1.2) and (S1.4) the shear stress τ_{xy} is given by

$$\tau_{xy} = \mu\left(\frac{\partial u}{\partial y} + \frac{\partial v}{\partial x}\right) \qquad (S1.6)$$

and the condition of isotropy implies that in a different coordinate system x', y' the corresponding stress must be

$$\tau_{x'y'} = \mu\left(\frac{\partial u'}{\partial y'} + \frac{\partial v'}{\partial x'}\right) \qquad (S1.6a)$$

invariance of stress formulae under coordinate axis rotation

where u' and v' are the velocity components in the new coordinate system. It will be shown in the Exercises below that the consequences of (S1.6a) is the following relationship between the coefficients μ, a and b:

$$2\mu = a - b \qquad (S1.7)$$

When (S1.7) is used to eliminate b from the stress formulas, the matrix of coefficients $|A|$ becomes

$$|A| = \begin{vmatrix} 0 & \mu & \mu & 0 \\ a & 0 & 0 & a - 2\mu \\ a - 2\mu & 0 & 0 & a \end{vmatrix} \quad (S1.8)$$

Finally it is assumed that the mean value of the three normal stresses acting in three mutually perpendicular directions is zero, which means that fluid is supposed to offer no viscous resistance to purely compressive rate of strain (kinetic theory justification of this step has been given for monatomic gases only). The normal stresses are obtained by substituting (S1.8) in (S1.1a) and by taking account of the fact that, in the expression for τ_{xx}, for example, the dependence on $\partial v/\partial y$ must also apply to the third (i.e., the z) direction of flow by including the term $\partial w/\partial z$ so that

$$\tau_{xx} = a\frac{\partial u}{\partial x} + (a - 2\mu)\left(\frac{\partial v}{\partial y} + \frac{\partial w}{\partial z}\right)$$

longitudinal stress, two parameters

$$= 2\mu \frac{\partial u}{\partial x} + (a - 2\mu)\theta \quad (S1.9)$$

where θ represents the divergence of the velocity as before

$$\theta \equiv \frac{\partial u}{\partial x} + \frac{\partial v}{\partial y} + \frac{\partial w}{\partial z} \quad (S1.10)$$

The normal stresses in the y- and z-directions are likewise expressible as

$$\tau_{yy} = 2\mu \frac{\partial v}{\partial y} + (a - 2\mu)\theta \qquad \tau_{zz} = 2\mu \frac{\partial w}{\partial z} + (a - 2\mu)\theta \quad (S1.9a)$$

so that the vanishing of the mean normal viscous stress gives

$$\tau_{xx} + \tau_{yy} + \tau_{zz} = 3(a - 2\mu)\theta + 2\mu\theta = 0$$

or

reduction effected with a physical assumption

$$a = \frac{4}{3}\mu \quad (S1.10)$$

The matrix of coefficients thus becomes

$$|A| = \mu \begin{vmatrix} 0 & 1 & 1 & 0 \\ \frac{4}{3} & 0 & 0 & -\frac{2}{3} \\ -\frac{2}{3} & 0 & 0 & \frac{4}{3} \end{vmatrix} \quad (S1.8a)$$

corresponding exactly to the formulas (2.1) and (2.2). The stresses are then

$$\tau_{xy} = \mu\left(\frac{\partial v}{\partial x} + \frac{\partial u}{\partial y}\right) \quad \tau_{yz} = \mu\left(\frac{\partial w}{\partial y} + \frac{\partial v}{\partial z}\right) \quad \tau_{zx} = \mu\left(\frac{\partial u}{\partial z} + \frac{\partial w}{\partial x}\right)$$

$$\tau_{xx} = 2\mu\left(\frac{\partial u}{\partial x} - \frac{\theta}{3}\right) \quad \tau_{yy} = 2\mu\left(\frac{\partial v}{\partial y} - \frac{\theta}{3}\right) \quad \tau_{zz} = 2\mu\left(\frac{\partial w}{\partial z} - \frac{\theta}{3}\right)$$

(S1.11)   final expressions, viscous stresses

in the general case of three-dimensional compressible fluid motion, while $\theta = 0$ gives the limit of incompressible flow.

Exercise 30 Axes x', y' form angle α with the set x, y. Consider a fluid element of triangular form as shown in Fig. IX-11, of unit length in the x'-direction, of length $\cos \alpha$ in the x-direction, $\sin \alpha$ in the y-direction, and show that

$$\tau_{x'y'} = \tau_{xy} \cos 2\alpha - (\tau_{yy} - \tau_{xx}) \frac{\sin 2\alpha}{2}$$

$$= \mu\left(\frac{\partial v}{\partial x} + \frac{\partial u}{\partial y}\right) \cos 2\alpha - \frac{(a-b)}{2}\left(\frac{\partial u}{\partial x} - \frac{\partial v}{\partial y}\right) \sin 2\alpha \quad (S1.12)$$

Exercise 31 Express u' and v' in terms of u, v and the angle α, and also the coordinates x and y in terms of x' and y' and the same angle, in order to obtain the two derivatives appearing in (S1.6a) in terms of the x and y derivatives of the velocity components u and v, as well as the angle α.

Exercise 32 Substitute the result of the preceding exercise in (S1.6a) and equate coefficients of the various derivative terms with the corresponding terms (S1.12). Show that these lead to (S1.7) in order to complete the derivation of the viscous stress formulas (2.2) and, by extension, of (S1.11).

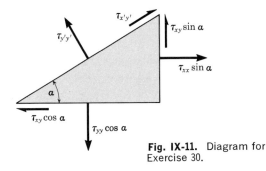

Fig. IX-11. Diagram for Exercise 30.

S2 GENERAL EQUATIONS OF UNSTEADY, VISCOUS, COMPRESSIBLE FLUID MOTION

The Navier-Stokes equations (2.4) are extended for compressible flows by taking the stress terms according to (S1.11), where $\theta \not\equiv 0$. The energy equation (8.3) is also modified, the complete set of equations of fluid mechanics being compiled and brought together in the present article.

The continuity equation, expressing a purely kinematical condition, is of course not affected by the presence of viscous friction, so that it remains

continuity equation: scalar form...

$$\frac{1}{\rho}\frac{D\rho}{Dt} + \left\{\frac{\partial u}{\partial x} + \frac{\partial v}{\partial y}\right\} = 0 \qquad (S2.1)$$

as seen, for example, by comparing with the one-dimensional compressible case (VIII.1.7) and the steady flow equation (6.4) for an incompressible fluid. To simplify the writing, the scalar equation (S2.1) is shown in two-dimensional form using cartesian coordinates, the corresponding three-dimensional form of which would include one additional term. The vector expression of the same equation, valid for one, two, or three dimensions, and any coordinate system, is

...and vector form

$$\frac{1}{\rho}\frac{D\rho}{Dt} + \nabla\cdot\mathbf{q} = 0 \qquad (S2.1a)$$

Momentum conditions (2.4) are now extended precisely as in Exercise 5, with the only difference that we do *not* require $\theta = 0$, leading to

Navier-Stokes equations

$$\frac{Du}{Dt} + \frac{1}{\rho}\frac{\partial p}{\partial x} = \nu\left\{\frac{\partial^2 u}{\partial x^2} + \frac{\partial^2 u}{\partial y^2} + \frac{1}{3}\frac{\partial \theta}{\partial x}\right\} - \frac{\partial U}{\partial x}$$

$$\frac{Dv}{Dt} + \frac{1}{\rho}\frac{\partial p}{\partial y} = \nu\left\{\frac{\partial^2 v}{\partial x^2} + \frac{\partial^2 v}{\partial y^2} + \frac{1}{3}\frac{\partial \theta}{\partial y}\right\} - \frac{\partial U}{\partial y} \qquad (S2.2)$$

In case of non-conservative forces, the last term on the right in each equation of (S2.2) is modified as in (IV.4.1), for example. The vector form of the viscous stress formulas is rather cumbersome (cf., (2.5)) and the Navier-Stokes equations of compressible fluid motion (S2.2) are not advantageously written in this notation. The pair of equations (S2.2) is our final expression of Newton's Second Law applied to a homogeneous fluid, the successive terms of which are seen to present the balance of inertia, pressure, viscous, and gravity forces. They may be compared with the system of equations (I.3.4a) of fluid statics taken as the starting point of the present development.

For later reference, we note that particular form of (S2.2) in steady, one-dimensional flow parallel to the x-axis. Writing the stress τ_{xx} as τ without subscripts gives, in the absence of external forces,

special case: steady rectilinear flow

$$\rho u \frac{\partial u}{\partial x} + \frac{\partial (p - \tau)}{\partial x} = 0$$

and since the continuity equation in the same case reduces to

$$\rho u = \text{const.} = m \qquad \text{(S2.1b)}$$

say, the preceding momentum equation is integrated at once to give

$$p + mu - \tau = \text{const.} = mu_1 B \qquad \text{(S2.2a)}$$

in terms of another constant B defined as shown and a reference velocity u.

The energy equation, likewise, is extended by proceeding as in (8.1), for unsteady, compressible fluid motion where for the sake of completeness heat addition will now also be included. If a fluid element receives heat only by molecular conduction (i.e., neglecting radiation, chemical reactions, electrical resistance, etc., any of which may be included in a manner analogous to the present development), the quantity of heat (measured in mechanical units) that crosses unit surface area in unit time is taken to be given by the product of the temperature gradient normal to the surface and the coefficient of thermal conductivity, denoted λ. Then in place of (8.3) we find directly

$$\frac{De}{Dt} + p \frac{D}{Dt}\left(\frac{1}{\rho}\right) = \frac{1}{\rho}\left\{\Phi + \lambda\left(\frac{\partial^2 T}{\partial x^2} + \frac{\partial^2 T}{\partial y^2}\right)\right\} \qquad \text{(S2.3)}$$

Energy equation, thermodynamic form . . .

if $\lambda = \text{const.}$, where Φ denotes the dissipation function appearing as the coefficient of $1/\rho$ in (8.3), also expressible in terms of the stresses as

$$\Phi = \tau_{xx} \frac{\partial u}{\partial x} + \tau_{yy} \frac{\partial v}{\partial y} + \tau_{xy} \frac{\partial u}{\partial y} + \tau_{yx} \frac{\partial v}{\partial x} \qquad \text{(S2.3')}$$

the complete expressions for three-dimensional flow obtainable immediately as before but not written explicitly in (S2.3) or (S2.3').

An alternate and sometimes useful form of the energy equation is

... and mechanical form

found directly from the balance (8.1) without the subtraction of momentum terms (cf., Exercise 25 and subsequent discussion), as

$$\rho \frac{D}{Dt}\left(h + \frac{u^2 + v^2}{2}\right) = \frac{\partial p}{\partial t} + \frac{\partial}{\partial x}\{u\tau_{xx} + v\tau_{yx}\}$$
$$+ \frac{\partial}{\partial y}\{u\tau_{yx} + v\tau_{yy}\}$$
$$+ \lambda\left\{\frac{\partial^2 T}{\partial x^2} + \frac{\partial^2 T}{\partial y^2}\right\} \quad \text{(S2.3a)}$$

where h represents the enthalpy as in (VIII.2.3). The latter equation, evidently, is generalized to account for unsteady flow, effects of fluid friction, and for heat conduction, in (S2.3a). For present purposes (S2.3) and (S2.3a) are regarded as equivalent and final forms of the energy equation, the most primitive form of which was developed as (II.5.5) and extended to non-viscous, compressible fluid in section VIII.2. In the special case of one-dimensional, steady, compressible, viscous flow, it is easily seen that in the absence of heat conduction

$$\rho u \frac{\partial}{\partial x}\left(h + \frac{u^2}{2}\right) = \frac{\partial}{\partial x}(u\tau_{xx})$$

When the stress is denoted more simply as τ, as before, and (S2.1b) is used, each of the terms of the preceding equation is recognized as an x-derivative. Integration with respect to x therefore gives

$$h + \frac{u^2}{2} - \frac{\tau}{\rho} = \text{const.} = u_1^2 A \quad \text{(S2.3b)}$$

in terms of a (constant) reference velocity u_1 and another constant A defined as shown. The equations of one-dimensional steady flow (S2.1b), (S2.2a), and (S2.3b) are studied in the following section, where particular meanings are given to the various constants introduced above.

S3 CONDITIONS WITHIN AND THICKNESS OF SHOCK WAVES

Viscosity effects are introduced into the argument leading to the normal shock wave relationships (VIII.3.3, 3.4) to analyze the finite shock transition zone analogous to the finite viscous boundary layer thickness determined in section 6 above.

One-dimensional steady flow is again assumed, and when heat conduction is neglected for the sake of simplicity, the mass, momentum, and energy conservation conditions in an ideal gas corresponding to (VIII.3.2) are extended as

$$\rho_2 u_2 = \rho_1 u_1 = \rho u = \text{const.} = m$$
$$p_2 + \rho_2 u_2^2 = p_1 + \rho_1 u_1^2 = p + mu - \tau = \text{const.} = m u_1 B \quad (S3.1)$$
$$\frac{\gamma}{\gamma - 1} \frac{p_2}{\rho_2} + \frac{u_2^2}{2} = \cdots$$
$$= \frac{\gamma}{\gamma - 1} \frac{p}{\rho} + \frac{u^2}{2} - \frac{\tau}{\rho} = \text{const.} = u_1^2 A$$

fundamental equations applied on both sides of shock wave, and within

according to (S2.1–3). As before, the constants m, A, and B are introduced as defined by (S3.1) and τ_{xx} is written more concisely as τ and given by (2.2) as

$$\tau = \frac{4}{3} \mu \frac{du}{dx} \quad (S3.2)$$

Terms ρ, u, and p without subscripts in (S3.1) refer to arbitrary points of the flow either within the transition zone ($\tau \neq 0$) or exterior to it ($\tau = 0$) where uniform conditions correspond to one or the other of the subscripts "1" and "2".

Exercise 33 Show that

$$A = \frac{1}{2} + \frac{1}{\gamma - 1} \frac{1}{M_1^2} \qquad B = 1 + \frac{1}{\gamma M_1^2} \quad (S3.3)$$

Exercise 34 Show that when velocities are normalized and written in dimensionless form by setting

$$v \equiv \frac{u}{u_1}$$

eliminating the pressures from (S3.1) yields the equation

$$\frac{1}{\gamma + 1} \frac{2\tau}{m u_1} = v^2 - \frac{2\gamma}{\gamma + 1} Bv + \frac{2(\gamma - 1)}{\gamma + 1} A \quad (S3.4)$$

When the right side of (S3.4) is denoted $f(v)$, the vanishing of the friction stress τ ahead of and behind the shock wave implies that the function f vanishes for the corresponding values $v = 1$ and $v = u_2/u_1$ as given by the normal shock relationship (VIII.3.3). Verify that this is the case by solving the quadratic equation $f(v) = 0$ and using the result of the preceding Exercise.

In terms of the physical velocities u, u_1, and u_2, (S3.4) can now be written as

$$\frac{1}{\gamma + 1} \frac{8}{3} \frac{\mu}{\rho} \frac{du}{dx} = (u - u_1)(u - u_2) \quad (S3.5)$$

velocity variation within the shock wave, differential form

which is integrable at once if the viscosity coefficient μ is regarded as constant. The variation of flow velocity is especially simple if the further restriction is imposed that the shock wave is weak so that the total velocity variation is small. In this case we have

$$\frac{\mu}{\rho} = \frac{\mu}{m} u$$

and the smallness of the shear viscosity in a gas like air justifies the use of an average value of u in the product with μ. Then the velocity dependence on location x within the shock wave is found in terms of the inverse function $x(u)$ by evaluating the indefinite integral

$$\int \frac{u \, du}{(u - u_1)(u - u_2)} \tag{S3.6}$$

where the coefficient of du in the numerator of (S3.6) can be brought outside of the integration.

Exercise 35 (i) Evaluate the integral (S3.6) *without* making use of the simplification just mentioned, i.e., integrate (S3.6) exactly as given, and simplify;

(ii) Integrate (S3.6) in the approximate manner discussed above, to obtain the indefinite integral as the bracketed term in the equation

$$x(u) = \frac{8}{3(\gamma + 1)} \frac{\mu}{m} \left\{ \frac{+u}{u_1 - u_2} \cdot \ln\left(\frac{u_1 - u}{u - u_2}\right) \right\} \tag{S3.6a}$$

What advantage is gained by the present approximation, as compared with the corresponding result in case (i)?

integrated form, inverse function: location vs velocity

An equivalent expression for the flow variation within the shock wave is obtained by replacing $\mu(u/m)$ in (S3.6a) by an average value of the kinematic viscosity ν, giving

$$x(u) = \frac{8}{3(\gamma + 1)} \frac{\nu}{u_1 - u_2} \ln\left(\frac{u_1 - u}{u - u_2}\right) \tag{S3.7}$$

It is readily seen that $x = 0$ is where the flow speed equals the mean of the upstream and downstream values, i.e., $u = (u_1 + u_2)/2$ and that each of these limiting flow speeds is strictly attained only at infinitely great (negative or positive) distances x. From the practical standpoint, as in the definition of the thickness of a viscous boundary layer, a sensible length can be easily assigned as a measure of the effective thickness by arbitrarily specifying a velocity tolerance to conform with any desired degree of accuracy. Thus if velocities ahead and behind the shock wave are defined respectively as

$$u_A = u_1 - \epsilon(u_1 - u_2) \quad \text{and} \quad u_B = u_2 + \epsilon(u_1 - u_2) \tag{S3.8}$$

then the value $\epsilon = 0.05$ corresponds to speed difference $u_A - u_B$ equal to 90 per cent of the total velocity change $u_1 - u_2$, etc.

Exercise 36 Substitute the values (S3.8) to find the distance $\delta \equiv x_B - x_A$ *practical estimation of shock thickness...*

$$\delta = \frac{8}{3(\gamma + 1)} \cdot \ln\left\{\frac{1 - \epsilon^2}{\epsilon^2}\right\} \cdot \frac{\nu}{u_1 - u_2} \quad (S3.9)$$

and from (VIII.3.3) to obtain

$$\frac{\delta \cdot u_1}{\nu} = \frac{4}{3} \cdot \ln\left\{\frac{1 - \epsilon^2}{\epsilon^2}\right\} \cdot \frac{M_1^2}{M_1^2 - 1} \quad (S3.9a) \quad \text{...in dimensionless form}$$

applicable when $0 < M_1^2 - 1 \ll 1$.

The particular interest of (S3.9) is in showing that the thickness determined by viscosity is proportional to the viscosity coefficient, and inversely proportional to the characteristic velocity $u_1 - u_2$ of the shock-generated flow (note that this is the flow speed behind a shock advancing into fluid at rest). The expression on the left side of (S3.9a), equivalent to a Reynolds Number based on shock thickness, is given by the dimensionless factors on the right, the numerical value of the product being small for all reasonable values of ϵ. The reader can easily verify that for air at standard conditions ($\nu \doteq 1.57 \times 10^{-4}$ ft^2 sec^{-1}) and speeds in excess of the acoustic velocity (of the order of 1120 ft sec^{-1}), shock wave thicknesses are of the order of a few microns, decreasing as shock strength increases. For strong shock waves, therefore, the characteristic length represented by a shock wave thickness approximates the mean free path of the molecules and a special study must be made of the validity of the continuum hypothesis. Although it is clear that effects of heat conduction and bulk viscosity must not be ignored in general, the influence of these corrections appears to be comparatively minor except in extreme cases.   *limits of validity of continuum analysis*

Exercise 37 Note that the shock thickness given by (S3.9a) varies inversely as the *first* power of the Reynolds Number formed with the appropriate characteristic quantities, while the thickness of the viscous boundary layer thickness (6.8) varies inversely as the *one-half* power of the corresponding parameter. Analyze the difference in the two cases by noting that the boundary layer thickness is determined by the balance of viscous stresses and momentum fluxes, each of which is dependent on δ in a different manner. HINT: Show that

$$\mu \frac{U}{\delta} \sim \rho U^2 \frac{\delta}{x}$$

X
Turbulence

1 TRANSITION TO TURBULENT FLOW; MEAN VALUES AND FLUCTUATIONS

In all the *laminar flows* studied in preceding chapters, streamlines curve gently around the boundary surfaces, with one fluid layer (or lamina) sliding smoothly over the next. The majority of practical flow problems also involves motions that are definitely no**t** laminar, however, the individual particle motions being highly irregular and rapidly fluctuating. In such regions of chaotically *turbulent flow*, intensive mixing normal to the principal flow direction is accomplished by disordered *velocity fluctuations* that destroy the familiar laminar flow streamline patterns. Special procedures are employed in calculations of turbulent flow velocities and resistance forces, some of the main ideas of which are traced in this chapter.

Steady laminar flow solutions of the Euler (or Navier-Stokes) equations are well confirmed observationally only for a limited range of small flow speeds. At speeds greater than some particular value, the character of the flow changes more or less abruptly from laminar to turbulent, and the same is generally true at such large distances from boundaries that any reasonably defined Reynolds number becomes large even when the velocities are small. Reynolds numbers that are sufficiently large either because of the largeness of the product of characteristic velocity and length, or because the kinematic viscosity is small, are invariably associated with turbulent flow that is always observed to be unsteady. Unsteady solutions of the flow equations present great mathematical difficulties only overcome in a few special cases. Practical flow solutions are then usually found only by recourse to drastic simplifications and semiempirical methods.

_{actual flows contain both laminar and turbulent regions}

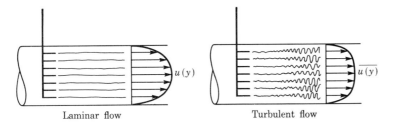

Fig. X-1 Reynolds' pipe flow experiments.

Experiments performed by the British physicist Osborne Reynolds nearly one hundred years ago provide the starting point for present understanding and calculations of turbulence (the inadequacies of which can still scarcely be understated). Reynolds injected colored dye into low-speed pipe flows in a manner that enabled him to observe sharply defined steady flow streamlines extending a certain distance from the point of injection (Fig. X-1). Beyond this distance the dye filaments first exhibited slight waviness, increasing with distance and ultimately becoming so severely irregular as to lose individual identity in the "transition" zone ahead of the fully turbulent flow regime associated with large values of the parameter now recognized as the Reynolds number. Although it is not possible to specify precisely one location where transition is properly considered to occur, fluid particles in the ensuing turbulent flow appear to move as if carried at a uniform speed (corresponding to the local mean flow) and simultaneously erratically buffeted about each instantaneous mean position. The physical situation is reminiscent of the atomistic description of a laminar fluid motion: uniform mean flow superimposed on random molecular agitations that are both far too complicated to analyze in detail and equally superfluous for most practical purposes.

Following the implicit suggestion of the kinetic theory of gases, early workers in turbulence accordingly established theoretical models that decompose the total motion of each fluid particle into a mean flow plus unsteady fluctuations assumed to be more or less random in character. Before details are presented, it is pointed out that the present state of knowledge does not permit the same degree of precision in dealing with turbulence that is customary with laminar flow, where the underlying physical principles are applied with comparatively fewer uncertainties and approximation errors. Results are typically imprecise as, for example, in specifying the "critical" values of the Reynolds number at which laminar flow ceases, or that at which turbulence is "fully established." Thus absence of sharp discontinuity between laminar and turbulent regions

Reynolds himself found values of the order of 2000 when he varied the kinematic viscosity of water in horizontal pipe flow by changing the temperature of the water, while well-rounded pipe inlets and carefully controlled supply conditions have led to values as great as 40,000 for the same flows. Boundary layers in high-speed flows, on the other hand, have been observed to remain laminar for values of the order of 10^6 and 10^7. It is fair to conclude that the Reynolds number is only one criterion affecting turbulence, and that others are either not known at all or only so poorly understood as to be of little general value. Questions about the dynamic stability of fluid motions are doubtless also relevant but also insufficiently developed to permit meaningful application to problems of turbulent flow.

Quantitative study begins with recognition that the complete description of a turbulent motion is, as already mentioned, hopelessly complicated by the continual agitation of turbulent "eddies" that can be likened to the actual molecules of a fluid in their unceasing fluctuations. These are so disorderly as to produce essentially zero displacements and velocities on any macroscopic scale. Since in every case only the *mean* motion of many neighboring fluid particles (or molecules) is of interest, the first requirement is to define a procedure for distinguishing between the mean velocities and the fluctuations at each point. When this is accomplished in a suitable manner for other flow quantities as well (e.g., the pressure), it will be seen in what manner the familiar conservation principles may be applied. Mean values (in time) are first related to instantaneous values as follows (Fig. X-2). If T denotes any time interval assumed to be great in comparison with the period of the flow irregularities, or fluctuations, then the mean value of the velocity component $u(x, y, z, t)$ is defined as

$$\bar{u} = \frac{1}{T}\int_0^T u(t)\,dt \qquad (1.1)$$

At any location $\bar{u} = \bar{u}(x, y, z)$ is independent of time if external (i.e., boundary) conditions are unchanging. The difference between

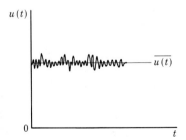

Fig. X-2 Mean value of turbulent flow quantity.

the instantaneous value u and the mean value $\bar{u}$, termed the *fluctuating component* of u and denoted u', is related to the former quantities by writing simply

$$u = \bar{u} + u'(t) \qquad (1.2)$$

decompose total turbulent motion: mean and fluctuating components

where the dependence on position coordinates x, y, z is suppressed. It is clear that the mean value of the fluctuation u' is zero:

$$\bar{u'} = \frac{1}{T}\int_0^T u'(t)\, dt = 0 \qquad (1.3)$$

and that the same rules apply to different velocity components, to the fluid pressure, and to all other quantities separable into mean and fluctuating components in the manner of (1.2).

Exercise 1 Establish the following rules for the algebra and calculus of mean values.
(a) The mean value of a constant multiplied by a turbulent quantity is given by the constant multiplied by the mean value of the quantity:

elementary properties of mean values

$$\overline{cu} = \frac{1}{T}\int_0^T \{cu(t)\}\, dt = c\bar{u} \qquad (1.4a)$$

(b) The mean value of the sum of two turbulent quantities equals the sum of their mean values:

$$\overline{u_1 + u_2} = \frac{1}{T}\int_0^T \{u_1(t) + u_2(t)\}\, dt = \overline{u_1} + \overline{u_2} \qquad (1.4b)$$

(c) The mean value of a mean value is the mean value itself:

$$\bar{\bar{u}} = \bar{u} \qquad (1.4c)$$

(d) The derivative of a turbulent quantity equals the sum of the derivatives of the mean value and the fluctuating component:

$$\frac{\partial u}{\partial y} = \frac{\partial \bar{u}}{\partial y} + \frac{\partial u'}{\partial y} \qquad (1.4d)$$

(e) The mean value of a spatial derivative is given by the derivative of the mean value:

$$\overline{\frac{\partial u}{\partial y}} = \frac{\partial \bar{u}}{\partial y} \qquad (1.4e)$$

(f) The time derivative of a mean value is zero:

$$\frac{\partial \bar{u}}{\partial t} = 0 \qquad (1.4f)$$

(g) The mean value of the product of two quantities is the sum of the product of their mean values plus the mean value of the product of the fluctuations:

$$\overline{uv} = \bar{u}\bar{v} + \overline{u'v'} \qquad (1.4g)$$

and the last quantity in (1.4g) is not necessarily zero.

248 TURBULENCE

mean velocity components satisfy usual continuity equation ...

(h) The mean values of the velocity components of an incompressible fluid in turbulent motion in two dimensions are interrelated by the equation

$$\frac{\partial \bar{u}}{\partial x} + \frac{\partial \bar{v}}{\partial y} = 0 \tag{1.4h}$$

2 EQUATIONS OF TURBULENT FLUID MOTION — REYNOLDS STRESSES

The mean velocities satisfy the continuity equation (1.4h) in a form identical with that of the continuity equation expressed in terms of the total velocity components, to which it reduces in fact when the flow is laminar, i.e., when the fluctuations are identically zero. It also follows that the instantaneous values of the fluctuations satisfy the same equation

... and the fluctuations also

$$\frac{\partial u'}{\partial x} + \frac{\partial v'}{\partial y} = 0 \tag{2.1}$$

as seen from the definition of flunctuations (1.2) and from the fact that the mean values and the total values separately satisfy the continuity equation [using relationships (1.4b,e)]. The relationship (1.4h) is more useful than (2.1) from the standpoint of determining the overall flow, because it furnishes an equation between two of the mean flow quantities:

$$\frac{\partial \bar{u}}{\partial x} + \frac{\partial \bar{v}}{\partial y} = 0 \tag{2.2}$$

Additional relationships between $\bar{u}$, $\bar{v}$, and $\bar{p}$ may be found from the Navier-Stokes equations (IX.2.4) by applying the averaging rules (1.4) to each sum of terms that add to zero in expressing the balance of forces and inertia.

When gravity is neglected and the viscous stresses of incompressible fluid motion are written as in (IX.2.1) and (IX.2.2), the total flow quantities satisfy the equation

total values

$$\frac{\partial u}{\partial t} + u\frac{\partial u}{\partial x} + v\frac{\partial u}{\partial y} + \frac{1}{\rho}\frac{\partial p}{\partial x} = \frac{1}{\rho}\left\{\frac{\partial \tau_{xx}}{\partial x} + \frac{\partial \tau_{xy}}{\partial y}\right\} \tag{2.3}$$

as well as the equation obtained from (2.3) by taking the mean values of the separate terms:

mean values

$$\overline{\frac{\partial u}{\partial t}} + \overline{u\frac{\partial u}{\partial x}} + \overline{v\frac{\partial u}{\partial y}} + \overline{\frac{1}{\rho}\frac{\partial p}{\partial x}} = \frac{1}{\rho}\left\{\overline{\frac{\partial \tau_{xx}}{\partial x}} + \overline{\frac{\partial \tau_{xy}}{\partial y}}\right\} \tag{2.3a}$$

according to (1.4b).

The determination of mean flow quantities could proceed in the same manner as in laminar flow if all the terms of (2.3a) could be reduced to exactly the same form as (2.3) with barred quantities replacing the total values, since this correspondence has already been established in (2.2) for the continuity equation. The rules (1.4) show that this is true only for the linear terms, while nonlinear terms like $u(\partial u/\partial x)$ require special attention. Thus, although the time mean value of each of the fluctuations u' and v' is separately zero, the mean value of their product is nonzero if they are not completely independent of each other at any point but are, instead, *correlated*. It is already apparent from (2.1) that fluctuations u' and v' are in fact interrelated by their spatial derivatives, and, since this type of correlation is still more obvious in the case of u' and u' itself at two nearby points, a separate investigation of the second and third terms of (2.3a) is required.

Applying the property indicated by (1.4g) to the product of u and $\partial u/\partial x$, for example, each factor being expressed as a sum of the form (1.2), gives

$$\overline{u\frac{\partial u}{\partial x}} = \bar{u}\frac{\partial \bar{u}}{\partial x} + \overline{u'\frac{\partial u'}{\partial x}} \qquad (2.4a)$$

while the third term of (2.3a) is similarly expressed as

$$\overline{v\frac{\partial u}{\partial y}} = \bar{v}\frac{\partial \bar{u}}{\partial y} + \overline{v'\frac{\partial u'}{\partial y}} \qquad (2.4b)$$

The last terms of (2.4a) and (2.4b) are written as derivatives of fluctuation products directly by noting that

$$\frac{\partial}{\partial x}(\overline{u'u'}) + \frac{\partial}{\partial y}(\overline{u'v'}) = \overline{u'\left(\frac{\partial u'}{\partial x} + \frac{\partial v'}{\partial y}\right)} + \overline{u'\frac{\partial u'}{\partial x}} + \overline{v'\frac{\partial u'}{\partial y}} \qquad (2.5)$$

The vanishing of the first pair of terms in parentheses on the right side of (2.5) according to (2.1) indicates that the sum of (2.4a) and (2.4b) is given by (2.5); hence (2.3a) becomes

$$\frac{\partial \bar{u}}{\partial t} + \bar{u}\frac{\partial \bar{u}}{\partial x} + \bar{v}\frac{\partial \bar{u}}{\partial y} + \frac{1}{\rho}\frac{\partial \bar{p}}{\partial x}$$
$$= \frac{1}{\rho}\left[\frac{\partial}{\partial x}\{\bar{\tau}_{xx} - \rho\overline{u'u'}\} + \frac{\partial}{\partial y}\{\bar{\tau}_{xy} - \rho\overline{u'v'}\}\right] \qquad (2.6)$$

(2.3) and (2.3a) differ only to the extent that fluctuations are correlated

which differs in form from (2.3) only by the presence of the two terms on the right side of (2.6) that contain the mean values of fluctuation products. These terms appear as additive to the viscous stress terms, and it is therefore possible to regard the problem of determining a turbulent flow as equivalent to that for laminar flow (mean values being understood in the present case), with the one difference that the stress formulae are not to be taken in the form in (IX.2.1,2) but as shown in the braces of (2.6).

Reynolds stresses account for the characteristic properties of turbulent flows

The additional "apparent" turbulent stress terms, denoted

$$T_{xx} = -\rho\overline{u'u'} \qquad T_{xy} = -\rho\overline{u'v'} \qquad (2.7)$$

and commonly termed Reynolds stresses, alone account for the greatly different character of laminar and turbulent flows. Hence, although it appears that if the fluctuations were uncorrelated there would be no difference between the two types of motion, the fact that the turbulent fluctuations are in fact usually both appreciable and correlated results in their total domination of the ordinary viscous stresses. When these terms are several orders of magnitudes greater than the viscous stresses, the latter are ignored completely in determining the mean flow from (2.6) and its companion equations. The main problem is then the expression of Reynolds stresses in terms of the measurable mean flow quantities. Wholly insufficient understanding of the physical mechanisms of turbulence at this juncture has led to a variety of approximations and conjectures aimed at the estimation of Reynolds stress terms like (2.7).

3 CORRELATION OF VELOCITIES: MIXING LENGTH HYPOTHESIS

seek expressions for Reynolds stresses in terms of mean flow velocities

An early approach to the question of estimating the Reynolds stresses was based on consideration of the correlation of velocity fluctuations in simple shear flow, $\bar{u}(y)$, $\bar{v} \equiv 0$. If it is supposed for definiteness that a fluid particle at $y = y_0$ experiences a fluctuation of velocity $v' > 0$ normal to the direction of mean flow (Fig. X-3), while its velocity in the flow direction is given by the mean value $\bar{u}(y_0)$ at the level y_0 then its velocity in the flow direction at a level $y > y_0$ will differ from that corresponding to the mean value $\bar{u}(y)$ there. Since

$$\bar{u}(y) = \bar{u}(y_0) + (y - y_0)\frac{d\bar{u}}{dy} \qquad (3.1)$$

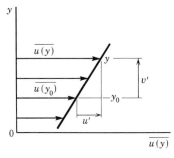

Fig. X-3 Turbulent velocity fluctuations in shear flow.

when the displacement $y - y_0$ is so small that the Taylor series expansion in $(y - y_0)$ can be truncated as shown, the fluid particle in its new location exhibits a velocity deficiency (fluctuation)

$$u' = -(y - y_0)\frac{d\bar{u}}{dy} < 0 \quad (3.2)$$

a basis for estimating correlation of fluctuations

in its x-component of velocity when $d\bar{u}/dy > 0$ as shown. Hence

$$u'v' < 0 \quad \text{when} \quad \frac{du}{dy} > 0 \quad (3.3)$$

The same conclusion is drawn from consideration of a particle displaced in the opposite direction, $v' < 0$, since the positive fluctuation $u' > 0$ is in this case attributed to the excess of mean velocity of the particle carried into a region of slower flow than at the "level of origin," y_0.

By supposing, in addition, that the fluctuation v' may also be expressed in the same form as u', i.e. as a distance multiplied by a velocity derivative as in (3.2), a single turbulent shear stress T can be written in analogy with the viscous shear stress τ of (IX.1.3) as

introduction of "mixing length"

$$T = \rho l^2 \left|\frac{d\bar{u}}{dy}\right| \frac{d\bar{u}}{dy} \quad (3.4)$$

where l is an appropriate length to be so determined as to provide agreement with actual stresses [it is noted that, by taking the absolute value of the velocity derivative in one of its powers in (3.4), the sign of the turbulent stress is positive when the shear $d\bar{u}/dy > 0$, etc.].

The *mixing length* l bears an obvious resemblance to the mean free path of molecules in the random thermal displacements considered in the kinetic theory of gases. The chaotic appearance of the turbulent velocity fluctuations has provided the basis for forming the analogy between turbulent *eddies* and gaseous molecules. From the standpoint of determining turbulent flows, however, (3.4) must be regarded as nothing more than a definition that shifts attention to the evaluation of the mixing length l instead of the turbulent fluctuations in (2.7). The concept of a mixing length was introduced by L. Prandtl, (3.4) being known as Prandtl's mixing length hypothesis, and has been exploited in numerous ways. Prandtl himself assumed that momentum is conserved as the turbulent eddy moves over a distance equal to the mixing length, whereas Sir Geoffrey Taylor, with better justification, imposed a condition of vorticity conservation. A somewhat more elegant development, due to Th. von Kármán, is based on the assumption that turbulence patterns exhibit a certain *similarity* at all points.

hypotheses for estimating mixing length
(i) momentum conservation
(ii) vorticity conservation
(iii) similarity

The usefulness of (3.4) lies in the fact that it has been successfully applied to flow inside pipes, along walls and plates, and also in jets surrounded by still fluid. The derivation of common design formulae are indicated in a few cases in the following sections.

4 UNIVERSAL RESISTANCE LAW FOR SMOOTH PIPES, LARGE REYNOLDS NUMBER

The frictional resistance at a flow boundary is again calculated in terms of the stress at the boundary surface. Examination of (2.6) at points of turbulent flow very close to the solid boundary reveals that the shear stress T_{xy} is constant, since only in this case is a balance established between the turbulent shear stress and the other small terms appearing in that equation. If the wall stress is denoted τ_0 as before and the mixing length hypothesis (3.4) is adopted, the mean velocity $\bar{u}(y)$ is found by integration when l is specified. Since fluid motion is considered to be most strongly inhibited by the presence of a boundary surface at points closest to it, a simple assumption for the mixing length takes l proportional to the distance y from the wall:

mixing length proportional to the local characteristic length

$$l = \kappa y \tag{4.1}$$

It is supposed that κ is constant, i.e., independent of y; therefore (3.4) may be written as

stress expressed as a function of mean flow — and an empirical constant

$$\tau_0 = \rho \kappa^2 y^2 \left(\frac{d\bar{u}}{dy}\right)^2 \tag{4.2}$$

in a form suitable for integration to find $\bar{u}(y)$ in terms of τ_0 and κ.

Exercise 2 Obtain the *universal velocity distribution law* for very large Reynolds number in the form

$$\bar{u}(y) = \frac{u_*}{\kappa} \ln \frac{u_* y}{\nu} + c \tag{4.3}$$

where c is an integration constant and the *friction velocity* u_* is defined by the relationship

"friction velocity" defined

$$u_*^2 \equiv \frac{\tau_0}{\rho} \tag{4.4}$$

by integrating (4.2).

The discharge Q through a pipe of diameter D furnishes the value of the average velocity V as in (IX.1.9), the integration of (4.3) being carried out between limits $y = 0$ (pipe wall) and $y = D/2$

X. 4 / UNIVERSAL RESISTANCE LAW FOR SMOOTH PIPES, LARGE REYNOLDS NUMBER

(pipe axis). When the Reynolds number is very large, the kinematic viscosity ν may be treated as very small, and it is thus found that V is expressible approximately as

$$V = \frac{u_*}{\kappa} \ln\left(\frac{u_*}{\nu}\frac{D}{2}\right) \tag{4.5}$$

pipe flow — average velocity in terms of mean flow-Reynolds number

Exercise 3 Carry out the integration of (4.3) to establish (4.5) and display the neglected terms. HINT: Introduce the integration variable $Y = (u_* y)/\nu$ and note that

$$\int Y \ln Y \, dY = \frac{Y^2}{2}(\ln Y) - \frac{Y}{2}$$

as well as the fact that

$$\lim_{Y \to 0} \left(\frac{Y^2}{2} \ln Y\right) = 0$$

Exercise 4 Define the friction factor as before, i.e., as the resistance force per unit length, $\tau_0 \pi D$, normalized with respect to the average flow quantities (IX.1.10), to show that

$$u_* = V\sqrt{\frac{f}{8}}$$

and establish from (4.5) the widely used logarithmic formula for the friction factor of turbulent flow in smooth pipes at very great Reynolds number:

$$\frac{1}{\sqrt{f}} = \frac{1}{\sqrt{8}\,\kappa} \ln\left(\frac{R}{2}\sqrt{\frac{f}{8}}\right) \tag{4.6}$$

friction factor — logarithmic law

The constant κ is not calculated theoretically, but has been estimated on the basis of experimental measurements of turbulent flow in circular pipes. Values $\kappa = 0.4$ and slightly different are used by different workers, sometimes with empirical adjustments in the form of an additive constant on the right side of (4.6).

Equation (4.6) is the counterpart of turbulent flow of the formula (IX.1.11) for friction factor in laminar pipe flow. Because in both cases the Reynolds number R is based on the average velocity and pipe diameter the two relationships may be compared to display the essential difference between laminar and turbulent friction effects. Although (4.6) is not easily solved explicitly for f as a function of R, its main features are evident from the sketch in Fig. X-4, which shows that turbulent friction generally is significantly greater than laminar at flow speeds corresponding to high values of the Reynolds number. On account of the wide range of values

turbulent stresses greater than laminar; diminish more slowly with increasing Reynolds number

254 TURBULENCE

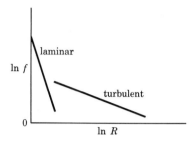

Fig. X-4 Frictional coefficient as function of Reynolds number; laminar and turbulent flows.

of R over which the two formulae are considered (roughly from 0 to 10^6 or even greater), logarithmic scales are employed.

The laminar case is represented by (IX.1.11) in logarithmic coordinates as a straight line, while the friction factor curve for turbulent flow is also nearly straight, having a smaller slope and hence decreasing more slowly with increasing Reynolds number.

Exercise 5 Set $y \equiv \ln f$, $x \equiv \ln R$ so that laminar flow is represented by a line having a slope given by $dy/dx = -1$, and show that when $y = -2$ the curve for turbulent flow has a slope roughly half as great as that for laminar flow. Turbulent flow resistance therefore decreases more slowly with increasing Reynolds number than does laminar pipe flow resistance. The characteristically higher resistance forces in turbulent flow are thereby demonstrated.

5 TURBULENT FLAT PLATE BOUNDARY LAYER; LOGARITHMIC FORMULA

The logarithmic variation of velocity (4.3) is applied to the calculation of boundary layer shear stresses in high-speed turbulent flow at high Reynolds Number. The procedure differs from that of Sections (IX.6,7) for laminar flow in three basic respects. Instead of calculating the shear stress as a velocity derivative in the manner of Exercise IX.18, it is noted first from (4.2) and (4.4) that the friction velocity u_* provides directly an expression for the stress τ_0, so that substitution of (4.3) in (4.2) simply yields an identity. It is also seen that (4.3) is invalid right at the boundary surface $y = 0$, where the logarithm is not defined. This is explained physically by observing that very close to the boundary surface the fluid is nearly at rest and its motion is laminar (i.e., zero fluctuations u' and v') within a very narrow zone called the *laminar sublayer*. Finally, the turbulent boundary layer can only be well defined if it has an outer edge that separates it from an exterior uniform flow, this edge giving rise to a definite turbulent boundary layer thickness $\delta(x)$ and the choice of an

boundary layer: laminar sublayer between solid boundary and turbulent flow origin for x also presents a new difficulty. The reason is that the turbulent boundary layer is usually a continuation of a laminar boundary layer upstream of it, i.e., corresponding to the smaller Reynolds number values (IX.6.8) at which the flow is laminar, but the transition to turbulent flow occurs at ill-defined and generally unpredictable locations. Under these circumstances, estimates of turbulent frictional resistance in boundary layers can be pursued only in a rough qualitative manner, while corrections to final formulae, for improved accuracy, are mainly empirical. An enormous number of measurements have been made for this purpose.

ill-defined position of upstream limit of turbulent boundary layer

When the turbulent flow is confined to a narrow zone bounded externally by uniform mean motion at speed U, (4.3) provides a relationship between the stress [expressed as u_* with the aid of (4.4)] and the thickness $\delta(x)$ of the layer:

turbulent friction stress related to Reynolds number

$$U = \frac{u_*}{\kappa} \ln \frac{u_* \delta}{\nu} \qquad (5.1)$$

A second relationship between the same two quantities is commonly taken (on dimensional grounds for lack of better information) as

$$\frac{\delta}{x} \doteq \frac{u_*}{U} \qquad (5.2)$$

This equation (5.2) is not regarded as having precise quantitative significance.

The stress τ_0 is next obtained by elimination between (5.1) and (5.2), using (4.4), the equations being written in more compact form by defining the dimensionless shear stress σ as

$$\sigma \equiv \frac{\tau_0}{(\rho/2)U^2} \qquad (5.3)$$

dimensionless shear stress...

Then the dimensionless stress is found, analogously to (4.6), from (5.1) as

$$\frac{\sqrt{2}}{\sqrt{\sigma}} = \frac{1}{\kappa} \ln \sigma R \qquad (5.4)$$

...given by a logarithmic formula

with "logarithmic accuracy," where, as in (IX. 6.15a)

$$R = \frac{Ux}{\nu}$$

U is the constant external flow speed. Friction stresses are calculated from equations of the form (5.4) in practice, although various modifications are employed by different workers. One commonly used formula obtained by the addition of an additive constant on the right side of (5.4), based on experimental measurements, is

a commonly used stress formula

$$\frac{1}{\sqrt{\sigma}} = 1.7 \ln \sigma R + 3.0 \quad (5.5)$$

Exercise 6 (a) By eliminating between the preceding equations, evaluate the neglected quantity when (5.4) appears as above. (b) Determine the value of the constant κ and of the additive correction to (5.4) that brings this equation strictly into accord with (5.5).

6 ONE-SEVENTH POWER LAW — MOMENTUM INTEGRAL FOR TURBULENT FLOW

When the order-of-magnitude arguments used in boundary layer theory are applied to the mean motion in a turbulent flow, it is easy to see that (2.6) simplifies to the same form as Prandtl's boundary layer equation (IX.7.2) for laminar flow. The momentum considerations that led to (IX.7.10) are then also valid for the turbulent boundary layer, where the stress τ_0 may either again be taken in the form (4.2) or be chosen in the manner of Exercise IX.23. Momentum thickness and displacement thickness definitions remain unchanged but with the understanding that mean velocities are employed. The application of the momentum integral equation to turbulent flow is now indicated for the case when the velocity variation in the boundary layer is given by an experimentally well-confirmed and simple algebraic relationship instead of the logarithmic function (4.3). For simplicity it is again assumed that the external flow is uniform, so that the momentum integral equation is

application of boundary layer integral equations

$$\frac{\tau_0}{\rho U^2} = \frac{d\theta}{dx} \quad (6.1)$$

which is a special case of (IX.7.11).

The stress τ_0 is taken from a formula of Blasius based on pressure-drop measurements in smooth pipes at moderate Reynolds numbers, expressible as

Blasius' formula: stress proportional to inverse fourth root of R

$$\frac{\tau_0}{\rho U^2} = \frac{0.0228}{(U\delta/\nu)^{1/4}} \quad (6.2)$$

that is, as the inverse *one-fourth* power of a Reynolds number based on the boundary layer thickness δ. It can be shown that the stress (6.2) corresponds to a much favored velocity profile according to which the mean velocity increases as the one-seventh power of the distance from the boundary wall:

"one-seventh power law" of turbulent velocity variation

$$\frac{\bar{u}}{U} = \left(\frac{y}{\delta}\right)^{1/7} \quad (6.3)$$

Exercise 7 Show that the momentum thickness for the velocity profile (6.3) is given by

$$\frac{\theta}{\delta} = \frac{7}{72} \qquad (6.4)$$

Exercise 8 Substitute (6.2) and (6.4) in (6.1) to obtain an equation for the boundary layer thickness $\delta(x)$. Show that the boundary layer thickens according to the *four-fifths* power of x.

Ans. $\delta/x = 0.376/(R)^{1/5}$

XI
The Molecular Basis of Fluid Mechanics

1 IDENTIFICATION OF PRESSURE AND TEMPERATURE

The molecular structure of matter is a view so firmly established in modern science that the question naturally arises whether persistence in the patently fallacious representation of fluids as *continuous* media is justifiable. Against the aesthetic objection that the dual view conflicts with the basic goal of science to unify different branches of knowledge, the indisputably great practical successes of the continuum description in a very wide range of physical phenomena is the most eloquent testimony to its proper value. For as soon as it is recalled that the number of molecules in a gas at standard conditions is roughly 10^{19}cm^{-3}, and that each molecule suffers upwards of one billion collisions per second, the staggering proportions of the alternative detailed calculations for the individual molecules in even the simplest flow situation is entirely evident — not to speak of the hopelessly cumbersome quantity of data that the results would entail if they could indeed be obtained. There are compelling reasons, nonetheless, for establishing the connection between the continuum and molecular descriptions. Logical completeness is one motive not to be denied, while practical caution may be deemed no less important, since the continuum treatment itself furnishes no clues as to when its calculations will fail by reason of overtaxed assumptions and idealizations.

hopeless complexity of complete molecular motion — and its uselessness from practical standpoint for study of macroscopic flow features

As in all physical theory, avoidance of undue complication in the study of a gas from the molecular, or "kinetic," standpoint requires that only the most essential features of the molecules and their behavior should be specified at the outset. Thus the *model* of an ideal gas is taken to conform to the following properties:

(a) Because molecules are all identical in form and their velocities are *randomly* distributed in direction, half of them are moving in any given direction at any time, half in the opposite direction.

(b) Each molecule is considered to occupy negligible volume and is therefore regarded as a point mass.

simplified physical model

(c) Molecular impacts with the walls of a containing vessel are perfectly elastic (intermolecular collisions are precluded by the vanishing of their cross sections).

Fluid pressure is computed by considering the reaction to the momentum reversal as a molecule impinges on and is reflected from a solid boundary. A molecule of mass m and speed v normal to the reflecting wall experiences a change of momentum amounting to $2mv$. The number of molecules in unit volume is denoted N, half of them are moving toward the wall at any time. The number of molecules that strike one unit of its area in one unit of time is $Nv/2$ if each one has the same normal velocity v. The corresponding momentum change in the gas per unit time is accordingly Nmv^2, subject to correction to account for the fact that the molecular velocities are supposed randomly directed.

pressure interpreted in fundamental terms: momentum reversals

Exercise 1 Show that, when account is taken of the randomness of molecular velocity directions, the fluid pressure is expressed in terms of the mean square velocity $\overline{v^2}$ as

$$p = \frac{Nm\,\overline{v^2}}{3} \tag{1.1}$$

corrected expression for pressure in terms of mean square velocity of molecules

HINT: Replace v in the preceding argument by $v\cos\theta$, where θ measures the velocity direction from the surface normal, and represent the θ-dependence of molecular density by noting that the end points of the N velocity vectors drawn from a common center describe the surface of a sphere. If its radius is unity, the surface is 4π, the density per unit area is $N/4\pi$, and thus the number of molecules moving at angle θ from the surface normal is $(N/4\pi)2\pi \sin\theta\, d\theta$ (cf. Section I.6). Obtain (1.1) by an integration of

$$2mv\cos\theta\, \frac{N}{4\pi}\, 2\pi \sin\theta\, d\theta\, v\cos\theta$$

and allow for the variation of molecular speeds v by introducing the spatial mean square as shown.

According to the meaning of the fluid density as the mass contained within unit volume, this density is expressed in terms of the molecular mass m and number density N as

$$\rho = Nm \tag{1.2}$$

macroscopic fluid density related to molecular characteristics...

The temperature of an ideal gas is similarly expressed with the aid of (1.1) and (1.2) in terms of the specific gas constant R and the spatial mean squared velocity $\overline{v^2}$ in the thermodynamic equation of state:

...and temperature...

$$RT = \frac{\overline{v^2}}{3} \tag{1.3}$$

A further reduction is obtained by noting that R is related to the *universal gas constant* $\tilde{R}$ and the molar weight of a gas. Thus R is given by the product of the mass m of a single molecule and the number N_0 of molecules in one gram-mole:

...and the specific gas constant

$$R = \frac{\tilde{R}}{N_0 m}$$

The number N_0, commonly known as Avogadro's number (sometimes as Loschmidt's), is taken as

$$N_0 = 6.06 \times 10^{23}$$

while the numerical value of the universal gas constant is

$$\tilde{R} = 8.315 \times 10^7 \text{ erg } (°K)^{-1}$$

It may be noted that the ratio of the two preceding constants arises frequently in thermodynamics and in physics broadly. Said to be the most important number in all of physical theory, *Boltzmann's constant* k is found to have the value

Fundamental physical constants
$N_0 = 6.06 \times 10^{23}$
$\tilde{R} = 8.315 \times 10^7 \text{ erg } (°K)^{-1}$
$k = 1.371 \times 10^{-16} \text{ erg } (°K)^{-1}$

$$k = \frac{\tilde{R}}{N_0} = 1.371 \times 10^{-16} \text{ erg } (°K)^{-1} \tag{1.4}$$

Exercise 2 Adopt the value of R given in Section I.5 in cgs units, to determine the effective molecular weight of air considered as an ideal gas. What is the mass of one air "molecule"?

Ans. $m = 4.76 \times 10^{-23}$ gm.

Exercise 3 Show that the mean kinetic energy of one molecule is determined by its temperature alone as

$$\tfrac{3}{2}kT$$

independently of molecular weight and of molecular mean square velocity.

Exercise 4 Recalling that the standard atmospheric sea level pressure is 1.013×10^6 dynes cm^{-2}, use the preceding values to determine the number of molecules, N, in unit volume at standard temperature (0°C) and pressure. Find the root mean square velocity of a molecule of air under the same conditions (i.e., the square root of $\overline{v^2}$).

2 FLOW VELOCITY EXPRESSED IN TERMS OF MOLECULAR SPEEDS; BOLTZMANN'S EQUATION

If molecular velocities are randomly distributed with respect to directions, the same is certainly not true of their magnitudes, the molecular speeds. While detailed calculations for all the individual molecules is out of the question for practical reasons, a satisfactory representation of the molecular motions for many purposes is obtained in the form of statistical statements giving the fractional number of molecules in a given condition at any time and location within a fluid. A discussion of the particular form taken by the statistical description is deferred until Section 3, and the manner in which such familiar macroscopic quantities as the fluid velocity can be evaluated when the particle statistics is known is now indicated.

statistical approach

Because of the very large numbers of particles under discussion, it is supposed that their individual velocities cover a wide range of values in such profusion that the *fractional* number density at each speed varies continuously from one place to another. Then a *distribution function* $f(x, y, z, v_x, v_y, v_z)$ gives the number of molecules situated in unit volume around the point x, y, z and having velocities ranging from v_x, v_y, v_z to $v_x + dv_x, v_y + dv_y$, etc., where f is considered to be a continuous and differentiable function of each of the six arguments $x, y, \ldots, v_y, v_z$. In a slightly more compact vector notation that extends the notion of a distribution function to the case of unsteady motion,

distribution function for positions and velocities

$$(\mathbf{r}, \mathbf{v}, t)\, d\mathbf{r}\, d\mathbf{v} \tag{2.1}$$

vector expression: number density in physical space and velocity space

is written to indicate the fractional f number of molecules which are located at time t within the range of coordinates $dx\, dy\, dz$ of the values x, y, z, and moving with velocity components as indicated above.

Summing over all velocities gives the particle number density at any location and time as

$$N(\mathbf{r}, t) = \int f(\mathbf{r}, \mathbf{v}, t)\, d\mathbf{v} \tag{2.2}$$

density related to distribution function f

where the integration limits for each velocity component may be taken as $-\infty$ and $+\infty$. Multiplication of (2.2) by the appropriate particle mass m recovers the familiar macroscopic fluid density (1.2), also interpreted as the expected or probable average number of molecules per unit volume at location $\mathbf{r}$ and time t. It is also clear that the distribution function f provides a convenient averaging technique for different fluid properties. If the component of velocity v_x is considered, for example, the fractional number of molecules per unit volume having speeds in the range $d\mathbf{v}$ of this value is simply the integral of $f\,d\mathbf{v}$ with respect to the components v_y and v_z (one of the arguments of the function f being v_x). For all the N molecules the average of their v_x velocity components recovers the familiar macroscopic Cartesian velocity component denoted u in the preceding chapters:

$$u(x, y, z, t) = \frac{1}{N} \int v_x f(\mathbf{r}, \mathbf{v}, t)\,d\mathbf{v} \qquad (2.3)$$

macroscopic flow properties expressed as integrals of f

where the integration now extends over all the velocity components, i.e., $d\mathbf{v} = dv_x \cdot dv_y \cdot dv_z$. Equation (2.3) may be regarded as the fundamental definition of the quantity on the left side. The other velocity components are also obtained as averages over the total range of respective molecular velocities in the same manner.

In view of the importance of the distribution function f in relating the molecular characteristics to measurable macroscopic flow quantities, it is useful to indicate its most important property, which also permits a development of the flow equations of Chapter II, and their subsequent extensions, on a purely "microscopic" basis. For this purpose it is necessary to consider the "total" time derivative of the distribution function in the manner of the streamline derivative of Section II.1.

Although the total number of molecules occupying a given region of fluid at any time is, of course, equal to the number of molecules in the region subsequently occupied by the same molecules, it is not to be expected that all of the gross properties defining the state of the fluid (characterized by the distribution function f) will necessarily also be constant in time. The variation of f itself with time is seen from (2.1) to depend on coordinates $x(t)$, $y(t)$, etc., represented by $\mathbf{r}(t)$, as well as on the three velocity components denoted $\mathbf{v}(t)$, so that a total time derivative of f contains no fewer than seven terms. The typical dependence on the velocity is indicated by applying the chain rule of differentiation giving rise to terms like:

time-variation of f: the total time derivative in "phase" space

$$\frac{\partial f}{\partial v_x} \frac{\partial v_x}{\partial t} = \frac{\partial f}{\partial v_x} \frac{F_x}{m}$$

where the x-component of particle acceleration is replaced by the appropriate force component F_x divided by the molecular mass. In the absence of all molecular collisions the total time derivative of f vanishes, since even the possible accelerating effect of external forces is accounted for by virtue of the definition of f as a function of the particle velocities. When collisions between molecules and boundary walls and between a molecule and other molecules are considered, however the corresponding changes in f are equated to the total time derivative. The result is *Boltzmann's equation*

$$\frac{\partial f}{\partial t} + v_x \frac{\partial f}{\partial x} + v_y \frac{\partial f}{\partial y} + v_z \frac{\partial f}{\partial z} + \frac{F_x}{m} \frac{\partial f}{\partial v_x} + \frac{F_y}{m} \frac{\partial f}{\partial v_y}$$
$$+ \frac{F_z}{m} \frac{\partial f}{\partial v_z} = \left(\frac{\partial f}{\partial t}\right)_{\text{coll.}} \quad (2.4)$$

Boltzmann's equation: f varies only as molecular collisions are accounted

for the distribution function f. Equation (2.4) forms the starting point of statistical thermodynamics and the kinetic theory of gases. When external forces are neglected and the gas behavior is ideal, this equation reduces to the simple statement of the vanishing of the particle derivative of the distribution function (cf. II.1.11a)

$$\frac{Df}{Dt} = 0 \quad (2.5)$$

collisions neglected, $f =$ const

because collisions then do not occur within the gas, as already indicated.

3 DEDUCTION OF FLOW EQUATIONS FROM BOLTZMANN'S EQUATION

The molecular coordinates x, y, z and their velocity components v_x, v_y, v_z are all treated as independent variables of the distribution function $f(\mathbf{r}, \mathbf{v}, t)$ as well as the time itself, and it follows that averages obtained as velocity integrals in the form (2.3) can be treated in a particularly simple fashion when the property in question is a function of the velocities alone. Thus, if an unspecified velocity-dependent fluid property is denoted Q (obvious examples are provided by an arbitrary constant, by a velocity component, and by the kinetic energy of a molecule), the average value $\bar{Q}(\mathbf{r}, t)$ is obtained as the integral with respect to velocities:

$$\bar{Q}(\mathbf{r}, t) = \frac{1}{N} \int Q(\mathbf{v}) f(\mathbf{r}, \mathbf{v}, t) \, d\mathbf{v} \quad (3.1)$$

standard formula for defining average-flow quantities as integrals of f at any location $\mathbf{r}$

The (partial) time derivative of the average quantity $\bar{Q}$ is now related to a slightly differently defined average of the molecular property Q since

$$\frac{\partial (N\bar{Q})}{\partial t} = \frac{\partial}{\partial t} \int Q(\mathbf{v}) f(\mathbf{r}, \mathbf{v}, t) \, d\mathbf{v} = \int Q(\mathbf{v}) \frac{\partial f}{\partial t} \, d\mathbf{v} \quad (3.2)$$

derivative of f only

Equation (3.2) shows that the time rate of change of an averaged (hence macroscopically meaningful) quantity is determined by the rate of change of the distribution function f.

The partial derivative with respect to any space coordinate is evidently obtained in the same manner as (3.2), since the coordinate x, for example, is mathematically equivalent to the time t in an integration with respect to the velocity variables. For reasons that will be apparent shortly, the velocity-dependent function is now written as one of the velocity components multiplied by Q, so that, for example,

$$\frac{\partial}{\partial x}(N\overline{v_x Q}) = \int v_x Q(\mathbf{v}) \frac{\partial f}{\partial x} d\mathbf{v} \qquad (3.3)$$

If Q is now taken to be the constant unity, $Q = 1$ and $\bar{Q} = 1$, so that multiplication of (3.2) by the molecular mass gives simply the partial derivative of the density with respect to time:

$$\frac{\partial \rho}{\partial t} = \frac{\partial (Nm)}{\partial t} = m \int \frac{\partial f}{\partial t} d\mathbf{v} \qquad (3.4)$$

The density derivative (3.4) is identified with the velocity integral of the first term in the full expression of (2.5) multiplied by m, since

$$m\frac{\partial f}{\partial t} + mv_x \frac{\partial f}{\partial x} + mv_y \frac{\partial f}{\partial y} + mv_z \frac{\partial f}{\partial z} = 0 \qquad (3.5)$$

in this case. It is now seen that all the remaining terms in (3.5) are of the same form as (3.3) after it is multiplied by m. When it is seen that the flow component u given by (2.3) reappears in (3.3) when $Q = 1$, and the analogous definitions of the other velocity components are used, (3.5) integrates to the continuity equation of an unsteady three-dimensional compressible fluid flow:

integration of Boltzmann's equation with $Q = 1$ recovers continuity equation

$$\frac{\partial \rho}{\partial t} + \frac{\partial (\rho u)}{\partial x} + \frac{\partial (\rho v)}{\partial y} + \frac{\partial (\rho w)}{\partial z} = 0 \qquad (3.6)$$

(cf. VIII.6.1).

The development of the continuity equation (3.6) in the preceding manner, i.e., by multiplying the Boltzmann equation by the same factor in each of its terms and then integrating, is spoken of as calculating a particular "moment" of the Boltzmann equation. When the factor Q is chosen differently, a different moment is calculated, and it is interesting to note that the assignment of the respective velocity components v_x, v_y, v_z in turn as Q-values gives each of the components of the vector equation of momentum conservation, i.e., the Navier-Stokes equations. Furthermore, "higher" moments are also of interest physically, and one of them recovers the energy equation. That the Boltzmann equation contains so much "information," or that its various slices, so to speak, convey all the physical principles of fluid mechanics and more, demonstrates its great importance in the several aspects of large aggregates of molecules represented by thermodynamics, kinetic theory, and fluid mechanics.

"higher moments" yield momentum and energy equations

4 BOLTZMANN'S LAW AND THE MAXWELLIAN VELOCITY DISTRIBUTION

The calculation of the velocity distribution function $f(\mathbf{r}, \mathbf{v}, t)$ is simplified when all the constituent molecules are assumed to be identical. For then it follows that two physically distinct dynamical *states* (each corresponding to a particular spatial arrangement of the molecules and their motions) result from the simple interchange of any pair of molecules, but all observable macroscopic properties are unaffected by the change. While the number of different possible states of N molecules is enormous when common gas densities are considered, very many of these states are essentially indistinguishable from many others. Greatly simplifying approximations are possible in accordance with rules of the so-called mathematics of large numbers, especially Stirling's formula. The problem is still so complex even when idealized in this manner that simple solutions are available only for equilibrium states. For convenience molecular motions are considered to be devoid of macroscopic velocity so that the total velocity $\mathbf{v}$ composed of a uniform flow $\mathbf{q}$ (having components u, v, w) plus the random motions $\mathbf{c}$ giving total velocity $\mathbf{v} = \mathbf{q} + \mathbf{c}$ in general now reduces to $\mathbf{c}$ alone:

to find the analytical form of f

consider the molecular motions alone

$$\mathbf{v} = \mathbf{c} \quad \text{and} \quad v^2 = c^2 \tag{4.1}$$

The scalar notation suffices whenever speeds and kinetic energies are important without regard to velocity directions. For further simplicity a *homogenous* gas is considered so that f does not depend on position $\mathbf{r}$, and a steady state that eliminates explicit dependence on t. The problem then is to find the velocity distribution function that can be written symbolically as a function of molecular speed:

$$f(c) \tag{4.2}$$

steady state, homogenous gas: f depends on molecular speeds only

The classical calculation considers the *entropy* of the aggregate of molecules and infers the form of $f(c)$ from the condition that the entropy takes a stationary value; standard treatises may be consulted for details. Here a more heuristic procedure is followed; it takes its starting point in the equilibrium atmospheric density

266 THE MOLECULAR BASIS OF FLUID MECHANICS

energy distribution shown by Halley's law

stratification given by Halley's law (I.5.4). When the density is written in the molecular form (1.2) and the specific gas constant is noted from Section 1 to be expressible as the quotient of Boltzmann's constant divided by the molecular mass,

$$R = \frac{k}{m} \tag{4.3}$$

Halley's law furnishes the variation of particle density with vertical coordinate z in the form

$$N(z) = (\text{const}) \exp\left(-\frac{mgz}{kT_0}\right) \tag{4.4}$$

The constant in (4.4) is evidently the number density at the level $z = 0$, and it is useful to recall the interpretation of the atmospheric pressure at any level z as the weight of the atmospheric column of unit cross section extending upward from z. Alternatively, now it is possible to read (4.4) as the statement that the level at which the particle density decreases to $N(z)$ is given by the expression on the right side of the equation, hence proportional to the negative exponential of the *potential energy* mgz of a molecule at level z. The energy is normalized by the factor kT_0 and the significance of the present interpretation is related to the fact that similar expressions are obtained for entirely different force fields. Equation (4.4) illustrates a general and exceedingly important principle known as *Boltzmann's law:* the probability of finding molecules in a given arrangement in space varies exponentially with the negative of the potential energy of that arrangement divided by kT_0. The form of the distribution function $f(c)$ can be written down when Boltzmann's law is applied to the kinetic energies of the molecules, and it is useful first to establish a relationship between the average potential energy and the average of the kinetic energy values of a dynamical system.

Boltzmann's law: energy distribution follows an exponential variation

For this purpose an elementary motion is considered with a single degree of freedom, exemplified by a vibrating spring-supported mass. The periodic motion is fully represented by the function $z(t)$:

average values of kinetic and potential energies are interrelated . . .

$$z = A \cos \omega t \tag{4.5}$$

where the circular frequency is related to the mass M and the spring parameter K in the form

$$\omega^2 = \frac{K}{M}$$

The kinetic energy T is given by

$$T = \frac{M}{2}\left(\frac{dz}{dt}\right)^2 = \frac{M}{2}\omega^2 A^2 \sin^2 \omega t \qquad (4.6)$$

at any time t, and it is averaged over a complete cycle of motion by integration with respect to t between limits 0 and $2\pi/\omega$; the result is the average kinetic energy of the motion,

$$\overline{T} = \frac{M\omega^2 A}{4} \qquad (4.6a)$$

The potential energy is represented by the spring extension or compression as

$$U = \frac{K}{2}z^2 = \frac{K}{2}A^2 \cos^2 \omega t \qquad (4.7)$$

and the time average is likewise easily evaluated and found to be given by (4.6a) exactly:

$$\overline{U} = \overline{T} \qquad (4.8)$$

... equal to each other, in case of harmonic oscillator

The property that the average values of kinetic and potential energy are equal is characteristic of a large class of dynamical systems.

Exercise 5 A simple pendulum of length a executes angular motion described by

$$\theta(t) = A \sin\left(\sqrt{\frac{g}{a}}\, t\right)$$

where A is an amplitude constant and g is the ordinary gravitational acceleration. Note that the kinetic energy of the pendulum mass M is $(M/2)[a^2\,(d\theta/dt)^2]$, and determine its average value over one complete period. Show that, when the angular displacements are small, the potential energy is given by $Mg(\theta^2/2)$. How is its average value related to the average kinetic energy?

If the potential energy mgz in (4.4) is replaced by the kinetic energy $mc_x^2/2$ of the x-component of a molecular motion in a gas at temperature T, it may be expected that a distribution function for this motion could be expressed as

$$f(c_x) = (\text{const})\exp\left\{-\frac{mc_x^2}{2kT}\right\} \qquad (4.9)$$

distribution of x-velocities

with similar expressions for the two orthogonal components.

268 THE MOLECULAR BASIS OF FLUID MECHANICS

Exercise 6 Evaluate the constant in (4.9) for a fictitious "one-dimensional" gas having only velocity components c_x by requiring the particle density to be N:

$$\int_{-\infty}^{\infty} f(c_x)\, dc_x = N \tag{4.10}$$

normalization

HINT:

$$\int_{0}^{\infty} e^{-x^2}\, dx = \frac{\sqrt{\pi}}{2} \tag{4.11}$$

Of even greater interest than the distribution of velocity components is the distribution of molecular speeds c. When it is considered that the speed c is the sum of squares of three orthogonal components and that a given value may be obtained in a great variety of ways corresponding to different combinations of speeds, the distribution of speeds is found to be given in the form (4.9) multiplied by the area $4\pi c^2$ of the spherical surface of radius equal to the speed c. The final form of the *Maxwellian distribution* of molecular speeds is obtained thereby as

Maxwellian velocity distribution

$$f(c) = 4\pi c^2 N \left(\frac{m}{2\pi kT}\right)^{3/2} \exp\left\{-\frac{mc^2}{2kT}\right\} \tag{4.12}$$

It is readily seen that (4.12) is a distribution ranging over all values of molecular speeds from zero to infinity, but with vanishingly small concentrations at each of these limits. Positive values between the two limits indicate the existence of at least one maximum in the curve of $f(c)$ plotted as a function of c.

*Exercise 7 Show that (4.12) contains a single peak value, which may be termed the most probable speed, and determine its value from the condition

most probable speed

$$\frac{df(c)}{dx} = 0 \tag{4.13}$$

Compare the most probable speed, c_m for example, with the root mean square velocity.

Exercise 8 Calculate the mean velocity $\bar{c}$ from its definition as

$$\bar{c} = \frac{1}{N}\int_{0}^{\infty} cf(c)\, dc \tag{4.14}$$

mean molecular speed

Evaluate $\bar{c}$ for the conditions of Exercise 4.

Ans. $\bar{c} = \sqrt{8kT/\pi m} \doteq 1.128 c_m$.

*Exercise 9   Calculate the mean square velocity of (1.1) and (1.3) as

$$\overline{c^2} = \frac{1}{N}\int_0^\infty c^2 f(c)\, dc = \frac{3kT}{m} \quad (4.15)$$

mean square velocity

by substituting from (4.12) and integrating by parts repeatedly to recover (4.11) and hence the last expression on the right side of (4.15). Also find the *root mean square velocity* $\sqrt{\overline{c^2}}$ and compare with the mean velocity, the most probable speed, and the speed of sound found in Exercise VIII.5. For what range of flow speeds is it justified to employ the molecular speeds in place of the total particle velocity according to (4.1)?

root mean square velocity

5 FINITE MOLECULAR DIAMETER: REAL-FLUID PROPERTIES

The crudest molecular model that regards each individual molecule as a point mass (occupying zero volume) has permitted both the fluid pressure and density to be related to the molecular mass m and number density N in Section 1 above. With the further introduction of Boltzmann's universal constant k and the Maxwellian distribution of molecular velocities, the temperature and the pressure of a gas are more completely expressed in terms of the kinetic parameters of the swarm of particles of which the gas is comprised. Taking account of the actual finite dimensions of the molecules opens the further possibility of relating the distinctive dynamic *properties* of real fluids to the assumed molecular characteristics. This section and succeeding ones illustrate how viscosity coefficient and real-gas corrections to the thermodynamic equation of state are estimated by studying the dynamics of a large number of molecules idealized as rigid spheres.

The molecular model of Section 1 is now modified not only with respect to feature (b), but also to extend (c) to include intermolecular collisions that are again assumed to be "elastic," i.e., such that the total translational kinetic energy of a pair of colliding molecules is unaltered by the collision. The first consequence of admitting finite molecular dimensions is that the individual molecule does not travel a straight trajectory unobstructed from one wall of a containing vessel to another. If the molecules are considered to be rigid spheres like billiard balls, all having the same diameter d, for example, it is clear that a collision between two molecules occurs whenever the distance between their centers is reduced to the value d. The average distance traveled between successive collisions can therefore be calculated in the following manner. It is supposed first that all molecules are at rest except one

real-fluid properties determined by non-ideal gas characteristics moving at speed c. The volume it sweeps out in unit time is given by the product of the speed c and its projected area $\pi(d^2/4)$ in the plane normal to its motion, and a collision occurs when the center of another molecule falls within distance $d/2$ of the cylindrical volume just described. It follows that the incidence of intermolecular collisions is determined by considering the cylindrical volume of *twice* the diameter of the molecule itself, i.e., by calculating the number of molecules whose centers fall within the volume $\pi d^2 c$. Since there are N molecules in each unit of volume, the volume in question contains $N\pi d^2 c$ centers on the average, and this is therefore the number of collisions occurring in the time the moving molecule travels the distance c. The average distance traveled between successive collisions is therefore given by the quotient of these two quantities. When allowance is made for the fact that the struck molecules are also in motion, a factor $\sqrt{2}$ is introduced which then gives for the *molecular mean free path*

molecular diameter and mean free path

$$l = \frac{1}{\sqrt{2}\pi N d^2} \tag{5.1}$$

which is independent of the molecular speed c. It should be noted that molecules are not accurately represented as rigid spheres and consequently that the numerical factor appearing in (5.1) is not strictly accurate, although it will be retained in the calculations that follow.

If now for definiteness a fluid motion is considered parallel to the x-axis so that the mean flow velocity is given by $u(y)$ and $v \equiv 0$, uniform shear is represented by writing

$$u = u(y) = u(0) + y\frac{du}{dy} \qquad v = 0 \tag{5.2}$$

where y is measured from an arbitrarily chosen reference level and $du/dy = $ const. The total velocity of an individual molecule is given as the sum of the components (5.2) plus the random thermal motions, so that

$$v_x = c_x + u(0) + y\frac{du}{dy} \qquad v_y = c_y \tag{5.3}$$

It is now possible to calculate the coefficient of viscosity with the aid of (5.3) by considering the force between two layers of fluid of unit area in contact along the plane $y = 0$. Newton's second law indicates that the force exerted by the upper layer on the lower layer is equal to the rate of change of momentum of the latter. Because of the molecular motions of thermal agitation, individual molecules passing from upper to lower levels possess a surplus of forward momentum, and the reverse is true on the average for a molecule for which $c_y > 0$. Proceeding now as in the analysis of turbulent stresses in Section X.3, we note that a molecule of mass m possesses momentum mv_x in the flow direction, and this leads to an upward flux

$$mv_x v_y \tag{5.4}$$

of x-momentum into a region of greater x-momentum. The shear stress is therefore found by summing terms (5.4) over the different combinations of values of v_x and v_y, i.e., by taking account of the *correlation* of the molecular motion c_y with the associated deficit of x-momentum resulting from the shear indicated in the first equation of (5.3).

shearing flow and transverse molecular motion leads to momentum flux...

The stress is accordingly written in the manner of (3.1) as

$$\tau = \int mf(v_x, v_y)v_x v_y \, dv_x \, dv_y = mN\overline{v_x v_y} \tag{5.5}$$

... and shear stress

Substituting from (5.3), and noting that the averages of the thermal agitations are zero, $\overline{c_x} = \overline{c_y} = 0$, leads to

$$\tau = mN \frac{du}{dy} \overline{yc_y} \tag{5.6}$$

when it is also supposed that c_x and c_y are independent (uncorrelated). The correlation of the vertical displacements with the vertical velocities in (5.6) is estimated by considering the "level of origin" to be situated at the plane $y = 0$ and accounting for the inclination (angle θ) of the random motion from the y-direction by writing

viscous shear stress in terms of correlated transverse displacements and molecular speeds

$$y = l \cos \theta \quad \text{and} \quad c_y = \bar{c} \cos \theta \tag{5.7}$$

where $\bar{c}$ is the mean molecular speed. The θ-average, effected with a weighting factor $\sin \theta \, d\theta$ as in (I.6.5) integrates to $\frac{1}{3}$, while writing (5.1) for l gives the stress in terms of molecular mass, velocity, and diameter as

evaluation

$$\tau = \frac{m\bar{c}}{3\sqrt{2}\pi d^2} \frac{du}{dy} \tag{5.6a}$$

Comparison with the Newtonian stress (IX.1.3) now gives the viscosity coefficient as

shear stress as a function of molecular motion, molecular dimension, and shear flow

$$\mu = \frac{m\bar{c}}{3\sqrt{2}\pi d^2} \tag{5.8}$$

elementary expression for μ in terms of molecular motion and diameter alone

where, as before, the numerical coefficients are regarded as approximate. A remarkable feature of (5.8) is the absence of N, which implies that viscosity is not directly dependent on density. Experi-

viscosity coefficient dependent mainly on temperature, not on density mental determinations of μ confirm this behavior, as well as the $\frac{1}{2}$-power temperature dependence inferred, for example, from the form of (1.3) and the results of Exercise 9.

Exercise 10 To understand the meaning of (5.8), which indicates that viscosity coefficients are not directly dependent on the particle density N even though the number of momentum carriers is clearly proportional to N, it is necessary to identify another effect which acts in the opposite sense. Indicate which of the relationships already developed furnishes the required explanation.

A variety of other fluid properties also depends on the molecular parameters like particle density, mass, structure, and motion; quantitative evaluations are obtained in a fundamental fashion analogous to the viscosity calculation (5.8). They include coefficients of heat conduction, molecular diffusion, and electrical conductivity, all termed transport properties and dealt with by kinetic theory and physical chemistry.

<small>mixing length and mean free path; turbulent "eddy" and individual molecule; analogy between turbulence and kinetic theory</small>

The reader can scarcely fail to note the striking similarity between the preceding calculations and the methods for calculating turbulent flows. The great contributions of Boltzmann and Maxwell in developing the kinetic theory of gases one century ago illustrate the not infrequent fortuitous benefits derived in one branch of science as a consequence of advances in another.

Exercise 11 Take the viscosity of air at 273°K as $\mu = 1.7 \times 10^{-4}$ gm cm sec^{-1}, use the result of Exercise 8, and estimate the effective diameter d of san air "molecule" directly from (5.8). Compare the result with the commonly used rough value quoted as "of the order of 10^{-8} cm." Calculate the mean free path from (5.1), and divide $\bar{c}$ from Exercise 8 by l to determine the number of collisions experienced by each molecule every second (how does the number calculated compare with the total human population of the Earth?).

Exercise 12 Show that (5.6a) can be written as

$$\tau = \frac{1}{3}\rho\bar{c}l\frac{du}{dy} = \frac{1}{3}\rho l^2\left(\frac{\bar{c}}{l}\right)\frac{du}{dy}$$

to establish the formal dimensional analogy with the turbulent stress representation (X.3.4) according to the mixing length hypothesis.

6 VIRIAL THEOREM — MOLECULAR BASIS OF IDEAL GAS EQUATION OF STATE

Fluid on a molecular scale gives the appearance of a completely coarse-grained substance, while the macroscopic scale employed in preceding chapters presents the same substance as a perfect continuum, precisely the opposite extreme of the former. The seeming paradox is, of course, resolved by considering the extremely numerous and frequent collisions between pairs of molecules (roughly 10^{30} collisions every second in each cubic centimeter of air at standard conditions). A strong measure of "order within disorder" is established thereby, so that nonuniformities are very rapidly wiped out in a characteristic time determined by the interval between two successive collisions of a molecule. It follows that strong and very useful conclusions can be drawn concerning the time-averaged molecular energies. These are applied in the present section to the deduction of the thermodynamic equation of state of an ideal gas, and in the following section to more accurate equations of state.

to study the time-averaged values of molecular motion characteristics

The required relationships are found by observing that a molecular velocity component $c_x \, (= dx/dt)$, position coordinate x, mass m, and external force X are connected to the kinetic energy term c_x^2 by the equation

$$\frac{d}{dt}(xc_x) = c_x^2 + x\frac{dc_x}{dt} = c_x^2 + \frac{xX}{m} \qquad (6.1)$$

Different interpretations of the force X are to be considered. When the y- and z-components are treated similarly, the kinetic energy is formed by adding the two companion equations and (6.1) and multiplying by $m/2$

$$\frac{m}{2}c^2 = \frac{m}{2}(c_x^2 + c_y^2 + c_z^2) = \frac{m}{2}\left\{\frac{d}{dt}(\mathbf{r}\cdot\mathbf{c}) - (\mathbf{r}\cdot\mathbf{F})\right\} \qquad (6.2)$$

kinetic energy related to position, velocity and force acting

The right side of (6.2) is written in the compact vector notation where

$$\mathbf{r}\cdot\mathbf{c} = xc_x + yc_y + zc_z$$

and the still unspecified force components X, Y, and Z define the vector $\mathbf{F}$:

$$\mathbf{r}\cdot\mathbf{F} = xX + yY + zZ$$

in an obvious extension of notation already introduced. Thus the total kinetic energy of a molecule is related to the force acting, $\mathbf{F}$, and to the velocity and position vectors $\mathbf{c}$ and $\mathbf{r}$. Equation (6.2) is now averaged over an interval of time T long compared with the typical time between two successive collisions of the molecule.

Time averages are defined as before, so that, for example,

$$\overline{\frac{m}{2}c^2} = \frac{1}{T}\int_0^T \frac{m}{2}c^2 \, dt \qquad (6.3)$$

Integration of the first term on the right side of (6.2) yields the difference of the product $\mathbf{r}\cdot\mathbf{c}$ at times T and 0, divided by T:

$$\frac{1}{T}\int_0^T \frac{d}{dt}(\mathbf{r}\cdot\mathbf{c})\, dt = \frac{1}{T}\{(\mathbf{r}\cdot\mathbf{c})_T - (\mathbf{r}\cdot\mathbf{c})_0\}$$

A molecule confined to a closed volume V is located by a position vector $\mathbf{r}$ that must remain bounded, and, when the same is true of the velocity $\mathbf{c}$, division by arbitrarily large T makes the quantity in braces arbitrarily small, while the average value of the kinetic energy in (6.2) is not affected by the magnitude of T. It follows at once that the average value of the kinetic energy of a molecule is determined completely by the average value of the position-force product represented as the last term of (6.2). The same is true for every molecule of gas contained within volume V, and, when the averaged equations are added for all molecules, the result is

average kinetic energy expressed in terms of forces: the virial theorem

$$\overline{\Sigma \frac{m}{2} c^2} = -\tfrac{1}{2} \overline{\Sigma (\mathbf{r}\cdot\mathbf{F})} \qquad (6.4)$$

where Σ indicates summation over all the molecules. The term on the right side (6.4) is called the *virial of the system*, and the virial theorem of Clausius stated by (6.4) expresses the equality of the total kinetic energy of the gas and of its virial. Different hypotheses for the forces $\mathbf{F}$ permit the thermodynamic equation of state of the gas to be found with degrees of accuracy varying with the quality of the assumptions for $\mathbf{F}$.

In an ideal gas where there are no intermolecular forces or collisions, the force $\mathbf{F}$ consists only of the pressure p acting on the gas at the walls of the vessel of volume V. The force is directed toward the interior of the gas and can therefore be written as $-p\mathbf{n}$ in terms of the outward-directed normal vector. The sum on the right side of (6.4) is evaluated over the bounding surface S as

force evaluation: wall pressure only

$$-\tfrac{1}{2}\int_S (-p\mathbf{n}\cdot\mathbf{r})\, dS \qquad (6.5)$$

that is, as the surface integral of the normal component of the vector $-p\mathbf{r}$. Gauss' transformation (A.6) permits expression of this quantity as the volume integral of the divergence of the same vector, and, since the pressure is a constant and $\nabla\cdot\mathbf{r} = \partial x/\partial x + \partial y/\partial y + \partial z/\partial z = 3$, (6.5) is equal to

$$\tfrac{3}{2}pV \qquad (6.5a)$$

The mean square velocity of the gas is given by the left side of (6.4), and, since there are VN particles, this becomes

$$VN\frac{m}{2}\overline{c^2} \qquad (6.5b)$$

Furthermore, in the absence of mass motion, $\overline{v^2} = \overline{c^2}$ and the temperature is given by (1.3) as

$$RT = \frac{\overline{c^2}}{3}$$

Equating the virial (6.5a) and the total kinetic energy (6.5b) therefore gives

$$p = \rho RT \qquad (6.6)$$

deduction of ideal gas equation of state

where Nm is written as ρ and (6.6) is recognized as the familiar equation of state (I.3.8).

7 INTERMOLECULAR FORCES — THE SECOND VIRIAL COEFFICIENT AND VAN DER WAALS' VOLUME CORRECTION

The parameter b in the van der Waals equation of state (I.4.3) is of dimensions equal to the reciprocal of the fluid density, hence volume per unit mass of fluid. It is evaluated in terms of the molecular diameter d and number density N if a correction for total gas volume is made to account for the reduced mobility of an individual molecule owing to the volume occupied by the molecules themselves. Thus the rigid sphere model considered above ascribes volume $\frac{4}{3}\pi(d/2)^3$ to each molecule, while its "sphere of influence" with respect to collisions with identical molecules has a diameter twice this great, hence volume $\frac{4}{3}\pi d^3$. Since the center of a second molecule can approach the center of the first no closer than distance d, the space in which it moves is reduced by this amount. The average reduced volume for both molecules corresponds to half this amount for each molecule, i.e., $\frac{2}{3}\pi d^2$ or just four times the volume of the molecular sphere. As there are $N - 1$ such excluded regions in unit volume of the gas, these give very nearly

the excluded volume — reduced mobility of molecules

$$b = \frac{2}{3} N\pi d^3 \div Nm = \frac{2}{3}\frac{\pi d^3}{m} \qquad (7.1)$$

for the effective volume reduction in unit mass of gas, the distinction between $N - 1$ and N being ignored.

It is now of interest to relate (7.1) to the equation of state obtained by applying the virial theorem (6.4) to a gas composed of rigid-sphere molecules. For this purpose the force **F** is considered to represent both the wall pressures (6.5) and the forces of attrac-

276 THE MOLECULAR BASIS OF FLUID MECHANICS

intermolecular collisions — and the associated forces

tion (positive or negative) between molecules. Then, instead of (6.6), the virial theorem now gives

$$\frac{p}{\rho} = RT - \frac{1}{3}\overline{\frac{\Sigma(\mathbf{r}\cdot\mathbf{F})}{Vnm}} \tag{7.2}$$

where the same symbol $\mathbf{F}$ is retained for simplicity in referring to the intermolecular forces. If two molecules, located at $\mathbf{r}_1$ and $\mathbf{r}_2$ for example, are considered, the force $\mathbf{F}_1$ on the first is associated with the second molecule only, and conversely, so that, according to Newton's third law, $\mathbf{F}_2 = -\mathbf{F}_1$. The two terms in (7.2) corresponding to this pair of molecules can thus be written as

$$\mathbf{r}_1\cdot\mathbf{F}_1 + \mathbf{r}_2\cdot\mathbf{F}_2 = \mathbf{r}_{12}\cdot\mathbf{F}_1 = r_{12}F_1 \tag{7.3}$$

where $\mathbf{r}_{12}$ is the vector drawn from the first molecule to the second, and the final scalar form of (7.3) indicates that the force $\mathbf{F}_1$ is assumed to be directed along the line joining the molecule centers. When the forces are assumed to be conservative, i.e., derived from a potential in the same manner as in (I.7.1), the expression (7.3) becomes

conservative force field

$$-r\frac{\partial\epsilon(r)}{\partial r} \tag{7.4}$$

when the subscripts are dropped and the potential is denoted $\epsilon(r)$, a function having dimensions corresponding to energy. Terms of the form (7.4) are required in (7.2) for every pair of molecules which can be formed from the VN individual members contained in the volume V.

Exercise 13 (a) Find the number $P(N)$ of ways in which pairs can be formed by choosing among N objects. HINT: Consider that, if the N objects are increased in number by the introduction of the $(N + 1)$st, it can form pairs with any of the initial N members, so that

$$P(N + 1) = P(N) + N \tag{7.5}$$

Find the appropriate solution of the *difference equation* (7.5) by noting that P cannot be a linear function of N and trying therefore to find a solution in quadratic form, i.e., as

$$P(N) = AN + BN^2 \tag{7.5a}$$

(b) Find the number $T(N)$ of threesomes that can be formed by choosing among N objects and applying the preceding reasoning to show that

$$T(N + 1) = T(N) + P(N)$$

and solving the difference equation for $T(N)$.

census of molecular pairs

Ans. (a) $P(N) = \dfrac{N(N-1)}{2}$.

Since the number N is very large compared with unity in the present case, the total number of pairs that can be formed from them is very nearly $N^2/2$. When macroscopic equilibrium is again assumed, Boltzmann's law can be applied to give the fractional number of pairs at a given distance r. Within the spherical shell of thickness dr and volume $4\pi r^2\, dr$, in direct analogy with the Maxwellian distribution (4.12), the fractional number is

assume Boltzmann's law gives spatial separation statistics

$$\frac{4\pi r^2\, dr}{V} \exp\left\{-\frac{\epsilon(r)}{kT}\right\}$$

so that the sum in (7.2) can be written as the integral

$$\int_0^\infty \frac{4\pi r^2\, dr}{V} \frac{N^2 V^2}{2} \exp\left\{-\frac{\epsilon(r)}{kT}\right\} r \frac{\partial \epsilon}{\partial r} \quad (7.6)$$

ranging over all distances from 0 to ∞.

to account for all two-body interactions contributing to pressure in (7.2)

Exercise 14 Perform an integration by parts, assuming $\epsilon(\infty) = 0$, to show that (7.6) is equal to

$$\frac{kTN^2 3V}{2} \int_0^\infty 4\pi r^2 \left[\exp\left\{-\frac{\epsilon(r)}{kT}\right\} - 1\right] dr \quad (7.6a)$$

and hence that (7.2) becomes

$$\frac{p}{\rho} = RT\{1 + NB(T)\} \quad (7.7)$$

equation of state containing virial coefficient $B(T)$

where the *second virial coefficient $B(T)$* is defined as the integral

$$B(T) = -\tfrac{1}{2} \int_0^\infty 4\pi r^2 [e^{-\epsilon/kT} - 1]\, dr \quad (7.8)$$

of the energy variation given by Boltzmann's law. $B(T)$ is evidently a function of temperature alone evaluated when the form of $\epsilon(r)$ is specified.

The simplest assumption for the potential of the intermolecular forces is that of the rigid-sphere model: zero force (hence constant energy) when molecules are separated by more than one molecular diameter d, but infinite energy at smaller distances:

$$\epsilon(r) = \begin{cases} \infty & r < d \\ 0 & r > d \end{cases} \quad (7.9)$$

simple force mode

In this case the term in square brackets in (7.8) vanishes for all values of r greater than d, and the integral is evaluated by considering the range 0 to d alone:

$$-\tfrac{1}{2} \int_0^d 4\pi r^2 [-1]\, dr = 2\pi \frac{d^3}{3}$$

Thus $NB(T)$ in (7.7) becomes

virial evaluation
$$\frac{2N\pi d^3}{3} \qquad (7.10)$$

interpretation
equal exactly to the value of Nmb obtained by elementary reasoning in (7.1). This establishes the meaning of the volume correction constant of van der Waals' equation of state in accordance with the virial theorem and Boltzmann's law. It can also be shown that the constant a in (I.4.3) represents the effect of *cohesive* forces among the molecules: a molecule near the boundary surface is attracted inward toward the inner molecules more strongly than one which is situated in the interior of the gas.

When it is recalled from Exercise I.11 that the critical density ρ_c of a van der Waals substance is given by

$$\rho_c = \frac{1}{3b}$$

(7.10) permits the liquid state to be characterized by the condition

$$N > \frac{1}{2\pi d^3}$$

the liquid-gas distinction, molecular interpretation
corresponding to $\rho > \rho_c$, which, it is readily seen, implies considerably closer spacing of the molecules than occurs in the gaseous state.

More complicated functions than (7.9) have been introduced to represent with greater accuracy molecular attractions corresponding to positive attraction increasing at first as the molecules approach, then strongly repelling at very close range (negative attraction). A large number of different equations of state have been obtained in this manner, although none can claim wider acceptance than van der Waals' for the broad range of physical conditions of interest in the study of fluid mechanics.

Appendix A
Green's Theorem—
Transformations of Gauss and Stokes

In physical theory surface integrals are frequently encountered in which the integrand appears as the derivative of a certain quantity with respect to one of the variables of integration. The inverse character of the operations of differentiation and integration then reduces the practical calculation from one differentiation and two integrations to a single integration. Thus in the surface integral of $\partial N(x, y)/\partial x$

$$\iint_S \frac{\partial N(x, y)}{\partial x} \, dx \, dy \qquad (A.1)$$

surface integral of a derivative...

extended over a region S of the x, y plane, the x-integration is carried out immediately as

$$\int \frac{\partial N(x, y)}{\partial x} \, dx = \{N(x_2, y) - N(x_1, y)\}$$

so that (A.1) is reduced to

$$\int \{N(x_2, y) - N(x_1, y)\} \, dy \qquad (A.2)$$

...reduces to single integral of a difference...

which is an integral with respect to y only, the integrand being the difference between the values of $N(x_2, y)$ and $N(x_1, y)$ at the upper and lower limits, respectively, of the x-integration. Adopting a standard "positive" sense of integration around the boundary C of the area S, (A.2) is written more compactly as the *line integral* around C, so that (Fig. A-1)

$$\iint_S \frac{\partial N}{\partial x} \, dx \, dy = \oint_C N(x, y) \, dy \qquad (A.3)$$

...and to a *line integral*...

279

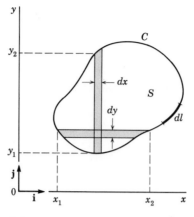

Fig. A-1

The distinctive integration symbol $\oint$ will be taken to imply the positive sense as just defined, and it is seen that $dy > 0$ when $x = x_2$, $dy < 0$ when $x = x_1$ in (A.3) in accord with (A.2); hence the integration around the complete circuit C is equivalent to (A.2), nad the positive sense of integration is anti-clockwise.

The integration of a function $\partial M(x, y)/\partial y$ over the same region S of the x, y plane leads to an analogous line integral of the function $M(x, y)$ itself around the bounding contour C. Because the upper limit of y-integration, y_2 for example, corresponds to a portion of the curve along which dx is negative when the standard sense of integration is followed, whereas dx is positive at the lower limit y_1, it follows that a reversal of sign is required and

positive sense: anti-clockwise

$$\iint_S \frac{\partial M(x, y)}{\partial y} dx\, dy = -\oint_C M(x, y)\, dx \qquad (A.4)$$

By adding the right sides of (A.3) and (A.4) it is established that

$$\iint_S \left\{ \frac{\partial M}{\partial y} + \frac{\partial N}{\partial x} \right\} dx\, dy = \oint_C \{N(x, y)\, dy - M(x, y)\, dx\} \qquad (A.5)$$

Green's theorem in general

which is known as Green's theorem.

Two cases of particular interest are seen when M and N are regarded as the Cartesian components of a vector field. Thus if

$$\mathbf{A} = \mathbf{i} A_x + \mathbf{j} A_y$$

and $M = A_y$, $N = A_x$, the integrand on the left side of (A.5) is the divergence of the vector field $\mathbf{A}$, i.e., $\nabla \cdot \mathbf{A}$, while the right side contains the integrand

$$A_x\, dy - A_y\, dx = \mathbf{A} \cdot \mathbf{n}\, dl$$

APPENDIX A GREEN'S THEOREM — TRANSFORMATIONS OF GAUSS AND STOKES

where **n** is the outward-directed unit vector normal to the boundary curve C at any point. Then Green's theorem assumes the form known as *Gauss' transformation* or theorem relating the surface integral of the divergence of a vector field to the line integral of its normal component around the boundary C of the surface S:

$$\iint_S \nabla \cdot \mathbf{A}\, dS = \oint_C \mathbf{A} \cdot \mathbf{n}\, dl \qquad (A.6) \qquad \text{special case: Gauss' theorem}$$

When the element of surface $dx\, dy$ is replaced by dS in the vector equation (A.6), the interpretation is no longer restricted to vectors referred to any particular base or coordinate system.

Another vector form of Green's theorem is obtained by setting $-M = A_x$ and $N = A_y$ so that the integrand on the left side of (A.5) is the normal component of the *curl* of the vector field $A(x, y)$, i.e., $\mathbf{k} \cdot (\nabla \times \mathbf{A})$, while the line integral becomes the sum of tangential components of **A** at all points of the curve C:

$$A_x\, dx + A_y\, dy = \mathbf{A} \cdot \boldsymbol{\tau}\, dl$$

Now Green's theorem takes the form known as *Stokes' transformation* or theorem; it expresses the surface integral of the normal component of the curl of a vector field as the line integral of its tangential component around the boundary curve C:

$$\mathbf{k} \cdot \iint_S (\nabla \times \mathbf{A})\, dS = \oint_C \mathbf{A} \cdot \boldsymbol{\tau}\, dl \qquad (A.7) \qquad \text{special case: Stokes' theorem}$$

The three-dimensional counterpart of Green's theorem (hence also those of the Gauss and Stokes transformations) is also obtained by the partial integration of derivative terms, which reduces a volume integral in each case to a surface integral (one less "dimension") of the boundary of the three-dimensional region.

Appendix B
On Partial Differential Equations—Gravitational Potential

The mutual gravitational force between a concentrated mass M and a second unit "test" mass at distance r from the first is given by Newton's law of gravitation as

Newtonian gravitation

$$= \frac{-k^2 M}{r^2} \tag{B.1}$$

where k^2 is the familiar universal constant and the minus sign indicates that the force is attractive, tending to diminish r. It is desired to express both the force magnitude and its direction in equation form, and it is convenient to introduce Cartesian coordinates so that $r^2 = x^2 + y^2 + z^2$ with M is taken as situated at the origin. Then the typical force component on the test mass is written in the form

$$f^{(x)} = \frac{-k^2 M x}{(x^2 + y^2 + z^2)^{3/2}} = \frac{-k^2 M}{r^2} \frac{x}{r} \tag{B.2}$$

force components

where the factor x/r, it will be recalled, is cosine of the angle subtended by the x-axis and the line joining the two masses, and similar equations give the components $f^{(y)}$ and $f^{(z)}$. From the set of three equations like (B.2) it is desired to infer the general properties of the field of forces at all points x, y, z.

The same constant $-k^2 M$ appears in all three force expressions for every set of values of x, y, and z, and it is readily verified that the constant is eliminated from a relationship between the separate force components by differentiating each of the three equations in the respective directions and adding:

vanishing divergence of force components

$$\frac{\partial f^{(x)}}{\partial x} + \frac{\partial f^{(y)}}{\partial y} + \frac{\partial f^{(z)}}{\partial z} = 0 \tag{B.3}$$

A more useful equation than (B.3) is obtained by noting that the force expression (B.1) is the *r*-derivative of the *potential U* given by

$$f = -\frac{dU}{dr} \quad \text{where} \quad U = -\frac{k^2 M}{r} \qquad (B.4)$$

introduction of gravitational potential

and consequently that the Cartesian force components are of the form

$$f^{(x)} = -\frac{\partial U}{\partial x} \qquad (B.4a)$$

as is readily verified by evaluating the respective derivatives. Substitution in (B.3) then gives

$$\frac{\partial^2 U}{\partial x^2} + \frac{\partial^2 U}{\partial y^2} + \frac{\partial^2 U}{\partial z^2} = 0 \qquad (B.5)$$

Laplace's equation satisfied by gravity potential

as the equation satisfied by the potential. Equation (B.5) is recognized as Laplace's equation in three dimensions and is identical with the equation satisfied by the velocity potential of an irrotational flow in incompressible fluid. In this case it replaces the statement (B.3) of the vanishing of the divergence of the three distinct components of gravitational force by an equation containing the gravitational potential alone. Since the potential U is of the same dimensions as energy, U is recognized as the common potential energy of the configuration consisting of the two masses M and unity when they are separated by a distance r. The variation of potential energy as the test mass is carried from one point to another is given by the solution (B.4) of the equation. It is a common feature of *continuously* varying properties like gravitational force and fluid velocity that the underlying equation contains derivatives with respect to more than one independent variable, hence the appearance of *partial* differential equations.

The appearance of U in the same linear (i.e., first power) form in each term of (B.5) assures that the sum of any two solutions is also a solution of the equation. In physical terms, two masses M_1 and M_2 exert a total force on unit test mass at x, y, z equal to the resultant of their separate attractions. The potential can then be written as $U = U_1 + U_2$, where

$$U_1 = -\frac{k^2 M_1}{r_1} \quad \text{and} \quad r_1{}^2 = (x - x_1)^2 + (y - y_1)^2 + (z - z_1)^2$$

and x_1, y_1, z_1 are the coordinates of M_1; a similar expression gives U_2. It follows that the potential of any number of attracting masses is expressible as a sum of terms equal to the number of masses.

from a fundamental solution, by superposition . . .

When the attracting mass is continuously distributed, the summation is replaced by an integration expressed as

$$U(x, y, z) = -k^2 \int \frac{\rho(\xi, \eta, \zeta)\, d\xi\, d\eta\, d\zeta}{\{(x - \xi)^2 + (y - \eta)^2 + (z - \zeta)^2\}^{1/2}} \quad (B.6)$$

...to a formula for the potential of an arbitrary distributed mass

where the integration is carried out over the volume occupied by the attracting mass. The density of the attracting mass, i.e., the mass per unit volume at location ξ, η, ζ is denoted $\rho(\xi, \eta, \zeta)$. The representation (B.6) is preferable to (B.1) both because the mass density is in reality finite, so that a finite mass M is not accurately regarded as concentrated at a point, and also because the potential U is a scalar quantity and no resolution of components of vector forces is required in its evaluation. It is also clear that (B.1) corresponds to a limiting case of (B.6) when the dimensions of the attracting body are small compared with its distance from the test mass: (B.6) is nearly $(-k^2 M)/r$, where M is the total mass, and the force is given by (B.4), approximately.

An important case is furnished by a spherical mass, and the potential will be calculated at a point on the x-axis, i.e., at distance $r = x$ from its center, by evaluating the integral (B.6) over one of the thin concentric spherical shells in its interior. When the shell radius is taken as a and the *surface density*, or mass per unit area of shell surface, is denoted τ, ring-shaped elements of mass are determined by the polar angle θ as shown in Fig. B-1, so that

$$dU_a = -\frac{k^2\, dm}{r}$$

surface distribution

where
$$dm = \tau a^2\, d\theta\, 2\pi \sin\theta \quad \text{and} \quad r^2 = a^2 + x^2 - 2ax \cos\theta$$

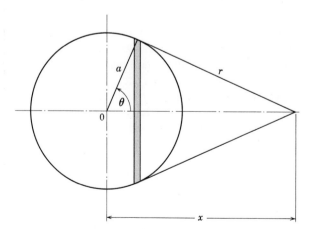

Fig. B-1 Sphere potential calculation.

The potential of the shell of radius a is obtained by summing θ over the interval $(0, \pi)$ as

$$U_a = -k^2 \int_0^\pi 2\pi\tau a^2 \frac{\sin\theta\, d\theta}{\sqrt{a^2 + x^2 - 2ax\cos\theta}}$$
$$= -k^2 2\pi\tau a^2 \left[\frac{\sqrt{a^2 + x^2 - 2ax\cos\theta}}{ax}\right]_0^\pi \quad \text{(B.7)}$$

When the point $(x, 0, 0)$ is outside the shell, i.e., when $x > a$, the term under the radical sign in (B.7) takes values $x + a$ and $x - a$ at the upper and lower integration limits, respectively, and the term in square brackets is

$$\frac{x + a - (x - a)}{ax} = \frac{2}{x} \quad \text{(B.8a)}$$

For an interior point, on the other hand, $x < a$, the positive distances give

$$\frac{a + x - (a - x)}{ax} = \frac{2}{a} \quad \text{(B.8b)}$$

Hence, if the shell mass is denoted $M_a = \tau 4\pi a^2$,

$$U_a = -\frac{k^2 M_a}{x} \quad \text{if } x > a \quad \text{(B.9a)}$$ shell potential, exterior point...

or

$$U_a = -\frac{k^2 M_a}{a} \quad \text{if } x < a \quad \text{(B.9b)}$$ ... and interior

The meaning of (B.9a) is that the attraction at an exterior point is exactly the same as if the entire mass M_a were concentrated at the center of the shell; differentiation with respect to the distance x shows that the force varies as the inverse square of x. For interior points the situation is entirely different, according to (B.9b), since in the absence of x the potential is a constant and the force is zero: an interior point anywhere inside a uniform spherical shell is attracted equally strongly in each pair of opposite directions by the mass that surrounds it.

The gravitational potential of a sphere of radius A at an exterior point $x > A$ is now obtained from (B.9a) by expressing the surface density τ as the product of an infinitesimal thickness da and the ordinary density ρ, $\tau = \rho\, da$, and, summing over all radii from 0 to A:

$$U(x) = \frac{-k^2 \int_0^A 4\pi\rho(a)a^2\, da}{x} = \frac{-k^2 M}{x} \quad \text{(B.10)}$$ sphere potential, exterior

where M is the total mass of the sphere, given by the integral in (B.10) and the density is not necessarily constant. It also follows from (B.9b) that a particle situated at an interior point of a sphere is attracted only by the mass of the concentric sphere that passes through the point, and not by more distant mass.

the sphere is *centrobaric*

The special property of the attraction of a radially symmetric spherical mass as given by (B.10) is that the potential, and hence the force also, is equivalent to the attraction of the entire mass concentrated at the center of the sphere, which is its center of inertia (also called the centroid or center of mass). This *centrobaric* property is not shared by other simple distributions in general.

an alternative calculation...

The attraction of a uniform sphere may also be calculated by taking it to consist of adjacent disks of varying radii and distances from the test mass at x. Thus the potential of a disk of surface density τ and radius H, at a point distant ξ from its center is found to be

$$U_{\text{disk}} = -k^2 2\pi\tau(\sqrt{H^2 + \xi^2} - \xi) \qquad (B.11)$$

... shows that circular disk is not centrobaric

and integration with respect to ξ from $x - A$ to $x + A$, when H is regarded as a function of ξ, recovers the result (B.10). Differentiation of (B.11) with respect to ξ, H being kept constant, gives the attraction of a circular disk at a point on its axis; this is different from the force of attraction of the same mass concentrated at the center of the disk.

Appendix C
Properties of Jacobian Determinants

A systematic procedure for changing independent variables in calculations requiring the evaluation of partial derivatives is furnished by the multiplication property of Jacobian determinants, employed in Section II.7 and elsewhere. This property is established by multiplying the determinant

$$\frac{\partial(u, v)}{\partial(x, y)} = \begin{vmatrix} \dfrac{\partial u}{\partial x} & \dfrac{\partial u}{\partial y} \\ \dfrac{\partial v}{\partial x} & \dfrac{\partial v}{\partial y} \end{vmatrix} \quad (C.1)$$

definition of symbol for Jacobian determinant

of derivatives with respect to x and y by the transformation determinant that relates these to the new variables ξ and η:

$$\frac{\partial(x, y)}{\partial(\xi, \eta)} = \begin{vmatrix} \dfrac{\partial x}{\partial \xi} & \dfrac{\partial x}{\partial \eta} \\ \dfrac{\partial y}{\partial \xi} & \dfrac{\partial y}{\partial \eta} \end{vmatrix} \quad (C.2)$$

according to the usual rule for determinant multiplication. Thus the element in the first row and first column of the product determinant is given by the sum of the products of the elements of the first row of (C.1) by the elements of the first column of (C.2); that is,

$$\frac{\partial u}{\partial x}\frac{\partial x}{\partial \xi} + \frac{\partial u}{\partial y}\frac{\partial y}{\partial \xi} \quad (C.3)$$

The sum (C.3), however, is recognized as

$$\left.\frac{\partial u}{\partial \xi}\right|_\eta$$

and a similar identification of the other elements shows that the product of (C.1) and (C.2) is the determinant

$$\frac{\partial(u, v)}{\partial(\xi, \eta)} = \begin{vmatrix} \frac{\partial u}{\partial \xi} & \frac{\partial u}{\partial \eta} \\ \frac{\partial v}{\partial \xi} & \frac{\partial v}{\partial \eta} \end{vmatrix} \quad (C.4)$$

In the Jacobian notation, therefore, the direct evaluation of product elements shows that

$$\frac{\partial(u, v)}{\partial(\xi, \eta)} = \frac{\partial(u, v)}{\partial(x, y)} \frac{\partial(x, y)}{\partial(\xi, \eta)} \quad (C.5)$$

If $v = \eta$, for instance,

partial derivative written as a Jacobian determinant, and transformed

$$\left.\frac{\partial u}{\partial \xi}\right|_\eta = \frac{\partial(u, \eta)}{\partial(\xi, \eta)} = \frac{\partial(u, \eta)}{\partial(x, y)} \frac{\partial(x, y)}{\partial(\xi, \eta)} \quad (C.5a)$$

so that the partial derivative with respect to the new variable ξ is expressed in terms of the derivatives of u with respect to the variables x and y of the original set.

If it happens that ξ and η are identified with u and v, respectively, the left side of (C.5) is equal to one, and the property

inverse property, like ordinary derivative

$$\frac{\partial(u\ v)}{\partial(x, y)} = \frac{1}{\frac{\partial(x, y)}{\partial(u, v)}} \quad (C.6)$$

is established; it is analogous to the familiar property of the ordinary derivative of a function of one variable.

transformation of multiple integrals

The Jacobian transformation determinant appearing as the last factor on the right side of (C.5a) relates partial derivatives with respect to one set of variables x, y to those in which a different set ξ, η is employed. An equally important and useful role of the same Jacobian transformation determinant appears when it is desired to change integration variables in multiple integrals. With only two independent variables x and y again for simplicity, it is required to transform a definite integral of a function $f(x, y)$ over a region S of these variables represented as an area in the plane having x and y as rectangular coordinates. If the new variables are again

APPENDIX C PROPERTIES OF JACOBIAN DETERMINANTS

denoted ξ and η, they are related to the first set by the pair of transformation equations

$$x = x(\xi, \eta) \qquad y = y(\xi, \eta) \qquad (C.7)$$

parametric formulae

and the associated Jacobian determinant of transformation is again given by (C.2) when the derivatives are evaluated from the definite functions of ξ and η represented on the right sides of (C.7). Then the integral of a function $f(x, y)$ over a region S of the x, y plane is expressed by substituting for x and y in terms of ξ and η from (C.7) and integrating with respect to ξ and η, after multiplying the integrand by the transformation Jacobian (C.2):

$$\iint_S f(x, y)\, dx\, dy = \iint_{S'} f(x[\xi, \eta], y[\xi, \eta]) \frac{\partial(x, y)}{\partial(\xi, \eta)}\, d\xi\, d\eta \qquad (C.8)$$

multiple integrals transform with a Jacobian determinant factor

where the region S' in the plane of ξ and η coordinates corresponds to the region S in the plane of x and y according to the transformation (C.7).

The proof of (C.8) is found by expressing the element of area dS in the x, y plane in terms of the differentials of ξ and η, instead of in terms of x and y as $dx\, dy$. For this purpose it is supposed that the transformation (C.7) is inverted to express ξ and η in terms of x and y:

$$\xi = \xi(x, y) \qquad \eta = \eta(x, y) \qquad (C.7')$$

inverse of (C.7) gives a curvilinear grid

so that the families of lines $\xi = $ const and $\eta = $ const form a curvilinear grid in the x, y plane. At each point x, y where the

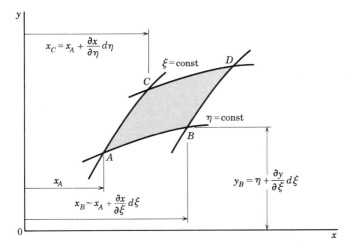

Fig. C-1

corresponding ξ-line intersects with a definite η-line, another pair of lines is obtained by increasing the values of ξ and η to $\xi + d\xi$ and $\eta + d\eta$ (Fig. C-1). When the increments $d\xi$ and $d\eta$ are small, the four-sided figure thus determined is closely approximated as a paralleleogram. Its area is given by the product of the lengths AB and AC of two adjacent sides multiplied by the sine of the included angle (expressed vectorially as the magnitude of the vector product of the corresponding vectors AB and AC). This can be written as the determinant

$$\begin{vmatrix} x_B - x_A & y_B - y_A \\ x_C - x_A & y_C - y_A \end{vmatrix} \quad (C.9)$$

or more symmetrically as the 3×3 determinant

$$\begin{vmatrix} 1 & 1 & 1 \\ x_A & x_B & x_C \\ y_A & y_B & y_C \end{vmatrix}$$

If the abscissa of vertex A is denoted x_A and the vertex B is situated on the line $\eta = $ const passing through A, at a distance $x_B - x_A$ from it in the x-direction, then

$$x_B - x_A = \frac{\partial x}{\partial \xi} d\xi$$

while

$$y_B - y_A = \frac{\partial y}{\partial \xi} d\xi$$

and the remaining elements of (C.9) are evaluated in the same manner. The area of the element $ABCD$ is thus given by (C.9) as

interpretation of Jacobian determinant: area distortion

$$\begin{vmatrix} \frac{\partial x}{\partial \xi} d\xi & \frac{\partial y}{\partial \xi} d\xi \\ \frac{\partial x}{\partial \eta} d\eta & \frac{\partial y}{\partial \eta} d\eta \end{vmatrix} = \frac{\partial(x, y)}{\partial(\xi, \eta)} d\xi \, d\eta \quad (C.9a)$$

The area element represented by (C.9a) is suitable for integration with respect to ξ and η, the determinant factor multiplying $d\xi \, d\eta$ being already expressed as a function of ξ and η. The variables ξ and η may, of course, also be regarded as Cartesian axes in the ξ, η plane, and, if the function $f(x, y)$ is taken as unity, the integral on the left side of (C.8) gives the area of the region of integration S in the x, y plane. The area of the corresponding region S' is in general different from that of S, and the interpretation of (C.8) is then that the transformation determinant $\partial(x, y)/\partial(\xi, \eta)$ gives the magnitude of the *area distortion* of the transformation (C.7) at any point.

A simple example is furnished by the transformation from Cartesian coordinates x and y to polar coordinates r and θ, when (C.7) becomes

$$x = r\cos\theta \quad y = r\sin\theta \qquad \text{familiar example}$$

ξ and η being interpreted as radial distance r and polar angle θ, respectively. From (C.2) it is found that the transformation determinant has the value r, the differential area expression (C.9a) becoming $r\,dr\,d\theta$.

Index

Acoustic propagation, 172, 230
Adiabatic relationship, 176, 184
Aerodynamics, 2
Aerostatics, 6, 10
   fundamental equation of, 9
Apparent mass, 127
Apparent stresses, 250
Approximations, linearized, 125, 185
   special intuitive, 201, 225
   hydrostatic, 161
   boundary, 164, 187
   employing homologous scaling, 191, 223
Archimedes' Principle, 31
Atmosphere, standard, 17
   barotropic, 18
   adiabatic, 18
Atmospheric stability, 32
Avogadro's number, 260

Barotropic relationship, 190
   and non-barotropic fluids, 95, 173
Bernoulli's Theorem, 47, 52, 62, 67, 73, 80, 116, 163
   for irrotational flow, 100
   not applicable, 84
Bessel functions, 214
Biot-Savart Law of vorticity induction, 150
Blasius' calculation, 219
Boltzmann's constant, 260
   equation, 261
   Law, 265
Boundary condition, 54, 124, 164, 169, 207, 209, 215, 216
Boundary layer, laminar, 219 et seq.
   turbulent, 254 et seq.
Buoyancy, 31
   ideal gas, 177

Centrobaric property, 286
Characteristic length, 158, 220
Characteristic velocity, 70, 158, 220
   time, 160
Chimney flow, 64
Circulation, 91, 92, 105
Coefficient, thrust force, 75, 83
   pressure, 83
Complex functions, method of, 121, 137
Complex potential, 154
Complex variable representation, 114
Compressibility, 3, 49
   isothermal, 12
   isentropic, 180
   liquid-gas distinction, 12
   isentropic, 15
Compression waves, 174
Condensation, 193
Conservation principles:
   fluid mass, 39
   fluid momentum, 45

fluid energy, 51
  summarized, 53
Conservative forces, 26, 46, 97, 276
Continuity equation
  steady flow, 40, 73
  vector form, 42
  integral form, 42, 138
  differential form, 44, 124, 131, 198, 264
  Lagrangian form, 57, 58, 173
  streamline form, 62
  of vorticity, 146
Continuous fluid medium, 2, 258
  limits of validity, 243
Coordinates, rotating, 25
  polar, 37, 44, 51, 141
  transformation, 38, 58, 107, 133, 191, 200, 223, 291
  cylindrical, 44, 141
  streamline, 60, 197
Coriolis acceleration, 76, 79, 216
Critical pressure and density, 15
  temperature, 14
Cyclindrical coordinates, *see* Coordinates, cylindrical
Cyclonic motion, 78
Cylinder flow, without circulation, 116
  with circulation, 124, 128

Decay of sound wave, 229
Density, 3, 259
Dimensionless equations, 219, 223
Dimensionless parameters, 159, 181, 210
Dimensionless variables, 158, 193
Dipole flow (doublet), 116, 135
Dispersion, 167, 192
Displacement thickness, 226, 257
Dissipation of energy, 228, 231, 233
  function, 229, 239
Distribution function of velocities, 261

Energy, conservation of, 1, 51, 176
Energy, potential, 26
  flux, 83
  minimum in irrotational flow
Energy loss in stars, 189
Entropy, 85, 181, 202, 265
  kinetic, 102
Euler's equations of fluid motion

49, 51, 98, 131, 163
  cylindrical coordinates, 143
  vector form, 53, 139, 148
  polar form, 140
Eulerian description, 37
  derivative, 38

Fluid, definition, 1
  perfect, 2, 45
  real, 2
Fluid statics, planetary and stellar, 18
Flow representations
  uniform, 89, 104, 107, 112, 116
  shear, 89
  dilatation, 89
  radial (source), 89, 108, 113
  rotation (vortex), 89, 105, 112
  doublet, 116
Forces, intermolecular, 275
Force expressed as momentum flux, 81
  obtained by integration of pressures, 117, 120, 189
Friction coefficient, 224, 227
Friction factor, 210, 253
Friction velocity, 252
Frictional cooling, 218
Frictional wind, 216
Froude number, 159
Fundamental principles of fluid mechanics, 1
  solutions, method of, 120, 283
Galilean invariance, 129
  counterexample, 132
Gas constant, 17
  for air, 17
Gases, 2
Gauss' transformation, 281
Geostrophic winds, 76, 216
Gravitational collapse, 192
Green's theorem, 23, 89, 91, 279
Ground effect, 153

Halley's Law, 11, 16, 17, 266
  isothermal atmosphere, 34
Heat addition, 182, 183
Heat shock, 182
Hodograph equations, 197
Hodograph representation of motion, 218
Hodograph variables, 62, 106, 197

Homogeneous fluid, 2
Hooke's Law, 234
Hugoniot's relationship, 179, 204
Hydrodynamics, 2
Hydrostatics, 6, 9
Hydrostatic force formulae, 31, 32
Hyperbolic functions, 126, 165, 166

Image vortex, 153
Impact Loss, 84, 85
Incompressible fluid, 3, 47
Infinitesimal analysis, 46
 higher order, 46
Integral, 26
Internal flow, 82
Invariance, 129, 235
Irrotational flow, methods for determining, 120
Irrotationality, 88, 198
 streamline and hodograph forms, 106, 108
Isobaric surfaces, 95
Isopycnic surfaces, 95

Jacobian determinant, 57, 106, 108, 130, 287
 partial derivative expression, 288
 fundamental interpretations, 288, 289
Jet, liquid, impingement of, 79

Kelvin's Theorem, 91, 94
 breaks down, 98
Kinetic energy of flow, 102, 127, 128
 molecular, 267
 related to virial, 274
Kutta-Joukowski Theorem, 118

Lagrangian description, 37, 172
 form of equations of fluid motion, 57
Laminar flow, 244
Laplace's Equation, 101, 107, 108, 113, 124, 164, 283
 in polar form, 111, 116, 122, 134
 solutions, 120, 122, 125
Lapse rate, defined, 17
 standard, 17
 adiabatic, 33
Legendre functions, 135, 169
Lift force, 118, 126
Line integrals, 92, 109, 279

 related to surface integrals (Green's, Gauss', Stokes' theorems), 279 et seq.
Linearized approximation, 125, 184
Liquid, 2

Mach Number, 180
Maxwellian distribution, 265
Mixing length hypothesis, 250
Molecular speeds, 269
 mean free path, 270
 momentum, 271
Momentum balance, 69
Momentum flux, 47, 49, 81, 226
Momentum integral equations, 226, 256
Momentum thickness, 226, 257

Natural convection, 64
Navier-Stokes equations, 212, 238
Newton's law of gravitation, 7, 282
 elastic collision, 85
 laws of mechanics, 4, 45
 iterative technique
 viscosity experiments, 206
No-slip condition, 207, 216, 220
Nonconservative forces, 97
Nonlinearity, 56, 190

Ocean Tide, *see* Tide
Order-of-magnitude analysis, 220

Partial derivative transformation, 288
Particle displacements, 175, 233
Particle paths (trajectories), 37, 170
Pascal's Principles, 3, 245
Phase velocity, 163
Pipe flow, 208, 252, 256
Pitot tube, 68
Planetary and stellar statics, fundamental equation of, 21
Poiseuille's law, 209
Poisson's equation, 22
Polytropic index, 18
 fictitious, 199
Potential, gravitational force, 26, 29, 283
 velocity, 101
 in complex form, 122, 154
Potential of perturbation flow, 186

Prandtl-Glauert compressibility correction, 195
Pressure, 1, 259
  gradient, 25, 53, 224
  recovery, 82
Product-type solutions, 121, 135, 215
Propeller flow, 72
Propulsive efficiency, 70, 73, 86

Quasi-one-dimensional flow, 84

Relative equilibrium, 24
Restitution, Newton's coefficient of, 85
Reynolds' experiments, 245
Reynolds' Number, 210, 219, 243, 245
Reynolds stresses, 248
Rocket propulsion, 69
  mass ratio, 70
  performance, fundamental equation of, 70
Rotating reference frame, 24, 76
  coordinate system, equilibrium in, 30, 131
Rotational flattening, 26

Sea breeze, 96
Shock waves, 178, 240
Similarity transformations, 191, 196, 223, 251
Sink flow, 114
Solar nebula hypothesis, 30
Sound wave, 175
Source flow, 113, 134
Specific heats, ratio of, 16
Sphere potential, 285
Spherical coordinates, *see* Coordinates
Spherical fluid masses, 18, 168
Stability, 32
Stagnation point, 68
  pressure, 68
Standing waves, 163, 168
Statics, fluid, 2
Steady motion, 36
Stokes' Theorem, 91, 147, 281
Stream surface, 38
  function, quantity of flow, 109, 110, 220
  axial symmetric flow, 143
  tube, 39
  filament, 39
Streamline, defined, 36
  equation of, 37, 104
  related to stream function, 110
Streamline derivative, 38
  coordinates, 60, 107
Stress, normal, 1
  tangential (viscous) 2, 206, 211, 228, 231, 234, 237
Substantial derivative, 38
Superposition of flows, 116, 120, 145
Supersonic flow, 181, 186, 219
Surface waves, 157
Swirling flow, 213

Temperature, molecular interpretation, 260, 271
Temperature lapse rate, 17
  scales, 17
Thermodynamic equation of state, 10
Three-dimensional flow, 81, 133 et seq.
Tide height, 136
  universal coefficient of, 137
Torricelli's Theorem, 66
Total derivative, 38, 55
Trapezoidal rule of integration, 204
Turbulent motion, 244 et seq.

Unit vectors, 28, 35, 54, 111, 140, 213
Units and dimensions, 6
Universal gas constant, 260
Universal gravitational constant, 7
Unsteady flow, 127, 129, 163
Upwelling, ocean, 78

van der Waals, equation of state, 14, 275
Variables, dependent, 4
  independent, 3
Velocity, molecular, 259
Velocity distribution, Maxwellian, 268
Velocity fluctuations, 244, 247 et seq.
Velocity perturbations, 123
Velocity potential, *see* Potential, velocity

Cartesian components, 37
Velocity vector, 35, 142
Virial coefficients, 275
Virial theorem, 272
Viscosity, 3, 88, 208, 271
   numerical values, 213
Viscous dissipation, 231
Viscous stresses, 207, 234, 237
Vortex, 113
   dynamics, 145
   lines, surfaces and tubes, 146
   rings (circular vortices), 150, 152
Vortices, rectilinear, 145

Vortices, lifting and starting, 155
Vorticity, 89, 91, 133, 141, 143, 251
   in polar form, 93
   solenoid, 95
   generation of, 148

Wave drag, 188
Wave equation, 162, 166, 175
   solutions, 162, 187
Wavy wall, incompressible flow past, 122
   supersonic flow, 186

About the author . . .

IRVING MICHELSON is Professor of Mechanical and Aerospace Engineering at the Illinois Institute of Technology in Chicago and has also held a concurrent appointment at the Faculté des Sciences de Nancy in France. He received a B.S. from Georgia Tech and his M.S. and Ph.D. in aeronautics and mathematics from the California Institute of Technology. He was formerly professor and department chairman at Penn State, and has also taught at Caltech, U.C.L.A., and the University of Southern California.

Dr. Michelson has published fundamental papers in several branches of theoretical and applied mechanics, including ocean and tide dynamics, high-speed flows, jets, satellite dynamics and celestial mechanics, cosmological aerodynamics and meteorology. Numerous examples discussed in the present volume are drawn from his original contributions, appearing for the first time in textbook form. Without exception, all of the unusual examples have been included in undergraduate courses he has taught.

Dr. Michelson has served as consultant to the Argonne National Laboratory, the Smithsonian Astrophysical Observatory, various Navy and Air Force agencies as well as numerous industrial organizations. His professional affiliations include the Society of Sigma Xi, the American Institute of Aeronautics and Astronautics, the American Astronomical Society, and the American Society for Engineering Education.